Authenticity of Honey: Characterization, Bioactivities and Sensorial Properties

Authenticity of Honey: Characterization, Bioactivities and Sensorial Properties

Editors

Olga Escuredo
M. Carmen Seijo

MDPI • Basel • Beijing • Wuhan • Barcelona • Belgrade • Manchester • Tokyo • Cluj • Tianjin

Editors
Olga Escuredo
University of Vigo
Spain

M. Carmen Seijo
University of Vigo
Spain

Editorial Office
MDPI
St. Alban-Anlage 66
4052 Basel, Switzerland

This is a reprint of articles from the Special Issue published online in the open access journal *Foods* (ISSN 2304-8158) (available at: https://www.mdpi.com/journal/foods/special_issues/Authenticity_Honey_Characterization_Bioactivities_Sensorial).

For citation purposes, cite each article independently as indicated on the article page online and as indicated below:

LastName, A.A.; LastName, B.B.; LastName, C.C. Article Title. *Journal Name* **Year**, *Volume Number*, Page Range.

ISBN 978-3-0365-4507-3 (Hbk)
ISBN 978-3-0365-4508-0 (PDF)

© 2022 by the authors. Articles in this book are Open Access and distributed under the Creative Commons Attribution (CC BY) license, which allows users to download, copy and build upon published articles, as long as the author and publisher are properly credited, which ensures maximum dissemination and a wider impact of our publications.
The book as a whole is distributed by MDPI under the terms and conditions of the Creative Commons license CC BY-NC-ND.

Contents

About the Editors vii

Preface to "Authenticity of Honey: Characterization, Bioactivities and Sensorial Properties" . ix

Olga Escuredo and M. Carmen Seijo
Authenticity of Honey: Characterization, Bioactivities and Sensorial Properties
Reprinted from: *Foods* **2022**, *11*, 1301, doi:10.3390/foods11091301 1

Eleni Tsavea, Fotini-Paraskevi Vardaka, Elisavet Savvidaki, Abdessamie Kellil, Dimitrios Kanelis, Marcela Bucekova, Spyros Grigorakis, Jana Godocikova, Panagiota Gotsiou, Maria Dimou, Sophia Loupassaki, Ilektra Remoundou, Christina Tsadila, Tilemachos G. Dimitriou, Juraj Majtan, Chrysoula Tananaki, Eleftherios Alissandrakis and Dimitris Mossialos
Physicochemical Characterization and Biological Properties of Pine Honey Produced across Greece
Reprinted from: *Foods* **2022**, *11*, 943, doi:10.3390/foods11070943 5

Saeed Mohamadzade Namin, Fatema Yeasmin, Hyong Woo Choi and Chuleui Jung
DNA-Based Method for Traceability and Authentication of *Apis cerana* and *A. dorsata* Honey (Hymenoptera: Apidae), Using the *NADH dehydrogenase* 2 Gene
Reprinted from: *Foods* **2022**, *11*, 9285, doi:10.3390/foods11070928 23

Kriss Davids Labsvards, Vita Rudovica, Rihards Kluga, Janis Rusko, Lauma Busa, Maris Bertins, Ineta Eglite, Jevgenija Naumenko, Marina Salajeva and Arturs Viksna
Determination of Floral Origin Markers of Latvian Honey by Using IRMS, UHPLC-HRMS, and ^{1}H-NMR
Reprinted from: *Foods* **2022**, *11*, 42, doi:10.3390/foods11010042 35

Lua Vazquez, Daniel Armada, Maria Celeiro, Thierry Dagnac and Maria Llompart
Evaluating the Presence and Contents of Phytochemicals in Honey Samples: Phenolic Compounds as Indicators to Identify Their Botanical Origin
Reprinted from: *Foods* **2021**, *10*, 2616, doi:10.3390/foods10112616 49

Marinos Xagoraris, Foteini Chrysoulaki, Panagiota-Kyriaki Revelou, Eleftherios Alissandrakis, Petros A. Tarantilis and Christos S. Pappas
Unifloral Autumn Heather Honey from Indigenous Greek *Erica manipuliflora* Salisb.: SPME/GC-MS Characterization of the Volatile Fraction and Optimization of the Isolation Parameters

Reprinted from: *Foods* **2021**, *10*, 2487, doi:10.3390/foods10102487 67

Donát Magyar, Paulian Dumitrica, Anna Mura-Mészáros, Zsófia Medzihradszky, Ádám Leelőssy and Simona Saint Martin
The Occurrence of Skeletons of Silicoflagellata and Other Siliceous Bioparticles in Floral Honeys
Reprinted from: *Foods* **2021**, *10*, 421, doi:10.3390/foods10020421 83

Olga Escuredo, María Shantal Rodríguez-Flores, Laura Meno and María Carmen Seijo
Prediction of Physicochemical Properties in Honeys with Portable Near-Infrared (microNIR) Spectroscopy Combined with Multivariate Data Processing
Reprinted from: *Foods* **2021**, *10*, 317, doi:10.3390/foods10020317 99

Asma Ghorab, María Shantal Rodríguez-Flores, Rifka Nakib, Olga Escuredo, Latifa Haderbache, Farid Bekdouche and María Carmen Seijo
Sensorial, Melissopalynological and Physico-Chemical Characteristics of Honey from Babors Kabylia's Region (Algeria)
Reprinted from: *Foods* **2021**, *10*, 225, doi:10.3390/foods10020225 . . . . **115**

Marcela Bucekova, Veronika Bugarova, Jana Godocikova and Juraj Majtan
Demanding New Honey Qualitative Standard Based on Antibacterial Activity
Reprinted from: *Foods* **2020**, *9*, 1263, doi:10.3390/foods9091263 . . . . **133**

Li-Ping Sun, Feng-Feng Shi, Wen-Wen Zhang, Zhi-Hao Zhang and Kai Wang
Antioxidant and Anti-Inflammatory Activities of Safflower (*Carthamus tinctorius* L.) Honey Extract
Reprinted from: *Foods* **2020**, *9*, 1039, doi:10.3390/foods9081039 . . . . **147**

Mounia Homrani, Olga Escuredo, María Shantal Rodríguez-Flores, Dalache Fatiha, Bouzouina Mohammed, Abdelkader Homrani and M. Carmen Seijo
Botanical Origin, Pollen Profile, and Physicochemical Properties of Algerian Honey from Different Bioclimatic Areas
Reprinted from: *Foods* **2020**, *9*, 938, doi:10.3390/foods9070938 . . . . **163**

Daniela Pauliuc, Florina Dranca and Mircea Oroian
Antioxidant Activity, Total Phenolic Content, Individual Phenolics and Physicochemical Parameters Suitability for Romanian Honey Authentication
Reprinted from: *Foods* **2020**, *9*, 306, doi:10.3390/foods9030306 . . . . **181**

About the Editors

Olga Escuredo

Olga Escuredo has been a professor at the Faculty of Sciences (University of Vigo, Spain) since 2012. She received her PhD in Agri-Food Science and Technology from the University of Vigo focusing on research tasks linked to the composition and physicochemical characterization of honeys. Her main research lines are based on the physicochemical characterization and botanical origin of bee products, and the application of NIR technology to food products and food industries. Her research interests include the chemistry of natural products such as honey and pollen, analysis of compounds with biological activity and the relationships with the quality and the origin of food products. Olga Escuredo has authored more than 70 research articles published in peer-reviewed international journals and has presented many communications in national and international seminars and conferences, within the areas of Food Sciences and Technology, Nutrition and Dietetics, multidisciplinary Chemistry, Environmental Sciences, Agronomy and Agriculture. She has also participated in national and European projects and dissemination activities.

M. Carmen Seijo

M. Carmen Seijo has been a professor at the Faculty of Sciences (University of Vigo, Spain) since 1995. She completed her PhD in Biology with a focus on the characterization of different honey types. She specializes in the study of honeybee products, mainly regarding botanical and geographical origin, authentication, quality control as well as physicochemical composition and sensory evaluation. She is a recognized expert in the field of melissopalynology and the evaluation of the botanical and geographical origin of honey. She has also led several research activities and projects dealing with the authentication of honeybee products using different analytical methodologies. M. Carmen Seijo has published more than 100 research articles in peer-reviewed journals and has delivered a number of presentations at some international and national conferences within the areas of Botany, Food Sciences and Technology, Nutrition and Dietetics, multidisciplinary Chemistry, Environmental Sciences, Agronomy and Agriculture.

Preface to "Authenticity of Honey: Characterization, Bioactivities and Sensorial Properties"

Honey is a natural food product with significant nutritional value, and several studies have proved its health-beneficial properties. The floral source of honey is one of the most significant factors contributing to the composition of honey, as it incorporates the nectar available to be transferred by bees during honey production. The chemical composition, stability and authenticity of honey are integral in determining the quality of this product. Hence, unifloral honey authentication requires different analytical procedures including sensorial profile, quality parameters and pollen analysis combined with statistical evaluation. There is a demand for this type of study because consumers request a greater variety of correctly characterized unifloral honeys in the market and with more information on their functional and healing properties. Therefore, establishing well-defined quality criteria and their origin guarantees the appreciation of honey in the market both locally and internationally.

This book aims to collect interdisciplinary approaches that combine physicochemical, palynological, sensory analyses and statistical treatments as tools for the authentication of honey as food. Finally, the editors appreciate the efforts of all the authors who participated in this Special Issue with their contributions.

Olga Escuredo and M. Carmen Seijo
Editors

Editorial

Authenticity of Honey: Characterization, Bioactivities and Sensorial Properties

Olga Escuredo * and M. Carmen Seijo

Department of Vegetal Biology and Soil Science, Faculty of Science, University of Vigo, As Lagoas, 32004 Ourense, Spain; mcoello@uvigo.es
* Correspondence: oescuredo@uvigo.es; Tel.: +34-988-387048

Citation: Escuredo, O.; Seijo, M.C. Authenticity of Honey: Characterization, Bioactivities and Sensorial Properties. *Foods* **2022**, *11*, 1301. https://doi.org/10.3390/foods11091301

Received: 26 April 2022
Accepted: 28 April 2022
Published: 29 April 2022

Publisher's Note: MDPI stays neutral with regard to jurisdictional claims in published maps and institutional affiliations.

Copyright: © 2022 by the authors. Licensee MDPI, Basel, Switzerland. This article is an open access article distributed under the terms and conditions of the Creative Commons Attribution (CC BY) license (https://creativecommons.org/licenses/by/4.0/).

Honey is a natural product well known for its beneficial properties, which depend on its composition. This composition of honey is related to its botanical and geographical origin, as well as its environmental and management conditions. Honey contains a wide spectrum of components, many of which are phytochemicals found in plants; these include compounds with highly demonstrated antimicrobial, antioxidant and anti-inflammatory properties. Studies about the composition, properties, and botanical and geographical origins of different honey types are relevant in society; they are especially relevant in the honey market, as they provide knowledge to guarantee honey authenticity.

For years, the determination of the geographical and botanical origin of honey has been based on the microscopic examination of its pollen profile. However, due to a lack of specialists in palynology, there is a tendency to replace pollen analysis with instrumental methods of finding biochemical markers for honey discrimination. Nevertheless, studies on unifloral honey, particularly if they have similar physicochemical characteristics, need to be complemented with palynological information to successfully discriminate its botanical origin. Research based on botanical and geographical origin that delves into the chemical, physical and health qualities—as well as the pollen profile—of honey as food is needed, to valorize the product in the market both locally and internationally.

In this Special Issue, our aim is to publish research papers on the authenticity, characterization, and biological properties of honey. Thus, various aspects of its physicochemical composition, quality parameters, sensorial profile, functional properties, healthy compounds, palynological characteristics, and the relationships of these factors with their botanical and geographical origins, are discussed. Methods to identify the entomological origin and possible adulteration of honey are also considered. The papers selected for this Special Issue were subject to a rigorous peer review procedure, with the aim of rapid and wide dissemination of research results, and their transfer to different stakeholders and the scientific community.

The paper presented by Labsvards et al. [1] assessed the chemical profiles of 78 honeys from Latvia (buckwheat, clover, heather, linden, rapeseed, willow, and polyfloral honey) using light stable-isotope-ratio mass spectrometry (IRMS), ultra-high-performance liquid chromatography coupled with high-resolution mass spectrometry (UHPLC-HRMS), and nuclear magnetic resonance (NMR) methods. The researchers validated the floral origin of honeys using melissopalynology. The study showed a combination of multiple analytical and statistical methods to differentiate between various types of monofloral honey, and to find adequate indicators for the classification of their botanical origins. Therefore, this study reports on the botanical authentication and quality of honey from this geographical origin.

The study carried out by Vázquez et al. [2] showed that the combination of chromatographic analysis with mass spectrometry detection and principal component analysis are adequate to investigate the botanical authentication of honey from Northwest Spain. A miniaturized, fast and environmentally friendly experimental procedure (VE-UAE) based on liquid chromatography–tandem mass spectrometry (LC-MS/MS) was successfully developed. Specifically, the statistical treatment—based on concentrations of total phenols, antioxidant activity and individual phenolic compounds—revealed significant differences according to the studied honey type, demonstrating that phenolic compounds can be used as indicators to identify botanical origins.

A melissopalynological analysis and some physicochemical parameters of 45 unifloral honeys (raspberry, mint, sunflower, thyme and rape) and polyfloral honeys from different regions of Romania were analyzed by Pauliuc et al. [3]. Principal component analysis confirmed the possibility of the botanical authentication of rape, sunflower and thyme honey samples based on the following physicochemical parameters: moisture content, pH, free acidity, electrical conductivity, hydroxymethylfurfural content, color, total polyphenol content, flavonoid content, DPPH radical scavenging activity, phenolic acids, flavonols, sugars and organic acids. However, the mint honey and raspberry honey were not satisfactorily separated. The researchers highlighted sugars, individual phenolic compounds and organic acids as the least influential compounds in the botanical origin of the studied honey types. It is important to note that for better classification of honey, it is necessary to study a large number of samples from different botanical origins.

Homrani et al. [4] characterized 62 honey samples produced in different bioclimatic areas of Algeria using palynology and their physicochemical properties. This paper highlighted the great diversity of honey production in Algeria, evidencing the importance of honey characterization to guarantee authenticity and to valorize local production. Some botanical taxa important for honey production, such as *Eucalyptus*, *Brassica napus*, *Hedysarum* and *Citrus*, are common honey plants in Mediterranean areas. However, others such as *Capparis spinosa*, *Asparagus*, *Ziziphus lotus* and some Apiaceae plants, are representative of the honey from this country, and are useful as markers to guarantee their authenticity. Based on the combination of palynological characteristics, physicochemical parameters and sensorial properties with statistical analysis, Ghorab et al. [5] differentiated 30 honey samples, produced in Babors Kabylia (North east of Algeria), by botanical origin. The pollen spectrum performed revealed a great diversity in pollen types (96 pollen types), with the main pollen types being the spontaneous species Fabaceae (*Hedysarum*, *Trifolium*, Genisteae plants), Asteraceae plants, Ericaceae (mainly *Erica arborea*), *Myrtus* and *Pistacia*. This is the first study to focus on sensory properties and their relationships with the botanical origin and the physicochemical properties of honey from Algeria. Considering that this area is a hotspot for biodiversity due to its high number of endemic plants, this study evidenced the wide variety of honey types that can be obtained in the area. Further studies could contribute to an increase in knowledge about these honey types and the most relevant plant species for honey yield, and could contribute to the valorization of local beekeeping.

Bucekova et al. [6] reported the need to specify some additional standards on the biological properties of honey. This study focused on evaluating the antibacterial activity, the content of hydrogen peroxide (H_2O_2) and the protein profile of 36 commercial honeys purchased in Slovakia, both from local beekeepers and medical-grade (recommended for treating infected wounds). More than 40% of the commercial honeys purchased presented low antibacterial activity, which is indicative of artificial honey (only sugars). In general, the honey samples exhibited high antibacterial activity, while generating low levels of H_2O_2. On the other hand, the honey samples from local beekeepers showed superior antibacterial activity compared to the medical-grade honeys. Tsavea et al. [7] demonstrated the ability of pine honeydew samples produced across Greece to generate high levels of H_2O_2, even higher than other types of honeydew honey. Furthermore, due to their high polyphenol content, a strong antioxidant activity of the honey samples was also denoted. The breakdown of H_2O_2, using a catalase treatment, into a honey solution resulted in a significant decrease in antibacterial activity. Similarly, the digestion of honey proteins by proteinase K resulted in lower antibacterial efficacy among the studied honey samples, depending, again, on the specific bacteria. Thus, the results suggest multiple underlying mechanisms of the antibacterial activity of pine honeydew honeys. These researchers demonstrated that the antibacterial activity of honey can be easily altered using adulteration, thermal treatment or prolonged storage; therefore, it fulfils strict criteria to be a suitable new quality standard [6,7].

Sun et al. [8] investigated the chemical composition and the biological and anti-inflammatory activities of Safflower honey from Northwest China, produced using nectar from the medicinal plant *Carthamus tinctorius*. The extract of this medicinal plant was studied as it has very interesting biological activity properties, but the chemical and biological composition of honey from this botanical origin has been poorly evaluated. The results of the study revealed a great capacity to capture DPPH and ABTS+ free radicals of Safflower honey in vitro [8]. The special antioxidant and anti-inflammatory properties of safflower honey revealed its potential as a novel functional food. Despite being a little-studied honey, the future of this type of honey is encouraging due to its biological qualities and excellent nutritional value.

Xagoraris et al. [9] analyzed the main volatile compounds of 25 autumn heather honeys from the indigenous Greek *Erica manipuliflora* using solid-phase microextraction (SPME) methodology, followed by gas chromatography–mass spectrometry (GC-MS). This is the first study to identify a volatile profile of honey from *E. manipuliflora*. Optimal method conditions were proposed for all the dominant volatile compounds, and predictive models were provided to evaluate each volatile compound separately. The objective of this study was to reinforce the characterization of this honey type.

Non-destructive characterization tools for quality control and the physicochemical compounds of honey, based on the principles of spectroscopy, are considered in this Special Issue [10]. Prediction models using portable near-infrared (MicroNIR) spectroscopy and reference physicochemical parameters of honey were developed to authenticate 100 honey samples from Northwest Spain. Using multivariate and partial least square regressions, the authors developed excellent models for moisture, hydroxymethylfurfural, color and total flavonoids, as well as acceptable statistics for electrical conductivity, pH and total phenols. Nevertheless, further experiments are proposed, to build a robust database that could support the use of this equipment as a quick alternative for honey authentication.

In addition to the identification of pollen grains, a great variety of particles in honey sediment can be found using microscopic analysis. Magyar et al. [11] identified some airborne particles present in honey from Poland and Tunisia, which provided information for the identification of its geographical origin. The study was based on the high percentage of anemophilous pollen grains and spores, and the great variety of particles found in the honey sediment. The presence of siliceous marine microfossils was related to the characteristics of the areas in which the honey was produced. The silicoflagellates are deposited from the air onto nectareous flowers and, consequently, bees transport them to their hives. These authors concluded that the silicoflagellates could be used as complementary indicators of the geographical origin of honeys collected in areas characterized by diatomite outcrops. This is the first document that provides a record of microfossils in honey, represented by silicoflagellates, diatoms and endoskeletal dinoflagellates.

Finally, this Special Issue includes a study on the importance of differentiating the entomological origin of honey [12]. *Apis mellifera* honey, *A. cerana* honey and *A. dorsata* honey are the three dominant honey types in the Asian market. Some studies indicate that *A. dorsata* honey has higher antioxidant properties and medical values than *A. mellifera* honey. *A. dorsata* honey and *A. cerana* honey are vulnerable to adulteration (as a result of false geographical and botanical origin, or from mixing their composition with sugars or syrups), due to the higher price in the market compared to *A. mellifera* honey [12]. Therefore, it is important to develop fast, reliable and cost-effective identification methods to determine the honey's origin. Mohamadzade et al. [12] proposed a rapid and accurate PCR-based method. In this study, three species-specific primers were designed to amplify the short part of the *NADH dehydrogenase 2* (*ND2*) region of mtDNA. At the same time, with this method, the authors detected a mixture of *A. mellifera* honey with *A. dorsata* and *A.*

cerana honey. Additionally, honey samples from different countries were used to evaluate the accuracy of the developed method.

In summary, the papers published in this Special Issue cover different analytical procedures for the identification of the botanical and geographical origins of honey. Authentication of the botanical and geographical origin of honey requires multiple analytical, statistical and mathematical methods for the determination of chemical compounds, and also requires the characterization of its sensory properties and pollen profile. The papers aim to report the sensory, chemical, nutritional, health and physical qualities of honey produced in different regions worldwide. New technologies looking for different purposes were even successfully adapted to the matrix of honey. The global trend towards healthy foods with high nutritional value arouses curiosity and sparks multidimensional and multidisciplinary approaches, as shown in the documents described above. The authors who have contributed to this issue inform the scientific community about the problem of, and the interest in, the authentication of honey as a food, which will undoubtedly help to deepen future studies.

Author Contributions: O.E. and M.C.S. contributed equally to the writing and editing of the editorial note. All authors have read and agreed to the published version of the manuscript.

Funding: This work did not receive external funding.

Conflicts of Interest: The authors declare no conflict of interest.

References

1. Labsvards, K.D.; Rudovica, V.; Kluga, R.; Rusko, J.; Busa, L.; Bertins, M.; Eglite, I.; Naumenko, J.; Salajeva, M.; Viksna, A. Determination of Floral Origin Markers of Latvian Honey by Using IRMS, UHPLC-HRMS, and 1H-NMR. *Foods* **2022**, *11*, 42. [CrossRef] [PubMed]
2. Vazquez, L.; Armada, D.; Celeiro, M.; Dagnac, T.; Llompart, M. Evaluating the Presence and Contents of Phytochemicals in Honey Samples: Phenolic Compounds as Indicators to Identify Their Botanical Origin. *Foods* **2021**, *10*, 2616. [CrossRef] [PubMed]
3. Pauliuc, D.; Dranca, F.; Oroian, M. Antioxidant Activity, Total Phenolic Content, Individual Phenolics and Physicochemical Parameters Suitability for Romanian Honey Authentication. *Foods* **2020**, *9*, 306. [CrossRef]
4. Homrani, M.; Escuredo, O.; Rodríguez-Flores, M.S.; Fatiha, D.; Mohammed, B.; Homrani, A.; Seijo, M.C. Botanical Origin, Pollen Profile, and Physicochemical Properties of Algerian Honey from Different Bioclimatic Areas. *Foods* **2020**, *9*, 938. [CrossRef]
5. Ghorab, A.; Rodríguez-Flores, M.S.; Nakib, R.; Escuredo, O.; Haderbache, L.; Bekdouche, F.; Seijo, M.C. Sensorial, Melissopalynological and Physico-Chemical Characteristics of Honey from Babors Kabylia's Region (Algeria). *Foods* **2021**, *10*, 225. [CrossRef] [PubMed]
6. Bucekova, M.; Bugarova, V.; Godocikova, J.; Majtan, J. Demanding New Honey Qualitative Standard Based on Antibacterial Activity. *Foods* **2020**, *9*, 1343. [CrossRef] [PubMed]
7. Tsavea, E.; Vardaka, F.-P.; Savvidaki, E.; Kellil, A.; Kanelis, D.; Bucekova, M.; Grigorakis, S.; Godocikova, J.; Gotsiou, P.; Dimou, M.; et al. Physicochemical Characterization and Biological Properties of Pine Honey Produced across Greece. *Foods* **2022**, *11*, 943. [CrossRef]
8. Sun, L.-P.; Shi, F.-F.; Zhang, W.-W.; Zhang, Z.-H.; Wang, K. Antioxidant and Anti-Inflammatory Activities of Safflower (*Carthamus tinctorius* L.) Honey Extract. *Foods* **2020**, *9*, 1039. [CrossRef]
9. Xagoraris, M.; Chrysoulaki, F.; Revelou, P.-K.; Alissandrakis, E.; Tarantilis, P.A.; Pappas, C.S. Unifloral Autumn Heather Honey from Indigenous Greek Erica manipuliflora Salisb.: SPME/GC-MS Characterization of the Volatile Fraction and Optimization of the Isolation Parameters. *Foods* **2021**, *10*, 2487. [CrossRef] [PubMed]
10. Escuredo, O.; Rodríguez-Flores, M.S.; Meno, L.; Seijo, M.C. Prediction of Physicochemical Properties in Honeys with Portable Near-Infrared (microNIR) Spectroscopy Combined with Multivariate Data Processing. *Foods* **2021**, *10*, 317. [CrossRef] [PubMed]
11. Magyar, D.; Dumitrica, P.; Mura-Mészáros, A.; Medzihradszky, Z.; Leelőssy, Á.; Saint Martin, S. The Occurrence of Skeletons of Silicoflagellata and Other Siliceous Bioparticles in Floral Honeys. *Foods* **2021**, *10*, 421. [CrossRef] [PubMed]
12. Mohamadzade Namin, S.; Yeasmin, F.; Choi, H.W.; Jung, C. DNA-Based Method for Traceability and Authentication of *Apis cerana* and *A. dorsata* Honey (Hymenoptera: Apidae), Using the NADH dehydrogenase 2 Gene. *Foods* **2022**, *11*, 928. [CrossRef] [PubMed]

Article

Physicochemical Characterization and Biological Properties of Pine Honey Produced across Greece

Eleni Tsavea [1], Fotini-Paraskevi Vardaka [2], Elisavet Savvidaki [2], Abdessamie Kellil [3], Dimitrios Kanelis [4], Marcela Bucekova [5], Spyros Grigorakis [3], Jana Godocikova [5], Panagiota Gotsiou [3], Maria Dimou [4], Sophia Loupassaki [3], Ilektra Remoundou [3], Christina Tsadila [1], Tilemachos G. Dimitriou [1], Juraj Majtan [5,6], Chrysoula Tananaki [4], Eleftherios Alissandrakis [2,7,*] and Dimitris Mossialos [1,*]

1. Laboratory of Microbial Biotechnology–Molecular Bacteriology–Virology, Department of Biochemistry & Biotechnology, University of Thessaly, Biopolis, 41500 Larissa, Greece; elenats89@hotmail.com (E.T.); tsadila@bio.uth.gr (C.T.); tidimitr@bio.uth.gr (T.G.D.)
2. Laboratory of Quality and Safety of Agricultural Products, Landscape and Environment, Department of Agriculture, Hellenic Mediterranean University, Stavromenos PC, 71410 Heraklion, Greece; foteinivardak@hmu.gr (F.-P.V.); elisavvidaki@hmu.gr (E.S.)
3. Food Quality & Chemistry of Natural Products, Mediterranean Agronomic Institute of Chania, International Centre for Advanced Mediterranean Agronomic Studies, 73100 Chania, Greece; kellilabdessamie@gmail.com (A.K.); grigorakis@maich.gr (S.G.); yiota@maich.gr (P.G.); sofia@maich.gr (S.L.); hlektra@maich.gr (I.R.)
4. Laboratory of Apiculture-Sericulture, Faculty of Agriculture, Forestry and Natural Environment, School of Agriculture, Aristotle University of Thessaloniki, 54124 Thessaloniki, Greece; dkanelis@agro.auth.gr (D.K.); mdimou@agro.auth.gr (M.D.); tananaki@agro.auth.gr (C.T.)
5. Laboratory of Apidology and Apitherapy, Department of Molecular Genetics, Institute of Molecular Biology, Slovak Academy of Sciences, Dubravska Cesta 21, 845 51 Bratislava, Slovakia; marcela.bucekova@savba.sk (M.B.); jana.godocikova@savba.sk (J.G.); juraj.majtan@savba.sk (J.M.)
6. Department of Microbiology, Faculty of Medicine, Slovak Medical University, Limbova 12, 833 03 Bratislava, Slovakia
7. Institute of Agri-Food and Life Sciences Agro-Health, Hellenic Mediterranean University Research Center, Stavromenos PC, 71410 Heraklion, Greece
* Correspondence: ealiss@hmu.gr (E.A.); mosial@bio.uth.gr (D.M.); Tel.: +30-28-1037-9409 (E.A.); +30-24-1056-5270 (D.M.)

Citation: Tsavea, E.; Vardaka, F.-P.; Savvidaki, E.; Kellil, A.; Kanelis, D.; Bucekova, M.; Grigorakis, S.; Godocikova, J.; Gotsiou, P.; Dimou, M.; et al. Physicochemical Characterization and Biological Properties of Pine Honey Produced across Greece. *Foods* **2022**, *11*, 943. https://doi.org/10.3390/foods11070943

Academic Editors: Olga Escuredo and María Carmen Seijo

Received: 6 March 2022
Accepted: 23 March 2022
Published: 25 March 2022

Publisher's Note: MDPI stays neutral with regard to jurisdictional claims in published maps and institutional affiliations.

Copyright: © 2022 by the authors. Licensee MDPI, Basel, Switzerland. This article is an open access article distributed under the terms and conditions of the Creative Commons Attribution (CC BY) license (https://creativecommons.org/licenses/by/4.0/).

Abstract: Pine honey is a honeydew honey produced in the East Mediterranean region (Greece and Turkey) from the secretions of the plant sucking insect *Marchalina hellenica* (Gennadius) (Coccoidea: Marchalini-dae) feeding on living parts of *Pinus* species. Nowadays, honeydew honey has attracted great attention due to its biological activities. The aim of this study was to study unifloral pine honey samples produced in Greece regarding their physicochemical parameters and antioxidant and antibacterial activity against five nosocomial and foodborne pathogens. These honeys showed physicochemical and microscopic characteristics within the legal limits, except for diastase activity, a parameter known to be highly variable, depending on various factors. Substantially higher levels of H_2O_2 were estimated compared to other types of honeydew honey, whereas protein content was similar. The total phenolic content was 451.38 ± 120.38 mg GAE/kg and antiradical activity ranged from 42.43 to 79.33%, while FRAP values (1.87 to 9.43 mmol Fe^{+2}/kg) were in general higher than those reported in the literature. Various correlations could be identified among these parameters. This is the first attempt to investigate in depth the antibacterial activity of pine honey from Greece and correlate it with honey quality parameters. All tested honeys exerted variable but significant antibacterial activity, expressed as MIC and MBC values, comparable or even superior to manuka honey for some tested samples. Although honey antibacterial activity is mainly attributed to hydrogen peroxide and proteins in some cases (demonstrated by elevated MICs after catalase and Proteinase K treatment, respectively), no strong correlation between the antibacterial activity and hydrogen peroxide concentration or total protein content was demonstrated in this study. However, there was a statistically significant correlation of moisture, antioxidant and antibacterial activity against *Klebsiella pneuomoniae*, as well as antioxidant and antibacterial activity against *Salmonella* ser. Typhimurium. Interestingly, a statistically significant negative correlation has been observed between diastase activity and *Staphylococcus aureus* antibacterial activity. Overall, our data indicate multiple mechanisms of antibacterial activity exerted by pine honey.

Keywords: honeydew; pine honey; physicochemical parameters; melissopalynology; sensory evaluation; bioactivity; antimicrobial; antioxidant

1. Introduction

Honey is a sweet, supersaturated solution of carbohydrates, composed of glucose, fructose, oligo- and polysaccharides, water, and other substances, such as proteins, enzymes, vitamins, minerals, phenolic compounds and amino acids that are of nutritional and health significance [1]. The chemical composition of honey and physicochemical parameters are variable and related to the botanical origin, geographic area and environmental conditions [2,3].

Pine honey is a honeydew honey produced by honeybees from the secretions of the plant sucking insect *Marchalina hellenica* (Gennadius) (Coccoidea: Marchalinidae) feeding on living parts of *Pinus* species. It is a typical honeydew honey, with high ash content, pH value and electrical conductivity. Additionally, the fructose and glucose content are low; therefore, its tendency to crystallize is low. This type of honey is produced in the Mediterranean region and specifically in Greece and Turkey. It constitutes 60–65% and 50% of the total annual honey production in Greece and Turkey, respectively [4]. It is of high nutritional value due to, among others, its high content of minerals such as potassium, calcium, iron, phosphorus, magnesium, sodium and zinc [5].

Greek legislation has adopted the Council Directive 2001/110/EC [6] regarding the physicochemical properties of honey. In addition, the Government Gazette B-239/23-2-2005 [7] describes the required properties for monofloral Greek honey varieties, according to which the electrical conductivity of Greek pine honey must be higher than 0.9, while also the presence of honeydew elements must be significant.

In recent decades, scientific interest has been focused on the antibacterial activity of diverse types of honey against clinical and foodborne pathogens [8–16] as well as on antioxidant activity [12,17–20]. Recently, honey antibacterial activity has been proposed as a valuable parameter determining its quality which takes into account the biological properties of honey [21].

Hydrogen peroxide (H_2O_2) is often considered as the major antibacterial compound of honey and it is produced by the enzyme glucose oxidase, which converts glucose into gluconic acid [22]. Several studies have clearly demonstrated the strong correlation between the antibacterial activity and the presence of H_2O_2 in certain types of blossom honeys [23–27]. However, a recent study by Farkasovska et al. [25] reported a weak or no correlation between antibacterial activity against particular bacteria and H_2O_2 concentration in linden honey samples. Similarly, despite the high level of H_2O_2 measured in Slovak honeydew honeys, no significant correlation was found between their overall antibacterial activity and the level of H_2O_2 [28]. In addition to H_2O_2, peptides and proteins such as bee defensin-1 and MRJP glycoproteins have been isolated from various honeys exhibiting antibacterial activity through cell lysis [29,30]. Additionally, phytochemicals such as phenolic compounds may significantly contribute to the antibacterial and antioxidant activity of honey, in particular honeydew honey [31].

A small fraction of honey's composition (2–5% of honey dry weight) contains compounds responsible for a plethora of biological properties, such as anti-inflammatory, antimicrobial antimutagenic, antioxidant, antiproliferative and antithrombotic [32,33]. The antioxidant activity involves the deactivation of free radicals, and it is classified into two mechanisms: hydrogen atom transfer (HAT) and electron transfer (ET). In the former, a hydrogen atom is donated to the radical, while in the latter, a single electron is transferred [34]. Phenolics are a class of phytochemicals that are primarily responsible for the antioxidant activity exerted by honey. Other non-phenolic compounds with the same activity are enzymes (catalase and peroxidase), ascorbic acid and carotenoids. Phenolic compounds that have been reported in honey include phenolic acids (coumaric, caffeic, ellagic, ferulic and chlorogenic acids) and flavonoids (chrysin, kaempferol, pinosembrin, quercetin, galangin, hesperetin, and myricetin) [35]. Very often, the presence of these compounds is expressed as total phenolic content, and it is positively correlated with the antioxidant capacity as well as antibacterial activity of the tested honey [27,34].

The aim of this study was to characterize unifloral pine honeys produced in Greece regarding their physicochemical parameters, antioxidant activity as well as their antibacterial activity against nosocomial and foodborne pathogens. To the best of our knowledge, this is the first in-depth attempt to investigate the antibacterial activity of pine honey from Greece and correlate it with various honey properties.

2. Materials and Methods

2.1. Honey Samples

Twenty-seven pine honey samples from diverse locations in Greece (Figure 1) were selected from samples collected in the framework of the national Emblematic Action "The Honeybee Routes". The classification as pine honey was based on the organoleptic, microscopic (honeydew elements) and electrical conductivity (>0.9 mS/cm) measurement (Table S1, Figure S1).

Figure 1. Locations of collected pine honey samples across Greece.

All samples were stored in glass containers at −18 °C until analysis. Before all assays, the samples were homogenized by stirring thoroughly for at least 3 min. Crystalized samples were liquefied in gentle heat of less than 40 °C for 5 min. Manuka honey UMF 24+ (MGO 1122) (Steens™, New Zealand, LOT 20NZH18) was used as reference honey to compare antibacterial activity of pine honey samples.

2.2. Microscopic Examination

Microscopic examination was performed according to von der Ohe et al. [36]. Honeydew elements (HDE) have been counted and the ratio HDE/P (Pollen) is given as the number of honeydew elements over the number of pollen grains from nectariferous plants. Quantitative melissopalynological analysis was based on the method of Yang et al. [37] and results are expressed as total number of all pollen grains (PG) in 10 g honey.

2.3. Physicochemical Parameters

All measurements of the physicochemical parameters were performed in duplicate ($n = 2$) unless otherwise stated.

2.3.1. Reagents

Glycerol standard ($\geq$99%) and starch soluble (for analysis) were purchased from Merck (Darmstadt, Germany). Glacial acetic acid ($\geq$99.8%), potassium iodide ($\geq$99.5%), sodium chloride ($\geq$99.5%), sodium hydroxide ($\geq$98%) and buffer solution (pH 4, 7 and 10) were purchased from Honeywell (Charlotte, NC, USA). Sodium acetate trihydrate (99%) was purchased from PENTA chemicals (Prague, Czech Republic) and conductivity standard (1413 µS/cm, 20 °C) was purchased from LLG International (Meckenheim, Germany). Iodine (99.5+) and zinc acetate dihydrate (for analysis) were from Fisher Chemical (Waltham, MA, USA). Potassium hexacyanoferrate (II) trihydrate (98+%) was purchased from Alfa Aesar (Ward Hill, MA, USA). Sodium bisulfite was purchased from Acros Organics (Geel, Antwerp, Belgium).

2.3.2. Moisture Content

Moisture content at 20 °C was determined from the refractive index of honey using a honey refractometer (Hanna, HI, USA). Initially, 2 g of honey was weighed and liquefied in an ultrasonic bath for a few minutes at 35 °C. Then, a small amount of liquefied, homogenized sample was spread on the surface of the prism and we marked the measurement on a percentage scale.

2.3.3. pH and Free Acidity

The determination of pH and free acidity was held by titration to pH 8.3 at 20 °C, according to the 'Harmonised Methods of the International Honey Commission (IHC)' [38]. A pH/conductivity device (Hanna, HI 9811-5, USA) was employed. The free acidity of honey measured the content of free acids, expressed in milliequivalents/kg honey (meq/kg).

2.3.4. Electrical Conductivity

Electrical conductivity at 20 °C was determined with the aforementioned portable pH/conductivity device, following the method of the IHC, except for the amount of honey which was reduced to 5 g (from 20 g in the initial method) as it was found to give the same results (data not shown). Results are expressed in mS/cm [38].

2.3.5. Color Analysis

Color measurements were performed using a color photometer (Hanna, HI 96785, USA), that measures the light transmittance of honey compared to analytical-grade glycerol. Honey samples of about 2 g were warmed gently in an ultrasound bath for a few minutes at 35 °C. The liquid honeys without air bubbles were transferred into a plastic cuvette and the color was read. Color grades were expressed in millimeters (mm) Pfund scale [24].

2.3.6. Hydroxymethylfurfural (HMF) after White

For the estimation of HMF content after White, a UV-VIS Spectrophotometer (Shimadzu, UV-1700, Japan) was employed according to the 'Harmonised Methods of the IHC' [38].

The method determines the concentration of 5-(hydroxymethyl-)furan-2-carbaldehyde. Results were expressed in milligrams per kilogram (mg/kg).

2.3.7. Diastase Activity (DN) after Schade

The procedure followed was in accordance with the respective method of IHC. For each sample of honey, a repeater was prepared to achieve accuracy in the result and calculate the repeatability (r). The unit of measurement of the enzyme is the unit DN (diastase number) which corresponds to the amount of enzyme needed for breaking down 0.01 g of starch in 1 h at 40 °C [38].

2.3.8. Sugars

For the determination of fructose, glucose and sucrose, a High-Performance Liquid Chromatography with Refractive Index Detector (HPLC-RID) technique was used. The preparation of the samples was achieved according to the literature [38], while for the separation and detection, the following parameters were used: column: Zorbax Carbohydrate Analysis (4.6 mm ID × 150 mm × 5 µm), temperature 35 °C, mobile phase: acetonitrile/water (75/25, v/v), flow rate: 1.8 mL min^{-1}. For the quantification, a five-point calibration curve was created and evaluated for each sugar.

2.4. Sensory Evaluation

For the sensory evaluation of samples, a two-round procedure was applied. At the first one the trained testers checked the botanical origin of the sample answering with Yes/No to the question: is the sample pine honey? At the second round, applied only for pine honeys, evaluating the visual, taste and aromatic characteristics, samples were classified in three levels: 3. very good, 2. medium, 1. good evaluated pine honey [39].

2.5. Bacterial Strains and Growth Conditions

All tested bacterial strains were isolated, identified and characterized by standard laboratory methods (kindly provided by Professor Spyros Pournaras, School of Medicine, University of Athens, Athens, Greece). Antibacterial activity of pine honeys was tested against *Acinetobacter baumannii*, *Klebsiella pneumoniae*, *Pseudomonas aeruginosa*, *Salmonella* ser. Typhimurium and *Staphylococcus aureus*. The bacteria were routinely grown in Mueller Hinton Broth or Mueller–Hinton Agar (Lab M, Bury, UK).

2.6. Determination of Minimum Inhibitory Concentration (MIC)

The determination of minimum inhibitory concentration (MIC) of tested honeys was carried out in sterile 96-well polystyrene microtiter plates (Kisker Biotech GmbH & Co. KG, Steinfurt, Germany) using a spectrophotometric bioassay as described by Tsavea and Mossialos [40]. Approximately 5×10^4 CFUs in 10 µL Mueller–Hinton broth was added to 190 µL of twofold diluted test honey (including manuka honey) at different concentrations which ranged from 25 to 0.78% (v/v). As control, Mueller–Hinton broth inoculated with bacteria was used. Optical density (OD) was determined at 630 nm using an ELx808 absorbance microplate reader (BioTek Instruments, Inc., Winooski, VT, USA), at t = 0 (prior to incubation) and after 24 h of incubation (t = 24) at 37 °C. The OD for each replicate well at t = 0 was subtracted from the OD of the same replicate well at t = 24. The following formula was used to determine the growth inhibition at each honey dilution: % inhibition = 1 − (OD test well/OD of corresponding control well) ×100. The MIC was determined as the lowest honey concentration which results in 100% growth inhibition [40]. MICs were determined in triplicates in at least two independent experiments.

2.7. Determination of Minimum Bactericidal Concentration (MBC)

Minimum bactericidal concentration (MBC) is the lowest concentration of any antibacterial agent that could kill tested bacteria. In order to determine the MBC, a small quantity of sample contained in each replicate well of the microtiter plates was transferred to Mueller–Hinton agar plates by using a microplate replicator (Boekel Scientific, Feasterville, PA, USA). The plates were incubated at 37 °C for 24 h. The MBC was determined as the lowest honey concentration at which no grown colonies were observed [41].

2.8. Determination of H_2O_2 Accumulation in Honey Samples

The ability to generate H_2O_2 in the diluted honey samples was determined using a Megazyme GOX assay kit (Megazyme International Ireland Ltd., Wicklow, Ireland) with some modification [27], which is based on the release of H_2O_2 after GOX catalysis of the oxidation of β-D-glucose to D-glucono-δ-lactone. As a standard, 9.8–312.5 µM diluted H_2O_2 was used. A total of 40% (w/w) of the honey solutions in 0.1 M potassium phosphate buffer (pH 7.0) were prepared and incubated for 24 h at 37 °C. Each honey sample and H_2O_2 standard were tested in duplicate in a 96-well microplate. The resultant H_2O_2 reacts with p-hydroxybenzoic acid and 4-aminoantipyrine in the presence of peroxidase to form a quinoneimine dye complex, which has a strong absorbance at 510 nm. The absorbance of reaction was then measured at 510 nm using a Synergy HT microplate reader (BioTek Instruments, Winooski, VT, USA).

2.9. Total Protein Content

Total protein content was measured using Quick Start™ Bradford reagent (Bio-Rad, Hercules, CA, USA) according to manufacturer's instructions.

2.10. Determining the Protein Profile of Honey Samples

For protein determination, 15 µL aliquots of diluted honey samples (50% w/w in distilled water) were loaded on 12% SDS-PAGE gels [Acrylamide/Bis solution, 37.5:1 (40% w/v), 2.6% C] and separated using a Mini-Protean II electrophoresis cell (Bio-Rad, Hercules, CA, USA). Protein content was assessed after gel staining with Coomassie Brilliant Blue R-250 (Sigma-Aldrich, Darmstadt, Germany).

2.11. Determination of the Antibacterial Activity Due to H_2O_2 and Proteinaceous Compounds

The MIC of honey treated with bovine catalase or proteinase K was determined and compared with that of untreated honey [40]. Briefly, 50% (v/v) honey in Muller–Hinton broth containing 100 µg/mL proteinase K (Blirt, Gdansk, Poland) or 600 U/mL bovine catalase (Serva, Heidelberg, Germany) was incubated for 16 h at 37 °C and then tested after being diluted twofold.

2.12. Total Phenolic Content (TPC) and Antioxidant Activity

2.12.1. Reagents

All the solvents used were of analytical grade. Methanol was purchased from Sigma-Aldrich Co (St. Louis, MO, USA). Sodium carbonate, L-ascorbic acid, 2,2-diphenylpicrylhydrazyl (DPPH.), 2,4,6-tris(2-pyridyl)-s-triazine (TPTZ), gallic acid (97.5%), fructose (>99.5%), glucose (>99.5%), maltose monohydrate (>99%), sucrose (>99.5%), Sigma-Aldrich Co (St. Louis, MO, USA). Folin–Ciocalteu (2N) phenol reagent, Iron (III) chloride hexahydrate and Iron (II) Sulfate heptahydrate ($FeSO_4 \cdot 7H_2O$) were from Honeywell-Fluka (Harvey St., Muskegon, MI, USA).

2.12.2. Sample Preparation

Five grams of honey was dissolved in 50 mL of distilled water. To eliminate the interferences of reducing sugars, the blank measurements were performed using an artificial honey containing: 40% fructose, 30% glucose, 8% maltose, 2% sucrose and 20% water [42]. All measurements were performed in triplicate ($n = 3$).

2.12.3. Total Phenolic Content (TPC)

The total phenol content was determined by a modified colorimetric assay using the Folin–Ciocalteu reagent [43,44]. Briefly, honey samples were dissolved in distilled water (0.1 g/mL), until a clear solution was obtained. Then, 100 µL of the honey solution was added to 1000 µL of Folin–Ciocalteu reagent previously diluted 1:10 with distilled water. After one minute, 300 µL of saturated sodium carbonate (Na_2CO_3) solution was added. The mixture was vortexed for 2 min, and the content was transferred into a 1.5 mL cuvette; absorbance was determined after one hour of incubation in the dark at 750 nm against a blank solution. A calibration curve was constructed with gallic acid (0.02–0.2 mg/mL) and the results were expressed as mg of gallic acid equivalents (GAE) per kg of honey.

2.12.4. Ferric Reducing Antioxidant Power (FRAP)

The reducing capacity of honey was estimated by the FRAP assay. This method involves the reduction of the ferric 2,4,6-tripyridyl-s-triazine complex (Fe^{+3}-TPTZ) to its ferrous, colored form (Fe^{+2}-TPTZ) in the presence of antioxidants [45]. The FRAP reagent consists of 2.5 mL of 10 mM of TPTZ (2,4,6-tripyridyl-s-triazine) solution in 40 mM HCl, 2.5 mL of 20 mM of $FeCl_3$ and 25 mL of 0.3 mM of acetate buffer, pH 3.6. Aliquots of 100 µL of the honey solution (0.1 g/mL) were mixed with 900 µL of FRAP reagent and the absorbance of the reaction mixture was measured at 593 nm after incubation at 37 °C for 30 min. A standard aqueous solution of $FeSO_4 \cdot 7H_2O$ (0.05–0.5 mM) was used for the construction of the calibration curve and the results were expressed as mmol Fe^{+2}/Kg of honey.

2.12.5. Antiradical Activity (DPPH)

The radical scavenging activity of honey was estimated spectrophotometrically using the stable free radical 2,2-diphenyl-1-picryl-hydrazile (DPPH). The principle of the method is based on the discoloration (purple) of DPPH solution in the presence of an antioxidant [46]. In short, 0.75 mL of DPPH solution (136 µM) was mixed with 0.375 mL of the honey solution (0.1 g/mL). The mixture was shaken vigorously and then incubated for 60 min at 25 °C in the dark; the absorbance of the remaining DPPH was determined at 517 nm against a blank solution. The scavenging activity was expressed as a percentage of absorbance reduction (RSA%) according to the following equation: RSA % = $[(A_{t=0} - A_{t=60})/A_{t=0}] \times 100$, where $A_{t=0}$ is the absorbance of the solution at t = 0 min and $A_{t=60}$ is the absorbance of the DPPH solution after 60 min of incubation [47]. All measurements of standards and samples were performed in triplicate.

2.13. Statistical Analysis

Correlation analysis was conducted using Spearman's correlation analysis. Data from Table S1 were tested to study any correlation among the variables. Values of $p < 0.05$ indicated statistically significant differences. Statistical analysis was performed using the SPSS version 13.0 statistical package (SPSS, Inc., Chicago, IL, USA).

3. Results and Discussion

3.1. Microscopic Examination

A significant number of honeydew elements were detected in all honeys. The great majority of the samples also contained fungal spores which are present solely in pine honeys [48]. The HDE/P ratio showed a great variation and was on average 8.23 ± 16.98, while the total number of pollen grains in 10 g of honey was 57 871 ± 55 867 (Tables 1 and S1). In about half of the samples, the HDE/P ratio was below 3, which is the minimum value for honeydew honeys proposed in the past [49]. However, later studies have shown that the HDE/P ratio of honeydew honeys may vary and Greek pine honeys have relatively few honeydew elements and large numbers of pollen grains [39,48,50–53]. Thus, the samples were considered typical Greek pine honeys.

Table 1. Mean values of physicochemical parameters.

	Mean	SD	Min	Max
Moisture (%)	16.1	1.0	14.0	18.2
pH	4.7	0.2	4.0	5.0
Free Acidity (meq/kg)	25.5	4.6	18	38
Electrical conductivity (mS/cm)	1.11	0.13	0.91	1.36
Color (mm Pfund)	91.9	15.9	60	118
HMF (mg/kg)	1.00	1.25	0	4.42
Diastase activity (DN)	20.90	8.71	6.95	37.29
Fructose (%)	31.81	2.64	25.9	39.2
Glucose (%)	25.94	4.30	14.4	33.2
Fructose + Glucose (%)	57.75	6.15	45.7	69.8
Fructose/Glucose	1.26	0.23	1.02	2.17
HDE/P	8.23	16.98	0.13	83.78
PG/10gr	57,871	55,867	4994	232,443
H_2O_2 (μM)	3341	921	1911	5620
Protein (μg/g of honey)	552	128	301	827
TPC (mgGAE/kg)	451.38	120.38	277.07	693.32
FRAP (mmol Fe^{+2}/kg)	4.91	1.85	1.87	9.43
DPPH (RSA%) [1]	60.11	9.21	42.43	79.33

[1] RSA: Radical Scavenging Activity. SD: Standard Deviation.

3.2. Physicochemical Parameters

The average physiochemical parameters of pine samples are presented in Table 1 (full data are available as Supplementary Materials Table S1). Moisture content, pH value and free acidity averaged 16.1 ± 1.0%, 4.7 ± 0.2 and 25.5 ± 4.6 meq/kg, respectively, values in accordance with previous publications regarding pine honey produced in Greece and Turkey [54–58]. Electrical conductivity values ranged between 0.91 and 1.31 mS/cm, averaging 1.11 ± 0.13 mS/cm. Moisture, free acidity, and electrical conductivity values are within legal requirements [6,7] for unifloral Greek pine honey (less than 20%, less than 50 meq/Kg honey and more than 0.9 mS/cm, respectively). Color values averaged 91.9 mm Pfund but presented high variability among samples (SD = 15.9). HMF was as low as 1 ± 1.25 mg/Kg of honey, with seven samples having no HMF at all, with these values being well below the legal limit of 40 mg/Kg. It is known that Greek pine honey has a low tendency to form HMF [59], which may be attributed to the less acidic nature of pine honey compared to other honey varieties. Fallico et al. [60] reported that, among others, pH values were correlated with the formation of HMF in honey and high pH honeys had a lower formation of HMF. Diastase activity averaged 20.9 ± 8.71 DN which is in accordance with previous works with Greek pine honey [56,57], but visibly higher than values reported for Turkish pine honey [55,58].

It should be stated that a certain number of samples showed relatively low diastase activity, two of them being slightly lower than the legislation limit of 8 DN (see Table S1). Honey enzyme is affected by storage and exposure to high temperatures [61]; however, this is not the case here as the samples are fresh. Fluctuation within a certain honey variety of fresh honey samples is known and can be explained by poor processing of nectar by the bees during an abundant nectar flow or seasonal activity of the pharyngeal glands [62], age of the bees, environmental conditions and beekeeping practices [61].

Fructose concentration was higher than glucose, ranging from 25.87 to 39.22%, while glucose from 14.38 to 33.20%. The average value for the sum of fructose and glucose was found to be 57.75 ± 6.15%, characteristic for a honeydew honey, like pine honey is. In all samples this value (min = 45.65%, max 69.76%) covers the legislation level for honeydew honeys "more than 45 g/100 g honey" [6]. Fructose/glucose ratio averaged 1.27 proving that pine honey has a moderate tendency to crystallize [63].

3.3. H_2O_2 Concentration and Total Protein Content of Pine Honeys

Antibacterial activity of honey is mainly attributed to accumulated H_2O_2 in diluted honey. The average value of accumulated H_2O_2 in 40% of honey solutions after 24 h was 3341 ± 921 µM and ranged from 1911 to 5620 µM (Table 1). The observed average ability of pine honey samples to generate high levels of H_2O_2 is substantially higher than in other types of honeydew honey, whereas total phenolic content was similar in both groups of honeydew honey samples [28]. In our previous study, the average value of H_2O_2 of Slovak honeydew honeys and blossom honeys was 1800 and 743 µM, respectively [27,28]. We assume that polyphenolic compounds, including specific flavonoids, contribute to higher levels of H_2O_2 by their pro-oxidant activities and/or the ability to act as electron acceptors in enzymatic reactions [28].

Apart from H_2O_2, defensin-1, an antibacterial bee-derived peptide found in honey, can contribute to the overall antibacterial activity of honey [30,64]. However, the effective concentration of defensin-1 at MIC of honeydew honeys is rather low and its contribution seems to be significant only in blossom honey [28].

Although the overall protein content can vary from honey to honey, the profile of the honey's most abundant proteins is similar among natural honey of different botanical and geographical origin. In the present study, the average of total honey protein content was 552 ± 128 µg/g, ranging from 301 to 827 µg/g (Table 1). The SDS-PAGE analysis showed an identical protein pattern among pine honeydew honeys where MRJP1 was the most dominant protein (Figure S2). These observations are in accordance with other studies where MRJP1 protein was found to be the most prominent band in all tested honeys, including medical-grade ones [65–67]. In fact, all major proteins identified in honey are of bee origin [66,68] and are secreted by hypopharyngeal glands into the nectar during collection and processing [69]. It has been proposed that some of these bee proteins and peptides including MRJP1 and defensin-1 could be considered as markers of honey quality and authenticity [30,70].

3.4. Total Phenolic Content (TPC) and Antioxidant Activity

Antioxidant activity is the outcome of different pathways which are not fully elucidated. Numerous assays are commonly employed to determine the antioxidant effect of a substrate. The lack of validated and standardized antioxidant protocols poses a challenge when authors compare their data [42,71]. Therefore, data regarding the antioxidant activity of pine honey are limited and discrepancies are observed.

In the present work, the total phenolic content was estimated to be 451.38 ± 120.38 mg GAE/kg, which is in line with the majority of previously published data. Cavrar et al. [72] studied pine honeys from northern Turkey and reported TPC values of 496 ± 148 mg GAE/kg. Similarly, Nayik et al. [73] determined the TPC of pine honeys from the Kashmir valley and found values of 598.4 ± 3.3 mg GAE/kg. In another study of Turkish pine honeys, Can et al. [33] found similar TPC values of 614.2 ± 55.9 mg GAE/kg. Higher TPC values were found by Karabagias et al. [74] (1583 ± 338 mg GAE/kg) regarding pine honey from Greece. In the lower range were the values by Ozkok et al. [75], reported for pine honeys from Turkey (155.55 ± 2.04 mg GAE/kg).

The antiradical activity of pine honeys was expressed as the percentage of the absorbance reduction (RSA%) of the stable DPPH radical at 517 nm. The values we measured ranged from 42.43 to 79.33% which are in excellent agreement with previously published data from Ekici et al. [76] (57.49 ± 20.15%) and Nayik et al. [73] (55.37 ± 6.8%). The reducing power of pine honeys was determined with the FRAP reagent and expressed as mmolFe^{+2}/kg. The values ranged from 1.87 to 9.43 mmol Fe^{+2}/kg, and in general were higher than those reported by Can et al. [33] (1.48 ± 0.83 mmol Fe^{+2}/kg).

3.5. Correlations of Parameters

Correlations were looked into among physicochemical parameters using Spearman's rho test and the results are presented in Table S2. Moisture content was negatively correlated with H_2O_2 value (r = −0.471, p = 0.013). Free acidity and pH were negatively correlated (r = −0.447, p = 0.019), which is logical since free acids contribute to a more acidic nature of honey. In addition, free acidity was positively correlated with several parameters, namely color (r = 0.577, p = 0.002), protein content (r = 0.556, p = 0.003), TPC (r = 0.592, p = 0.001), DPPH (r = 0.410, p = 0.034) and FRAP (r = 0.493, p = 0.009). Color has been found to correlate with TPC and DPPH [77]. Apart from free acidity, color was positively correlated with the electrical conductivity (r = 0.482, p = 0.011), which was expected since minerals considerably contribute to the color of honey [78] and has been demonstrated in the literature [79]. In addition, positive correlations of color were found with H_2O_2 (r = 0.381, p = 0.050), TPC (r = 0.399, p = 0.039) and FRAP (r = 0.547, p = 0.003). Dark honeys are known to have higher antibacterial potential, which is partly connected to the H_2O_2 value [80], yet this was not shown in our work. In addition, phenolic compounds contribute to honey color [81] and FRAP has been related to color intensity [82–84]. Diastase activity was negatively related to fructose (r= −0.425, p = 0.027) for no apparent reason, and positively to protein content (r = 0.590, p = 0.001), which was expected due to the protein nature of enzymes. The HDE/P ratio was negatively correlated with fructose content (r = −0.422, p = 0.028) and DPPH (r = −0.417, p = 0.030). H_2O_2 value was strongly positively correlated with protein content (r = 0.501, p = 0.008). Higher amount of protein could relate to a higher amount of glucose oxidase which is involved in the production of hydrogen peroxide. As expected, TPC, DPPH and FRAP showed a very strong positive correlation with each other (r = 0.637 and 0.684, p < 0.001). In the literature, correlations among these three parameters depend on the honey type [17,81,85]. These data indicate that the antioxidant activity is mainly attributed to phenolic compounds. FRAP values were positively correlated with fructose/glucose ratio (r = 0.412, p = 0.033) and H_2O_2 value (r = −0.497, p = 0.008). Finally, HMF was not correlated with any other physicochemical parameter.

3.6. Antibacterial Activity of Pine Honey

Nosocomial infections are a major cause of high morbidity and mortality both in developing and developed countries. Most common nosocomial pathogens include A. baumannii, P. aeruginosa, S. aureus and K. pneumoniae. Infections caused especially by hypervirulent strains of those pathogens are very difficult to treat due to multidrug resistance. Therefore, alternative therapeutic approaches to combat nosocomial infections are urgently needed [86]. On the other hand, S. Typhimurium-related serotypes are implicated in salmonellosis, the second most common gastrointestinal infection in Europe, thus leading to serious public health issues and economic losses in the food industry [87].

It is generally acknowledged that MIC measurement in broth is a more sensitive and quantitatively accurate method to study honey antimicrobial activity in comparison to an agar-well diffusion assay due to slower diffusion rates of active substances in agar [9,33]. Therefore, MICs of honeys were determined in broth using a spectrophotometric-based assay.

Honey samples exerted antibacterial activity against all tested bacterial strains. MIC and MBC values are presented in Table 2. MIC values of tested honeys against S. *aureus* varied from 3.125% (v/v) to 12.5% (v/v). Nine honeys exhibited comparable MIC values to manuka honey with an MIC of 3.125% (v/v). Regarding K. pneumoniae, the MIC values of tested honeys varied from 6.25% (v/v) to 12.5% (v/v) while manuka's MIC value has been determined at 6.25% (v/v). Thirteen honeys demonstrated an MIC value equal to manuka honey. Similarly, for honeys that tested against A. baumannii and S. Typhimurium, the MIC values varied from 6.25% (v/v) to 25% (v/v), whereas the MIC value of manuka honey has been determined at 6.25% (v/v). Five and twelve honey samples, respectively, demonstrated MIC values equal to manuka honey. Interestingly, MIC values of honeys tested against P. aeruginosa ranged from 6.25% (v/v) to 25% (v/v) whereas the MIC value of manuka has been determined at 12.5% (v/v), meaning that two honey samples exerted superior antibacterial activity against this particular pathogen, while twenty-one honeys were comparable to that of manuka.

Table 2. Antibacterial activity of pine honeys (n = 27) compared to manuka honey expressed as MIC and MBC values.

Honey Number	S. aureus		A. baumannii		K. pneumoniae		S. Typhimurium		P. aeruginosa	
	MIC	MBC	MIC	MBC	MIC	MBC	MIC	MBC	MIC	MBC
1	3.125	3.125	12.5	12.5	6.25	6.25	6.25	6.25	25	ND
2	3.125	3.125	12.5	12.5	6.25	6.25	6.25	6.25	25	ND
3	3.125	3.125	12.5	12.5	6.25	6.25	6.25	6.25	12.5	ND
4	3.125	3.125	12.5	12.5	6.25	6.25	6.25	12.5	25	ND
5	6.25	6.25	12.5	12.5	6.25	6.25	12.5	12.5	12.5	ND
6	12.5	12.5	25	25	12.5	12.5	12.5	12.5	12.5	ND
7	6.25	6.25	12.5	12.5	6.25	6.25	6.25	12.5	12.5	ND
8	3.125	3.125	12.5	12.5	6.25	12.5	12.5	12.5	12.5	ND
9	6.25	6.25	12.5	12.5	12.5	12.5	6.25	6.25	12.5	ND
10	6.25	6.25	12.5	12.5	12.5	12.5	12.5	12.5	12.5	ND
11	12.5	12.5	6.25	6.25	6.25	6.25	6.25	6.25	12.5	ND
12	3.125	3.125	25	25	6.25	6.25	6.25	6.25	12.5	ND
13	6.25	6.25	25	25	6.25	6.25	6.25	6.25	12.5	ND
14	6.25	6.25	6.25	6.25	6.25	6.25	12.5	12.5	12.5	ND
15	6.25	6.25	12.5	12.5	12.5	12.5	12.5	12.5	12.5	ND
16	3.125	3.125	6.25	6.25	6.25	6.25	6.25	6.25	6.25	ND
17	6.25	6.25	12.5	12.5	12.5	12.5	12.5	12.5	12.5	ND
18	6.25	6.25	12.5	12.5	12.5	12.5	12.5	12.5	12.5	ND
19	3.125	3.125	12.5	12.5	6.25	6.25	6.25	6.25	6.25	ND
20	6.25	6.25	12.5	12.5	12.5	12.5	12.5	12.5	12.5	ND
21	6.25	6.25	12.5	12.5	12.5	12.5	12.5	12.5	12.5	25
22	12.5	12.5	12.5	12.5	12.5	12.5	12.5	12.5	12.5	12.5
23	12.5	12.5	12.5	12.5	12.5	12.5	12.5	12.5	12.5	12.5
24	3.125	3.125	6.25	6.25	12.5	12.5	25	25	25	25
25	6.25	6.25	12.5	12.5	12.5	12.5	6.25	6.25	12.5	12.5
26	6.25	6.25	12.5	12.5	12.5	12.5	12.5	12.5	12.5	12.5
27	6.25	6.25	6.25	6.25	12.5	12.5	12.5	12.5	12.5	12.5
Manuka	3.125	3.125	6.25	6.25	6.25	6.25	6.25	6.25	12.5	12.5

Values expressed as % (v/v). ND: Not Determined.

The variation in MICs could possibly reflect differential bacterial susceptibility due to distinct antibacterial mechanisms. Furthermore, it has been shown that *S. aureus*, a Gram-positive bacterium was, in general, more susceptible to honey and other bee products compared to Gram-negative bacteria [12,88], which is in accordance with the present study. However, in a recent study that tested blossom honeys from the Greek island of Lemnos, it was demonstrated that Gram-positive bacteria were more resistant compared to the Gram-negative bacteria [89].

In order to find out whether honey samples exert bacteriostatic or bactericidal activity, MBC was determined (Table 2). The MBC values of all tested honey, including manuka honey, against all tested bacterial strains were identical to the MIC values, demonstrating that pine honey kills bacteria, not just inhibits their growth.

Overall, the antibacterial activity exerted by pine honeys, especially of those honeys demonstrating superior or comparable activity to manuka, warrants further investigation.

3.7. Antibacterial Activity of Pine Honey Could Be Attributed to Multiple Mechanisms

The underlying mechanisms that could contribute to exerted antibacterial activity were further assessed in those honey samples demonstrating comparable or superior antibacterial activity to manuka. In that respect, catalase-treated honeys demonstrated higher MIC values up to 16-fold in some cases (some honeys against *S. aureus*, for instance) compared to untreated samples (Figure S3). Of note, two honey samples tested against *P. aeruginosa* did not demonstrate higher MICs after catalase treatment.

Similarly, after proteinase K treatment, 2 out of 14 honey samples tested against *P. aeruginosa*, 9 out of 12 against *S.* Typhimurium, 10 out of 14 against *S. aureus* and 6 out of 11 against *K. pneumoniae* demonstrated higher MICs up to 4-fold. Surprisingly, no increase in MIC value was observed for all tested honeys against *A. baumannii* after proteinase K treatment, indicating that proteins present in honey that might exert antibacterial activity have no effect on this certain pathogen (Figure S3).

Spearman's analysis was performed to assess the correlation between the physicochemical parameters and antibacterial activity. It is shown (Table 3) that no correlation between pH, HMF content, H_2O_2 concentration and the antibacterial activity was observed. Surprisingly, there is a statistically significant positive correlation of moisture and MIC and MBC values against *K. pneuomoniae*, indicating that moisture negatively affects the antibacterial activity against *K. pneuomoniae*. A statistically significant positive correlation of antibacterial and antioxidant activity was observed for *K. pneuomoniae* and *S.* Typhimurium. Interestingly, a statistically significant negative correlation was observed between DN and *S. aureus's* MIC and MBC values, indicating that higher diastase activity correlates with higher antimicrobial activity against *S. aureus*.

Table 3. Correlation coefficient (r) and significance (parenthesis) values calculated by Spearman's correlation analysis.

	S. aureus		*A. baumannii*		*K. pneumoniae*		*S.* Typhimurium		*P. aeruginosa*
	MIC	MBC	MIC	MBC	MIC	MBC	MIC	MBC	MIC
Moisture	0.363 (0.058)	0.363 (0.058)	0.148 (0.451)	0.148 (0.451)	0.481 ** (0.010)	0.408 * (0.031)	0.371 (0.052)	0.367 (0.054)	0.157 (0.424)
pH	−0.036 (0.855)	−0.036 (0.855)	0.137 (0.487)	0.137 (0.487)	0.117 (0.552)	0.127 (0.518)	−0.110 (0.579)	−0.213 (0.277)	−0.171 (0.385)
HMF	−0.092 (0.641)	−0.092 (0.641)	−0.065 (0.741)	−0.065 (0.741)	−0.327 (0.089)	−0.371 (0.052)	−0.345 (0.072)	−0.272 (0.161)	−0.135 (0.493)
DN	−0.584 ** (0.001)	−0.584 ** (0.001)	−0.097 (0.624)	−0.097 (0.624)	−0.155 (0.431)	−0.161 (0.414)	−0.194 (0.324)	−0.262 (0.178)	−0.010 (0.960)
H_2O_2	−0.242 (0.215)	−0.242 (0.215)	−0.297 (0.125)	−0.297 (0.125)	0.137 (0.486)	0.232 (0.234)	0.155 (0.430)	0.098 (0.620)	−0.216 (0.269)
Protein	−0.420 * (0.026)	−0.420 * (0.026)	−0.197 (0.316)	−0.197 (0.316)	−0.137 (0.486)	−0.089 (0.651)	−0.125 (0.528)	−0.035 (0.861)	−0.174 (0.377)
TPC	0.051 (0.795)	0.051 (0.795)	−0.285 (0.142)	−0.285 (0.142)	0.288 (0.137)	0.411 * (0.030)	0.316 (0.102)	0.346 (0.072)	−0.101 (0.609)
DPPH	0.306 (0.113)	0.306 (0.113)	−0.365 (0.056)	−0.365 (0.056)	0.554 ** (0.002)	0.554 ** (0.002)	0.381 * (0.045)	0.371 (0.052)	−0.190 (0.334)
FRAP	0.136 (0.490)	0.136 (0.490)	−0.202 (0.302)	−0.202 (0.302)	0.563 ** (0.002)	0.688 ** (0.000)	0.432 * (0.022)	0.421 * (0.026)	−0.216 (0.270)

* Correlation is statistically significant at the 0.05 level; ** Correlation is statistically significant at the 0.01 level.

Overall, our data indicate multiple mechanisms of antibacterial activity exerted by pine honey. This is further supported by our recent study, whereas RNA−sequencing analysis revealed that pine honey affected the transcriptomic profile of *P. aeruginosa* by inducing the expression of 189 genes and by suppressing the expression of 274 genes [90]. Pine honey treatment exerted a broad range of action on several pathways and biological processes, including oxidative stress, transmembrane transport and regulation of DNA-templated transcription, two-component regulatory systems, ABC transporters and SOS response. Interestingly, pine honey downregulates key physiological responses in *P. aeruginosa* such as quorum sensing, bacterial chemotaxis and biofilm formation [90].

4. Conclusions

In this study, 27 pine honeydew samples showed physicochemical and microscopic characteristics within the legal limits, except for diastase activity, a parameter known to be highly variable depending on many factors. The ability of pine honeydew samples to generate high levels of H_2O_2 is substantially higher than in other types of honeydew honey, whereas protein content was similar. Furthermore, due to their high polyphenol content, a strong antioxidant activity of honey samples was demonstrated. In addition, various correlations were identified among these parameters.

The antibacterial activity of pine honeydew honey samples was variable and MICs of honey solutions varied from 3.125 to 25% depending on the pathogen. The breakdown of H_2O_2 by catalase treatment into honey solution resulted in a significant decrease in antibacterial activity. Similarly, the digestion of honey proteins by proteinase K resulted in lower antibacterial efficacy among honey samples, again depending on specific bacteria. Interestingly, the antibacterial activity of proteinase K-treated honey samples against *A. baumannii* was not affected at all. Taken together, these observations suggest multiple underlying mechanisms of antibacterial activity of pine honeydew honeys.

Supplementary Materials: The following supporting information can be downloaded at: https://www.mdpi.com/article/10.3390/foods11070943/s1, Table S1: Full data of each tested honey, Table S2: Correlation coefficient (r, first value) and significance (second value) of physicochemical parameters of pine honey calculated by Spearman's correlation analysis, Figure S1: Sensory three−point scale evaluation of pine honey samples, Figure S2: Protein profile of pine honeydew honey samples ($n = 27$) from Greece, Figure S3: MIC values after proteinase K (MICp) and bovine catalase (MICc) treatment of honey samples.

Author Contributions: Conceptualization, D.M., P.G. and E.A.; methodology, D.K., S.G., M.B., J.G. and P.G.; validation, formal analysis, P.G., E.A., M.B., J.G., J.M., E.T., D.M. and T.G.D.; investigation, E.T., F.-P.V., E.S., A.K., D.K., M.D., S.L., I.R. and C.T. (Christina Tsadila); resources, D.M. and E.A.; data curation, P.G.; writing—original draft preparation, D.M., S.G., E.A., T.G.D. and J.M.; writing—review and editing, D.M., S.G., P.G., E.A., T.G.D. and J.M.; supervision, C.T. (Chrysoula Tananaki), S.G., P.G., E.A. and D.M.; project administration, C.T. (Chrysoula Tananaki); funding acquisition, C.T. (Chrysoula Tananaki), P.G., E.A., D.M. and J.M. All authors have read and agreed to the published version of the manuscript.

Funding: This research was supported by the General Secretariat for Research and Innovation (G.S.R.I.) under the national emblematic action "Honeybee Routes" and partially by the Scientific Grant Agency of the Ministry of Education of the Slovak Republic and the Slovak Academy of Sciences VEGA 2/0022/22.

Institutional Review Board Statement: Not applicable.

Informed Consent Statement: Not applicable.

Data Availability Statement: Not applicable.

Conflicts of Interest: The authors declare no conflict of interest. The funders had no role in the design of the study; in the collection, analyses or interpretation of data; in the writing of the manuscript; or in the decision to publish the results.

References

1. Machado De-Melo, A.A.; de Almeida-Muradian, L.B.; Sancho, M.T.; Pascual-Maté, A. Composition and properties of Apis mellifera honey: A review. *J. Apic. Res.* **2018**, *57*, 5–37. [CrossRef]
2. Pita-Calvo, C.; Vázquez, M. Honeydew Honeys: A Review on the Characterization and Authentication of Botanical and Geographical Origins. *J. Agric. Food Chem.* **2018**, *66*, 2523–2537. [CrossRef] [PubMed]
3. Tarapatskyy, M.; Sowa, P.; Zaguła, G.; Dżugan, M.; Puchalski, C. Assessment of the botanical origin of polish honeys based on physicochemical properties and bioactive components with chemometric analysis. *Molecules* **2021**, *26*, 4801. [CrossRef] [PubMed]
4. Bacandritsos, N.; Saitanis, C.; Papanastasiou, I. Morphology and Life cycle of Marchalina Hellenica (Gennadius) (Hemiptera: Margarodidae) on Pine (Parnis Mt.) and Fir (Helmos Mt.) Forests of Greece. In *Annales de la Société Entomologique de France*; Taylor & Francis Group: Abingdon, UK, 2004; pp. 169–176. [CrossRef]
5. Tananaki, C.; Thrasyvoulou, A.; Giraudel, J.L.; Montury, M. Determination of volatile characteristics of Greek and Turkish pine honey samples and their classification by using Kohonen self organising maps. *Food Chem.* **2007**, *101*, 1687–1693. [CrossRef]
6. European Council. Council directive 2001/110/EC of 20 December 2001 relating honey. *Off. J. Eur. Communities* **2002**, *45*, 47–52.
7. AXS Decision. Decision of the Greek Higher Chemical Commission Regarding the Characteristics of Monofloral Honey Pine, Fir, Castanea, Erica, Thymus, Citrus, Cotton and Helianthus. FEK 239/B/23-2-2005 2004, 127. Available online: http://www.et.gr/index.php/anazitisi-fek (accessed on 2 March 2022).
8. Combarros-Fuertes, P.; Estevinho, L.M.; Teixeira-Santos, R.; Rodrigues, A.G.; Pina-Vaz, C.; Fresno, J.M.; Eugenia Tornadijo, M. Antibacterial action mechanisms of honey: Physiological Effects of Avocado, Chestnut, and Polyfloral Honey upon Staphylococcus aureus and Escherichia coli. *Molecules* **2020**, *25*, 1252. [CrossRef]
9. Godocikova, J.; Bugarova, V.; Kast, C.; Majtan, V.; Majtan, J. Antibacterial potential of Swiss honeys and characterisation of their bee-derived bioactive compounds. *J. Sci. Food Agric.* **2020**, *100*, 335–342. [CrossRef]
10. Grabek-Lejko, D.; Słowik, J.; Kasprzyk, I. Activity of selected honey types against *Staphylococcus aureus* methicillin susceptible (MSSA) and methicillin resistant (MRSA) bacteria and its correlation with hydrogen peroxide, phenolic content and antioxidant capacity. *Farmacia* **2018**, *66*, 37–43.
11. Anthimidou, E.; Mossialos, D. Antibacterial Activity of Greek and Cypriot Honeys Against *Staphylococcus aureus* and *Pseudomonas aeruginosa* in Comparison to Manuka Honey. *J. Med. Food* **2013**, *16*, 42–47. [CrossRef]
12. Stagos, D.; Soulitsiotis, N.; Tsadila, C.; Papaeconomou, S.; Arvanitis, C.; Ntontos, A.; Karkanta, F.; Adamou-Androulaki, S.; Petrotos, K.; Spandidos, D.A.; et al. Antibacterial and antioxidant activity of different types of honey derived from Mount Olympus in Greece. *Int. J. Mol. Med.* **2018**, *42*, 726–734. [CrossRef]
13. Majkut, M.; Kwiecińska-Piróg, J.; Wszelaczyńska, E.; Pobereżny, J.; Gospodarek-Komkowska, E.; Wojtacki, K.; Barczak, T. Antimicrobial activity of heat-treated Polish honeys. *Food Chem.* **2021**, *343*. [CrossRef] [PubMed]
14. Piotrowski, M.; Karpiński, P.; Pituch, H.; van Belkum, A.; Obuch-Woszczatyński, P. Antimicrobial effects of Manuka honey on in vitro biofilm formation by Clostridium difficile. *Eur. J. Clin. Microbiol. Infect. Dis.* **2017**, *36*, 1661–1664. [CrossRef] [PubMed]
15. Sowa, P.; Grabek-Lejko, D.; Wesołowska, M.; Swacha, S.; Dżugan, M. Hydrogen peroxide-dependent antibacterial action of *Melilotus albus* honey. *Lett. Appl. Microbiol.* **2017**, *65*, 82–89. [CrossRef] [PubMed]
16. Velásquez, P.; Montenegro, G.; Leyton, F.; Ascar, L.; Ramirez, O.; Giordano, A. Bioactive compounds and antibacterial properties of monofloral Ulmo honey. *CyTA-J. Food* **2020**, *18*, 11–19. [CrossRef]
17. Akgün, N.; Çelik, Ö.F.; Kelebekli, L. Physicochemical properties, total phenolic content, and antioxidant activity of chestnut, rhododendron, acacia and multifloral honey. *J. Food Meas. Charact.* **2021**, *15*, 3501–3508. [CrossRef]
18. Bueno-Costa, F.M.; Zambiazi, R.C.; Bohmer, B.W.; Chaves, F.C.; da Silva, W.P.; Zanusso, J.T.; Dutra, L. Antibacterial and antioxidant activity of honeys from the state of Rio Grande do Sul, Brazil. *LWT-Food Sci. Technol.* **2016**, *65*, 333–340. [CrossRef]
19. Ramón-Sierra, J.M.; Villanueva, M.A.; Yam-Puc, A.; Rodríguez-Mendiola, M.; Arias-Castro, C.; Ortiz-Vázquez, E. Antimicrobial and antioxidant activity of proteins isolated from Melipona beecheii honey. *Food Chem. X* **2022**, *13*, 100177. [CrossRef]
20. Smetanska, I.; Alharthi, S.S.; Selim, K.A. Physicochemical, antioxidant capacity and color analysis of six honeys from different origin. *J. King Saud Univ.-Sci.* **2021**, *33*, 101447. [CrossRef]
21. Majtan, J.; Bucekova, M.; Kafantaris, I.; Szweda, P.; Hammer, K.; Mossialos, D. Honey antibacterial activity: A neglected aspect of honey quality assurance as functional food. *Trends Food Sci. Technol.* **2021**, *118*, 870–886. [CrossRef]
22. Kwakman, P.H.S.; Zaat, S.A.J. Antibacterial components of honey. *IUBMB Life* **2012**, *64*, 48–55. [CrossRef]
23. Bucekova, M.; Valachova, I.; Kohutova, L.; Prochazka, E.; Klaudiny, J.; Majtan, J. Honeybee glucose oxidase-Its expression in honeybee workers and comparative analyses of its content and H_2O_2-mediated antibacterial activity in natural honeys. *Naturwissenschaften* **2014**, *101*, 661–670. [CrossRef] [PubMed]
24. Ramos, O.Y.; Salomón, V.; Libonatti, C.; Cepeda, R.; Maldonado, L.; Basualdo, M. Effect of botanical and physicochemical composition of Argentinean honeys on the inhibitory action against food pathogens. *LWT-Food Sci. Technol.* **2018**, *87*, 457–463. [CrossRef]
25. Farkasovska, J.; Bugarova, V.; Godocikova, J.; Majtan, V.; Majtan, J. The role of hydrogen peroxide in the antibacterial activity of different floral honeys. *Eur. Food Res. Technol.* **2019**, *245*, 2739–2744. [CrossRef]
26. Grecka, K.; Kuś, P.; Worobo, R.; Szweda, P. Study of the Anti-Staphylococcal Potential of Honeys Produced in Northern Poland. *Molecules* **2018**, *23*, 260. [CrossRef] [PubMed]

27. Bucekova, M.; Jardekova, L.; Juricova, V.; Bugarova, V.; Di Marco, G.; Gismondi, A.; Leonardi, D.; Farkasovska, J.; Godocikova, J.; Laho, M.; et al. Antibacterial Activity of Different Blossom Honeys: New Findings. *Molecules* **2019**, *24*, 1573. [CrossRef] [PubMed]
28. Bucekova, M.; Buriova, M.; Pekarik, L.; Majtan, V.; Majtan, J. Phytochemicals-mediated production of hydrogen peroxide is crucial for high antibacterial activity of honeydew honey. *Sci. Rep.* **2018**, *8*, 1–9. [CrossRef] [PubMed]
29. Brudzynski, K.; Sjaarda, C. Honey glycoproteins containing antimicrobial peptides, jelleins of the Major Royal Jelly Protein 1, are responsible for the cell wall lytic and bactericidal activities of honey. *PLoS ONE* **2015**, *10*, e0120238. [CrossRef] [PubMed]
30. Valachová, I.; Bučeková, M.; Majtán, J. Quantification of bee-derived peptide defensin-1 in honey by competitive enzyme-linked immunosorbent assay, a new approach in honey quality control. *Czech J. Food Sci.* **2016**, *34*, 233–243. [CrossRef]
31. Ramlan, N.A.F.M.; Zin, A.S.M.; Safari, N.F.; Chan, K.W.; Zawawi, N. Application of heating on the antioxidant and antibacterial properties of malaysian and australian stingless bee honey. *Antibiotics* **2021**, *10*, 1365. [CrossRef]
32. Hau-Yama, N.E.; Magaña-Ortiz, D.; Oliva, A.I.; Ortiz-Vázquez, E. Antifungal activity of honey from stingless bee Melipona beecheii against Candida albicans. *J. Apic. Res.* **2020**, *59*, 12–18. [CrossRef]
33. Can, Z.; Yildiz, O.; Sahin, H.; Akyuz Turumtay, E.; Silici, S.; Kolayli, S. An investigation of Turkish honeys: Their physico-chemical properties, antioxidant capacities and phenolic profiles. *Food Chem.* **2015**, *180*, 133–141. [CrossRef] [PubMed]
34. Karadag, A.; Ozcelik, B.; Saner, S. Review of methods to determine antioxidant capacities. *Food Anal. Methods* **2009**, *2*, 41–60. [CrossRef]
35. Mărgăoan, R.; Topal, E.; Balkanska, R.; Yücel, B.; Oravecz, T.; Cornea-Cipcigan, M.; Vodnar, D.C. Monofloral Honeys as a Potential Source of Natural Antioxidants, Minerals and Medicine. *Antioxidants* **2021**, *10*, 1023. [CrossRef] [PubMed]
36. Von Der Ohe, W.; Persano Oddo, L.; Piana, M.L.; Morlot, M.; Martin, P. Harmonized methods of melissopalynology. *Apidologie* **2004**, *35*, S18–S25. [CrossRef]
37. Yang, Y.; Battesti, M.-J.; Djabou, N.; Muselli, A.; Paolini, J.; Tomi, P.; Costa, J. Melissopalynological origin determination and volatile composition analysis of Corsican "chestnut grove" honeys. *Food Chem.* **2012**, *132*, 2144–2154. [CrossRef]
38. Bogdanov, S.; Luellmann, C.; Martin, P. Harmonised methods of the International Honey Commission. *Int. Honey Comm.* **2009**, 1–63. Available online: https://www.ihc-platform.net/ihcmethods2009.pdf (accessed on 22 February 2022).
39. Piana, M.L.; Persano Oddo, L.; Bentabol, A.; Bruneau, E.; Bogdanov, S.; Guyot Declerck, C. Sensory analysis applied to honey: State of the art. *Apidologie* **2004**, *35*, S26–S37. [CrossRef]
40. Tsavea, E.; Mossialos, D. Antibacterial activity of honeys produced in Mount Olympus area against nosocomial and foodborne pathogens is mainly attributed to hydrogen peroxide and proteinaceous compounds. *J. Apic. Res.* **2019**, *58*, 756–763. [CrossRef]
41. Szweda, P. Antimicrobial Activity of Honey. In *Honey Analysis*; de Toledo, V.D.A., Ed.; InTech: Rijeka, Croatia, 2017; ISBN 978-953-51-2880-9.
42. Bertoncelj, J.; Doberšek, U.; Jamnik, M.; Golob, T. Evaluation of the phenolic content, antioxidant activity and colour of Slovenian honey. *Food Chem.* **2007**, *105*, 822–828. [CrossRef]
43. Beretta, G.; Granata, P.; Ferrero, M.; Orioli, M.; Facino, R.M. Standardization of antioxidant properties of honey by a combination of spectrophotometric/fluorimetric assays and chemometrics. *Anal. Chim. Acta* **2005**, *533*, 185–191. [CrossRef]
44. Lachman, J.; Orsák, M.; Hejtmánková, A.; Kovářová, E. Evaluation of antioxidant activity and total phenolics of selected Czech honeys. *LWT-Food Sci. Technol.* **2010**, *43*, 52–58. [CrossRef]
45. Benzie, I.F.F.; Strain, J.J. The ferric reducing ability of plasma (FRAP) as a measure of "antioxidant power": The FRAP assay. *Anal. Biochem.* **1996**, *239*, 70–76. [CrossRef] [PubMed]
46. Sharma, O.P.; Bhat, T.K. DPPH antioxidant assay revisited. *Food Chem.* **2009**, *113*, 1202–1205. [CrossRef]
47. Perna, A.; Simonetti, A.; Intaglietta, I.; Sofo, A.; Gambacorta, E. Metal content of southern Italy honey of different botanical origins and its correlation with polyphenol content and antioxidant activity. *Int. J. Food Sci. Technol.* **2012**, *47*, 1909–1917. [CrossRef]
48. Dimou, M.; Katsaros, J.; Klonari, K.T.; Thrasyvoulou, A. Discriminating pine and fir honeydew honeys by microscopic characteristics. *J. Apic. Res.* **2006**, *45*, 16–21. [CrossRef]
49. Louveaux, J.; Maurizio, A.; Vorwohl, G. Methods of Melissopalynology. *Bee World* **1978**, *59*, 139–157. [CrossRef]
50. Diez, M.J.; Andres, C.; Terrab, A. Physicochemical parameters and pollen analysis of Moroccan honeydew honeys. *Int. J. Food Sci. Technol.* **2004**, *39*, 167–176. [CrossRef]
51. Tsigouri, A.; Passaloglou-Katrali, M.; Sabatakou, O. Palynological characteristics of different unifloral honeys from Greece. *Grana* **2004**, *43*, 122–128. [CrossRef]
52. Karabagias, I.K.; Badeka, A.; Kontakos, S.; Karabournioti, S.; Kontominas, M.G. Characterisation and classification of Greek pine honeys according to their geographical origin based on volatiles, physicochemical parameters and chemometrics. *Food Chem.* **2014**, *146*, 548–557. [CrossRef]
53. Persano Oddo, L.; Piro, R. Main European unifloral honeys: Descriptive sheets. *Apidologie* **2004**, *35*, S38–S81. [CrossRef]
54. Xagoraris, M.; Lazarou, E.; Kaparakou, E.H.; Alissandrakis, E.; Tarantilis, P.A.; Pappas, C.S. Botanical origin discrimination of Greek honeys: Physicochemical parameters versus Raman spectroscopy. *J. Sci. Food Agric.* **2021**, *101*, 3319–3327. [CrossRef]
55. Beykaya, M. Determination of physicochemical properties of raw honey samples. *Prog. Nutr.* **2021**, *23*, e2021020. [CrossRef]
56. Thrasyvoulou, A.; Manikis, J. Some physicochemical and microscopic characteristics of Greek unifloral honeys. *Apidologie* **1995**, *26*, 441–452. [CrossRef]
57. Rodopoulou, M.; Tananaki, C.; Kanelis, D.; Liolios, V.; Dimou, M.; Thrasyvoulou, A. A chemometric approach for the differentiation of 15 monofloral honeys based on physicochemical parameters. *J. Sci. Food Agric.* **2022**, *102*, 139–146. [CrossRef]

58. Özcan, M.M.; Ölmez, Ç. Some qualitative properties of different monofloral honeys. *Food Chem.* **2014**, *163*, 212–218. [CrossRef] [PubMed]
59. Thrasyvoulou, A.T. The Use of HMF and Diastase as Criteria of Quality of Greek Honey. *J. Apic. Res.* **1986**, *25*, 186–195. [CrossRef]
60. Fallico, B.; Zappalà, M.; Arena, E.; Verzera, A. Effects of conditioning on HMF content in unifloral honeys. *Food Chem.* **2004**, *85*, 305–313. [CrossRef]
61. Belay, A.; Haki, G.D.; Birringer, M.; Borck, H.; Lee, Y.-C.; Kim, K.-T.; Baye, K.; Melaku, S. Enzyme activity, amino acid profiles and hydroxymethylfurfural content in Ethiopian monofloral honey. *J. Food Sci. Technol.* **2017**, *54*, 2769–2778. [CrossRef]
62. Persano Oddo, L.; Baldi, E.; Accorti, M. Diastatic activity in some unifloral honeys. *Apidologie* **1990**, *21*, 17–24. [CrossRef]
63. Dettori, A.; Tappi, S.; Piana, L.; Dalla Rosa, M.; Rocculi, P. Kinetic of induced honey crystallization and related evolution of structural and physical properties. *LWT* **2018**, *95*, 333–338. [CrossRef]
64. Proaño, A.; Coello, D.; Villacrés-Granda, I.; Ballesteros, I.; Debut, A.; Vizuete, K.; Brenciani, A.; Álvarez-Suarez, J.M. The osmotic action of sugar combined with hydrogen peroxide and bee-derived antibacterial peptide Defensin-1 is crucial for the antibiofilm activity of eucalyptus honey. *LWT* **2021**, *136*, 110379. [CrossRef]
65. Bucekova, M.; Bugarova, V.; Godocikova, J.; Majtan, J. Demanding New Honey Qualitative Standard Based on Antibacterial Activity. *Foods* **2020**, *9*, 1263. [CrossRef] [PubMed]
66. Mureșan, C.I.; Cornea-Cipcigan, M.; Suharoschi, R.; Erler, S.; Mărgăoan, R. Honey botanical origin and honey-specific protein pattern: Characterization of some European honeys. *LWT* **2022**, *154*, 112883. [CrossRef]
67. Bucekova, M.; Majtan, J. The MRJP1 honey glycoprotein does not contribute to the overall antibacterial activity of natural honey. *Eur. Food Res. Technol.* **2016**, *242*, 625–629. [CrossRef]
68. Di Girolamo, F.; D'Amato, A.; Righetti, P.G. Assessment of the floral origin of honey via proteomic tools. *J. Proteom.* **2012**, *75*, 3688–3693. [CrossRef]
69. Lewkowski, O.; Mureșan, C.I.; Dobritzsch, D.; Fuszard, M.; Erler, S. The Effect of Diet on the Composition and Stability of Proteins Secreted by Honey Bees in Honey. *Insects* **2019**, *10*, 282. [CrossRef]
70. Zhang, Y.; Chen, Y.; Cai, Y.; Cui, Z.; Zhang, J.; Wang, X.; Shen, L. Novel polyclonal antibody-based rapid gold sandwich immunochromatographic strip for detecting the major royal jelly protein 1 (MRJP1) in honey. *PLoS ONE* **2019**, *14*, e0212335. [CrossRef]
71. Badarinath, A.V.; Mallikarjuna Rao, K.; Madhu Sudhana Chetty, C.; Ramkanth, S.; Rajan, T.V.S.; Gnanaprakash, K. A review on In-vitro antioxidant methods: Comparisions, correlations and considerations. *Int. J. PharmTech Res.* **2010**, *2*, 1276–1285.
72. Cavrar, S.; Yıldız, O.; Şahin, H. Comparison of Physical and Biochemical Characteristics of Different Quality of Turkish Honey. *Uludag Bee J.* **2013**, *13*, 55–62.
73. Nayik, G.A.; Suhag, Y.; Majid, I.; Nanda, V. Discrimination of high altitude Indian honey by chemometric approach according to their antioxidant properties and macro minerals. *J. Saudi Soc. Agric. Sci.* **2018**, *17*, 200–207. [CrossRef]
74. Karabagias, I.K.; Karabagias, V.K.; Papastephanou, C.; Badeka, A.V. New insights into the typification of Hellenic monofloral honeys using selected physico-chemical and bio-chemical indicators coupled with z score analysis and chemometric models. *Eur. Food Res. Technol.* **2021**, *247*, 169–182. [CrossRef]
75. Özkök, A.; D'arcy, B.; Sorkun, K. Total Phenolic Acid and Total Flavonoid Content of Turkish Pine Honeydew Honey. *J. ApiProd. ApiMed. Sci.* **2010**, *2*, 65–71. [CrossRef]
76. Ekici, L.; Sagdic, O.; Silici, S.; Ozturk, I. Determination of phenolic content, antiradical, antioxidant and antimicrobial activities of Turkish pine honey. *Qual. Assur. Saf. Crops Foods* **2014**, *6*, 439–444. [CrossRef]
77. Hailu, D.; Belay, A. Melissopalynology and antioxidant properties used to differentiate Schefflera abyssinica and polyfloral honey. *PLoS ONE* **2020**, *15*, e0240868. [CrossRef]
78. Kędzierska-Matysek, M.; Teter, A.; Stryjecka, M.; Skałecki, P.; Domaradzki, P.; Rudaś, M.; Florek, M. Relationships Linking the Colour and Elemental Concentrations of Blossom Honeys with Their Antioxidant Activity: A Chemometric Approach. *Agriculture* **2021**, *11*, 702. [CrossRef]
79. Bodó, A.; Radványi, L.; Kőszegi, T.; Csepregi, R.; Nagy, D.U.; Farkas, Á.; Kocsis, M. Quality Evaluation of Light- and Dark-Colored Hungarian Honeys, Focusing on Botanical Origin, Antioxidant Capacity and Mineral Content. *Molecules* **2021**, *26*, 2825. [CrossRef]
80. Alygizou, A.; Grigorakis, S.; Gotsiou, P.; Loupassaki, S.; Calokerinos, A.C. Quantification of Hydrogen Peroxide in Cretan Honey and Correlation with Physicochemical Parameters. *J. Anal. Methods Chem.* **2021**, *2021*, 1–7. [CrossRef]
81. Becerril-Sánchez, A.L.; Quintero-Salazar, B.; Dublán-García, O.; Escalona-Buendía, H.B. Phenolic Compounds in Honey and Their Relationship with Antioxidant Activity, Botanical Origin, and Color. *Antioxidants* **2021**, *10*, 1700. [CrossRef]
82. Khalil, M.I.; Moniruzzaman, M.; Boukraâ, L.; Benhanifia, M.; Islam, M.A.; Islam, M.N.; Sulaiman, S.A.; Gan, S.H. Physicochemical and Antioxidant Properties of Algerian Honey. *Molecules* **2012**, *17*, 11199–11215. [CrossRef]
83. Islam, A.; Khalil, I.; Islam, N.; Moniruzzaman, M.; Mottalib, A.; Sulaiman, S.A.; Gan, S.H. Physicochemical and antioxidant properties of Bangladeshi honeys stored for more than one year. *BMC Complement. Altern. Med.* **2012**, *12*, 177. [CrossRef]
84. Anand, S.; Pang, E.; Livanos, G.; Mantri, N. Characterization of Physico-Chemical Properties and Antioxidant Capacities of Bioactive Honey Produced from Australian Grown Agastache rugosa and its Correlation with Colour and Poly-Phenol Content. *Molecules* **2018**, *23*, 108. [CrossRef] [PubMed]

85. Hunter, M.; Ghildyal, R.; D'Cunha, N.M.; Gouws, C.; Georgousopoulou, E.N.; Naumovski, N. The bioactive, antioxidant, antibacterial, and physicochemical properties of a range of commercially available Australian honeys. *Curr. Res. Food Sci.* **2021**, *4*, 532–542. [CrossRef] [PubMed]
86. Khan, H.A.; Baig, F.K.; Mehboob, R. Nosocomial infections: Epidemiology, prevention, control and surveillance. *Asian Pac. J. Trop. Biomed.* **2017**, *7*, 478–482. [CrossRef]
87. Antunes, P.; Mourão, J.; Campos, J.; Peixe, L. Salmonellosis: The role of poultry meat. *Clin. Microbiol. Infect.* **2016**, *22*, 110–121. [CrossRef]
88. Didaras, N.A.; Kafantaris, I.; Dimitriou, T.G.; Mitsagga, C.; Karatasou, K.; Giavasis, I.; Stagos, D.; Amoutzias, G.D.; Hatjina, F.; Mossialos, D. Biological Properties of Bee Bread Collected from Apiaries Located across Greece. *Antibiotics* **2021**, *10*, 555. [CrossRef]
89. Gkoutzouvelidou, M.; Panos, G.; Xanthou, M.N.; Papachristoforou, A.; Giaouris, E. Comparing the Antimicrobial Actions of Greek Honeys from the Island of Lemnos and Manuka Honey from New Zealand against Clinically Important Bacteria. *Foods* **2021**, *10*, 1402. [CrossRef]
90. Kafantaris, I.; Tsadila, C.; Nikolaidis, M.; Tsavea, E.; Dimitriou, T.G.; Iliopoulos, I.; Amoutzias, G.D.; Mossialos, D. Transcriptomic Analysis of Pseudomonas aeruginosa Response to Pine Honey via RNA Sequencing Indicates Multiple Mechanisms of Antibacterial Activity. *Foods* **2021**, *10*, 936. [CrossRef]

Article

DNA-Based Method for Traceability and Authentication of *Apis cerana* and *A. dorsata* Honey (Hymenoptera: Apidae), Using the *NADH dehydrogenase 2* Gene

Saeed Mohamadzade Namin [1,2,†], Fatema Yeasmin [3,†], Hyong Woo Choi [3] and Chuleui Jung [1,3,*]

1. Agricultural Science and Technology Institute, Andong National University, Andong 36729, Korea; saeedmn2005@gmail.com
2. Department of Plant Protection, Faculty of Agriculture, Varamin-Pishva Branch, Islamic Azad University, Varamin 3381774895, Iran
3. Department of Plant Medicals, Andong National University, Andong 36729, Korea; fatema.setudu@gmail.com (F.Y.); hwchoi@anu.ac.kr (H.W.C.)
* Correspondence: cjung@andong.ac.kr
† These authors contributed equally to this work.

Abstract: Honey is a widely used natural product and the price of honey from *Apis cerana* (ACH) and *A. dorsata* (ADH) is several times more expensive than the one from *A. mellifera* (AMH), thus there are increasing fraud issues reported in the market by mislabeling or mixing honeys with different entomological origins. In this study, three species-specific primers, targeting the *NADH dehydrogenase 2* (*ND2*) region of honeybee mitochondrial DNA, were designed and tested to distinguish the entomological origin of ACH, ADH, and AMH. Molecular analysis showed that each primer set can specifically detect the *ND2* region from the targeted honeybee DNA, but not from the others. The amplicon size for *A. cerana*, *A. dorsata* and *A. mellifera* were 224, 302, and 377 bp, respectively. Importantly, each primer set also specifically produced amplicons with expected size from the DNA prepared from honey samples with different entomological origins. The PCR adulteration test allowed detection of 1% of AMH in the mixture with either ACH or ADH. Furthermore, real-time PCR and melting curve analysis indicated the possible discrimination of origin of honey samples. Therefore, we provide the newly developed PCR-based method that can be used to determine the entomological origin of the three kinds of honey.

Keywords: honey; entomological origin; mitochondrial DNA; *NADH dehydrogenase 2*; PCR

Citation: Mohamadzade Namin, S.; Yeasmin, F.; Choi, H.W.; Jung, C. DNA-Based Method for Traceability and Authentication of *Apis cerana* and *A. dorsata* Honey (Hymenoptera: Apidae), Using the *NADH dehydrogenase 2* Gene. *Foods* **2022**, *11*, 9285. https://doi.org/10.3390/foods11070928

Academic Editor: Alessandra Bendini

Received: 28 February 2022
Accepted: 22 March 2022
Published: 23 March 2022

Publisher's Note: MDPI stays neutral with regard to jurisdictional claims in published maps and institutional affiliations.

Copyright: © 2022 by the authors. Licensee MDPI, Basel, Switzerland. This article is an open access article distributed under the terms and conditions of the Creative Commons Attribution (CC BY) license (https://creativecommons.org/licenses/by/4.0/).

1. Introduction

Honey is a sweet natural product produced by honey bees using the nectar, secretions of living parts, or honeydew of plants [1,2]. Due to the broader geographical distribution, *Apis mellifera* honey (AMH), *A. cerana* honey (ACH), and *A. dorsata* honey (ADH) are the three dominant types of honey in the Asian market. Giant honeybee (*Apis dorsata* F.) is distributed throughout South and Southeast Asia and China [3]. The colonies of *A. dorsata* are generally found in rainforests or on the cliffs, but they also can be occasionally found in building ledges of urban areas. Even though *A. dorsata* is not domesticated and cannot be maintained for honey harvest or pollination purposes, it plays an important role in the pollination of tropical rainforest plants and local crops [4–6]. Due to the fact that *A. dorsata* is considered the most defensive honeybee compared to other *Apis* spp., ADH is harvested by highly motivated experts (so-called honey hunters) [7]. Furthermore, ADH contains the highest concentration of phenolic compounds and flavonoids compared to other honeys, thus exhibiting high DPPH (2,2-diphenyl-1-picrylhydrazy) free radical-scavenging activity, FRAP (Ferric reducing-antioxidant power assay) values and the lowest AEAC (Ascorbic acid Equivalent Antioxidant Capacity) values, as well. This indicates that ADH has strong antioxidant properties and medical values [8,9]. Asian honeybee (*A. cerana* F.) is widespread in South, South East, and Eastern Asia from Afghanistan to Far East Russia and Japan [10]. It is one of the domesticated honeybees; however, due to its lower productivity compared to *A. mellifera*, most beekeepers prefer *A. mellifera* over *A. cerana* [11]. In addition, high interspecific competition between *A. mellifera* and *A. cerana* on the same niche resulted in the decline of *A. cerana* colonies in many countries such as China and Korea in the last decades [11–13]. This led to lower production of ADH and ACH compared to AMH.

Although honey is one of the most widely consumed natural products, it is one of the most counterfeited food products in the market [14]. Due to the fast growth of the human population and the rising demands toward the consumption of organic and local products, the entomological origin of honey has been taken into consideration. Therefore, the market price of ADH and ACH is several times higher than AMH. This situation makes ADH and ACH vulnerable to adulteration problems, either by mislabeling (claiming the false geographical, botanical, or entomological origin of honey) or by mixing (overfeeding the bees with sugars, adding sweeteners or syrups, and dilution with cheaper honey) in order to increase the economic profit [2,3,15–19]. Thus, it is important to develop rapid, reliable, and cost-effective identification methods for the entomological origin of honey to solve the adulteration problem in the market.

Molecular detection of the entomological origin of honey by using the set of specific primers is regarded as a rapid, accurate, and suitable tool for the identification of the origin of animal products and processed foods [20–23]. Considering the method of processing honey by honeybees, the bee cells can remain inside of the honey. Given the opportunity to extract bee DNA from honey, it is possible to use it for the identification of the entomological origin of honey. Compared to the other identification methods for entomological origin of honey, such as SDS-PAGE or chemical-based methods [12,24,25], the DNA-based method is more precise, quick, and suitable for analysis of a large sample size [26]. Recently, several studies were conducted using DNA-based methods to identify the entomological origin of honey. Zhang et al. [26] developed a gDNA-based method for the identification of two different major honeys, AMH and ACH, in the market. Two sets of primers were designed to amplify *Major royal jelly protein 2 (MRJP2)* gene, resulting in the different sizes of PCR product in the gel electrophoresis, making it useful to discriminate ACH from AMH. In addition, it is also possible to identify the honey samples through Real-Time PCR based on their melting temperature analysis [26].

Mitochondrial DNA (mtDNA) is present in most cells with high copy numbers. It is characterized by a high genetic variation between related species but a low intraspecific variation [27–29]. Therefore, it is suitable to use mtDNA for taxonomic and phylogenetic analysis. Targeting the *cytochrome oxidase I (COI)* gene of mtDNA, Kim et al. [30] designed species-specific primers to differentiate ACH and AMH. PCR with designed primer sets produced amplicons with a length of 133bp and 178 bp for *A. mellifera* and *A. cerana*, respectively. Although the size of the amplicons is distinguishable and even applicable for relatively old honey samples, Zhang et al. [26] reported the designed primers for *A. mellifera* was not species-specific as they made the same length of band from ACH-originated DNA extracts in China. Soares et al. [31] developed species-specific primers to amplify the intergenic region of tRNAleu-cox2, enabling the detection of *A. cerana* DNA using PCR. In addition, they discriminated ACH and AMH using high-resolution melting curve analysis targeting the 16S rRNA gene, making it possible to detect the entomological origin of ACH. However, the lack of species-specific primer designed for mtDNA of *A. mellifera* makes it difficult to use it for adulteration studies. The only species-specific primer set that is available to detect *A. mellifera* is provided by Zhang et al. [26]. However, the size of the PCR product (~560 bp) is largely applied to relatively old honey samples due to DNA degradation, and it is important to design species-specific primers targeting smaller regions. On the other hand, there is no species-specific primer available for reliable and cost-effective identification of the entomological origin of ADH.

In this study, we aimed to develop a rapid and accurate PCR-based method to recognize the entomological origin of ADH, ACH, and AMH. This method can also be applied to discriminate between pure and adulterated honey. We also aimed to provide species-specific primers targeting smaller parts of mtDNA to avoid the negative effect of possible DNA degradation, which may happen during the storage of honey. In this study, three species-specific primers for ADH, ACH, and AMH were designed to amplify the short part of the *NADH dehydrogenase 2 (ND2)* region of the mtDNA. Our experiment suggests that species-specific primer sets targeting *ND2* not only successfully distinguished ADH, ACH,s and AMH, but also detected mixed 1% AMH from ADH or ACH. Additionally, several honey samples from different countries were used to evaluate the accuracy of the developed method.

2. Material and Methods

2.1. Schematic Overview of the Experimental Design

In this study, 3 species-specific primer sets were designed to test the traceability of the entomological origin of honey. The specificity and sensitivity of the primers were tested first with the DNA extracts from honeybees with different geographical origins. Then, the DNA extracts from artificially mixed honey samples were used to evaluate the applicability of using designed primer sets in honey authentication. Subsequently, the developed method was used to check the entomological origin of honey provided by honey hunters and beekeepers (Figure 1).

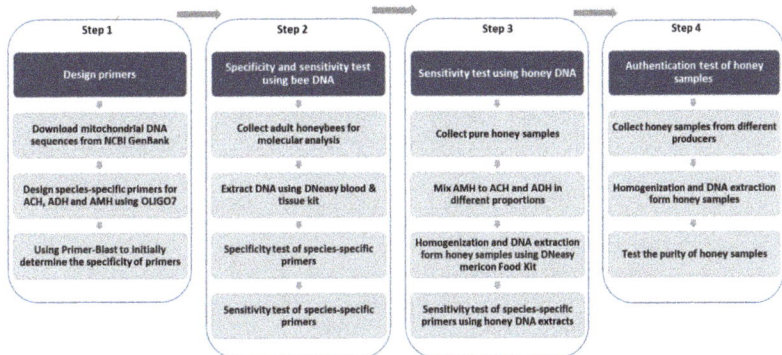

Figure 1. Schematic overview of the experimental design.

2.2. Designing Species-Specific Primers

NADH dehydrogenase 2 (ND2) region of mitochondrial DNA was used as a target area. The complete mitochondrial genome sequence of *A. cerana*, *A. dorsata* and, *A. mellifera* were obtained from NCBI (Table S2) and used for designing species-specific primers using OLIGO 7 primer analysis software (Table 1). The Primer-BLAST tool was initially used to determine primer specificity (http://www.ncbi.nlm.nih.gov/tools/primer-blast) accessed on 25 March 2020). Designed primers were synthesized by Macrogen (Daejan, Korea).

Table 1. Specific-primers for honey identification, nucleotide sequence, primer length, and the expected length of PCR product.

Species	Primer	5'-3'	Length	Target Fragment
A. cerana	AC-F	TCATTAGATTTTACAAAATCAGATCA	26	224 bp
	AC-R	CTTATAACTAAATATGTTAATGATCATA	28	
A. dorsata	AD-F	TATATTAATTGTTATAACTTACATAAATAA	31	302 bp
	AD-R	GGATTAAGAATATATAATATTCATATTTT	29	
A. mellifera	AM-F	CTATTAGATTTACTAAAACAGATACT	26	377 bp
	AM-R	ATAATTAAATGAATATAAAATAATTATAGCA	31	

2.3. Evaluate the Specificity and Sensitivity of Designed Primer Sets Using Bee DNA

- DNA extraction from honeybees

DNA extracts from adult or larvae samples of honeybees of *A. cerana* (5 adults from Nepal, 5 adults from Korea), *A. dorsata* (5 larvae from Thailand, 5 adults from Nepal) and *A. mellifera* (5 adults from Nepal, 5 adults from Korea) were used to test the specificity and sensitivity of the primer sets. Bee DNA was extracted from the left hind leg of adult honeybee or head and thorax part of larvae using DNeasy blood and tissue kit (Qiagen, Hilden, Germany) following manufacturer's instruction.

- Specificity and sensitivitytest of designed primers

The DNA extracts from honeybees with different geographical origins were used to examine the specificity of the designed species-specific primers. The PCR procedure was carried out in a 20 µL reaction volume mixture containing 100 ng of template DNA and 1µL of each primer (10 pmole/µL) using AccuPower PCR PreMix (Bioneer, Daejan, Korea). The thermal cycling procedure contained an initial pre-denaturation at 95 °C for 5 min, and 35 cycles of 95 °C for 30 s, 52 °C for 30 s, 72 °C for 40 s, and a final extension of 72 °C for 5 min in a BIOER thermal cycler. 7 µL of PCR products were analyzed using 2.5% agarose gel in TAE buffer and the bands were visualized by EcoDye (BIOFACT) and gel document system (GSD-200D).

To evaluate the detection limit of the species-specific primers, DNA extracts from different bee samples were serially diluted by 10-fold (100 to 0.01 ng/µL) and used for PCR analysis.

- Melting curve analysis by real-time PCR

DNA extract from honeybees were used to evaluate the possibility of using real-time PCR-based detection of adulteration of ACH and ADH. The real-time PCR was carried out using 10 µL of 2X Real-Time PCR Master Mix (BioFACT) including SYBR Green I, 100 ng of DNA template and 1 µL of each primer (10 pmole/µL) in 20 µL of total reaction mixture. PCR cycling was as follows: 95 °C for 15 min, following 35 cycles of 95 °C for 30 s, 52 °C for 30 s and 72 °C for 40 s. For analyzing melt curve, real-time PCR products were denatured at 95 °C for 15 s, annealed at 52 °C for 1 min then followed by melting curve ranging from 52 to 95 °C with temperature increments of 0.3 °C every 20 s. The data of real-time PCR and melt curve analysis were processed using FQD-96a V1.0.13 software (BIOER, Hangzhou, China).

2.4. PCR-Based Sensitivity Test of Honey Samples

- Preparation of honey samples

Three pure honey samples (ACH and AMH from Korea and ADH from Thailand) were used to test the sensitivity of species-specific primers to detect honey adulteration. AMH was mixed with ACH or ADH in different proportions (100:0, 99:1, 95:5, 50:50, 20:180, 0:100), and then used for DNA extraction.

- DNA extraction from honey samples

To extract DNA from honey samples, 40 mL distilled water was added to 15 g of honey, incubated at 45 °C for 30 min, vortexed and centrifuged at 15,000 rpm for 30 min. The supernatant was discarded, and the pellet was dissolved in 1 mL of distilled water and centrifuged at 15,000 rpm for 15 min. The supernatant was discarded, and the pellet was used for DNA extraction using DNeasy mericon Food Kit (Qiagen, Hilden, Germany) following manufacturer's protocol. The concentration and purity of the DNA extracts were evaluated using Nano Drop spectrophotometer (Life Real). Extracted DNA was used for the subsequent PCR analysis.

- Polymerase chain reaction (PCR)

For sensitivity (or adulteration) test, 2 rounds of PCR were performed with same primer sets. In the first round of PCR, 100 ng of DNA was used as a template DNA following the same protocol described above. Then, 5 µL of PCR product was used as a template DNA for the second round of PCR with the same protocol described above. A total of 7 µL of final PCR products were analyzed in 2.5% agarose gel. In addition, PCR using DNA extracts from honey samples (10 with ACH and 5 with ADH labels) was conducted for honey adulteration test using the species-specific primers. Two rounds of PCR were performed as described above. To confirm the amplified DNA sequence, PCR products were analyzed in 2.5% agarose gel, purified and sequenced by Macrogen (Daejan, Korea) using an ABI 3130xl capillary automated.

2.5. Adulteration Analysis of the Honey Samples

- Honey samples

The purity of 20 honey samples (10 ACH, 5 ADH and 5 AMH) from different localities (Nepal, Thailand and Korea) were evaluated. ACH samples were provided by beekeepers from Nepal ($n = 2$), Thailand ($n = 5$) and Korea ($n = 3$). ADH samples were provided by honey hunters from Thailand ($n = 5$). AMH samples were harvested directly by beekeepers from Korea ($n = 5$) (Table S1). All samples were collected in 2020. Honey samples were stored at -20 °C and 4 °C prior to DNA extraction, respectively.

- DNA extraction and PCR-based authentication of honey samples

DNA was extracted from all honey samples using the method that was described before. PCR using DNA extracts from honey samples (10 with ACH and 5 with ADH labels) was conducted to check honey adulteration using the species-specific primers. There were 100 ng of DNA used for the first PCR and 5 µL of PCR product used as a template DNA in the second PCR following the procedure described before. PCR products were analyzed in 2.5% agarose gel and sequenced by Macrogen (Daejan, Korea) using an ABI 3130xl capillary automated. All sequences were generated in both directions and the forward and reverse sequences were assembled in BIOEDIT v7.0.5.2 (Hall, 1999) to produce a consensus sequence for each sample and the assembled sequences generated in this study were used to confirm the identification through DNA barcoding and have been deposited in GenBank under accession numbers MW660861-MW660880.

- Data analysis

From melting curve analysis, melting temperatures for 3 species of honey bees were compared by one-way analysis of variance (ANOVA) followed by Tukey's post-hoc test. p values less than 0.05 were considered to be statistically significant. The statistical analysis was conducted using The R project software version 4.0.5 [32].

3. Results

3.1. Specificity Test of Species-Specific Primers

The DNA extracts from different honeybees with different geographical origins were used to examine the specificity of the designed species-specific primers. Each primer set successfully amplified *ND2* region from the DNA samples extracted from *A. cerana*, *A. dorsata*, and *A. mellifera* with an amplicon size of 224, 302, and 377 bp, respectively (Figure 2). None of the non-specific DNA amplification was observed with tested primer sets, suggesting these three species-specific primers can be successfully used to distinguish the origin of the honeybee at the DNA level.

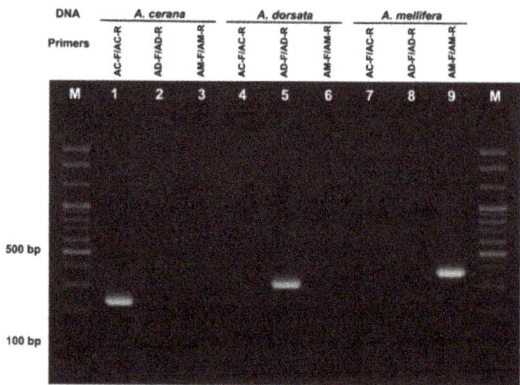

Figure 2. Agarose gel electrophoresis of PCR products amplified from DNA extracted of honeybees with species-specific primers. Bee DNA (Lanes 1–3: *A. cerana* DNA, Lanes 4–6: *A. dorsata* DNA, Lanes 7–9: *A. mellifera* DNA) Primers (Lanes 1, 4 and 7: *A. cerana* specific primers AC-F/AC-R, Lanes 2, 5 and 8: *A. dorsata* specific primers AD-F/AD-R, Lanes 3, 6 and 9: *A. mellifera* specific primers AM-F/AM-R), M: 100 bp ladder.

3.2. Sensitivity Test of Primers Using Bee and Honey DNA

To evaluate the sensitivity of PCR-based assay, DNA samples from 3 different bees (*A. cerana*, *A. dorsata* and *A. mellifera*) were serially diluted (100 to 0.01 ng/μL) and used for PCR. The PCR condition using primer sets showed that all primer sets are able to amplify the specific bands (Figure 3). From the *A. dorsata* DNA, the AD-F/AD-R primer set successfully amplified the band with the expected size (302 bp) (Figure 3A). The intensity of characteristic bands was gradually raised as the concentration of DNA template increased, and the band could be visible when the DNA template was as low as 0.1 ng. From the *A. mellifera* (Figure 3B) and *A. cerana* DNA, similarly, AM-F/AM-R and AC-F/AC-R primer sets were also able to amplify specific bands with a detection limit of total 0.1 ng template DNA in the PCR reaction. This suggests that our species-specific primers can be used to detect the origin of honeybee samples with a low amount of DNA.

In other to test the ability to detect the target DNA among pure and adulterated honey samples, AMH was mixed with either ADH or ACH in different proportions. DNA was extracted from pure and mixed honey and used for the subsequent PCR analysis. Importantly, although the same amount of DNA (100 ng) from honey and bee samples was used for PCR, we were not able to detect the specific band from the first round of PCR with DNA from honey, unlike with DNA from bees (Figures 2 and 3). This is likely due to actual amount of bee DNA being lower in DNA extracted from honey, as the honey sample contains biological tissues of other organisms (e.g., plant, microorganism, and other insect tissues). Thus, we performed another round of PCR by using 5 μL of PCR product as a template for analyzing the honey samples. In the second round of PCR, AC-F/AC-R and AD-F/AD-R primer sets produced a single band at the expected size with DNA from 100% ACD (Figure 4A, lane 1) and 100% ADH (Figure 4B, lane 1), respectively. On the contrary, the AM-F/AM-R primer set failed to amplify the band from the DNA extracted from 100% ACH (Figure 4A, lane 2) or 100% ADH (Figure 4B, lane 2). Neither AC-F/AC-R nor AD-F/AD-R primer sets amplified the specific bands from DNA extracted from 100% AMH (Figure 4A,B, lane 11).

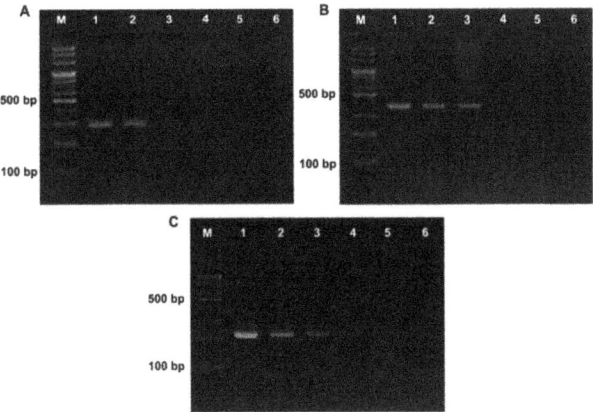

Figure 3. Sensitivity test of the designed species-specific primers using serially diluted DNA extract of *A. dorsata* (**A**), *A. mellifera* (**B**), and *A. cerana* (**C**). Lane M, DNA marker; Lane 1, 100 ng; lane 2, 10 ng; lane 3, 1 ng; lane 4, 0.1 ng; lane 5, 0.01 ng; lane 6, negative control.

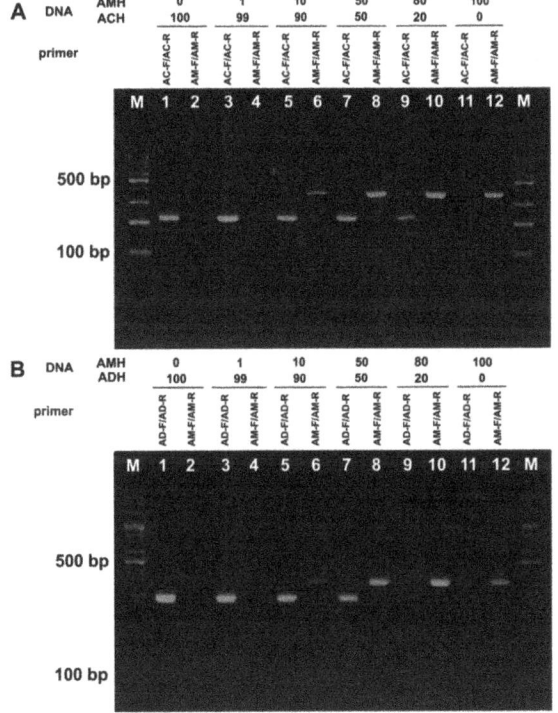

Figure 4. Adulteration test with artificially mixed honey samples. PCR products amplified from DNA extracted either from mixtures of ACH and AMH (**A**) or ADH and AMH (**B**) were analyzed by DNA gel electrophoresis. The proportions of AMH inside either ACH or ADH and usage of species-specific primer sets are shown. AC-F/AC-R, *A. cerana* specific primers; AD-F/AD-R, *A. dorsata* specific primers; AM-F/AM-R, *A. mellifera* specific primers; M, 1 kb ladder.

DNA form ACH and ADH with different concentrations of AMH were also tested with species-specific primers (Figure 4A,B, lanes 3–12). In both conditions, the species-specific band of for A. mellifera was visible when the concentration of AMH was as low as 1%. The intensity of A. mellifera species-specific band gradually increased when the DNA from mixed honey with a proportion of 1 to 50% AMH were used and remained constant up to 100% AMH. To examine the possibility of using species-specific primer sets in the practical adulteration assay, 20 honey samples labeled as ADH, ACH, and AMH from different localities were tested. Analysis of the sequences of PCR products indicated that the primer sets are specific enough to detect the entomological origin of honey from different geographical localities.

3.3. Melting Curve Analysis by Real-Time PCR

To evaluate the possibility of use of melting curve analysis for detecting ACH or ADH adulteration, a real-time PCR experiment was conducted using the same PCR condition and primer sets and DNA extracted from honeybees. The result was confirmed using agarose gel electrophoresis and sequencing. Melting curve analysis of real-time PCR products demonstrated two distinct curves allowing the discrimination of A. dorsata from A. mellifera (Figure 5A) and of ADH from AMH. The melting temperature (Tm) of amplicons generated from A. dorsata (69.2 ± 0.1 °C) was distinct from the A. mellifera (72.4 ± 0.1 °C); hence, the detection of Tm could be an alternative method to detect the origin of ADH in addition to standard PCR method. Melting curve analyses of PCR products between A. cerana and A. mellifera were also performed (Figure 5B). Tm of amplicons of A. cerana (71.9 ± 0.2 °C) was distinct from A. mellifera (72.4 ± 0.1 °C) but very similar; hence, the use of Tm for distinguishing A. cerana and A. mellifera need more caution. The results of one-way ANOVA indicated that there was a significant difference between the Tm values of all three species (F value = 325.2, p-value < 0.001).

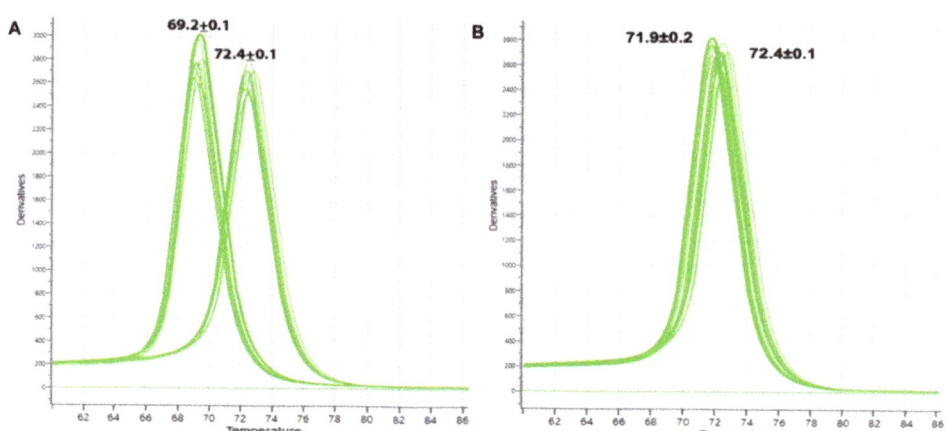

Figure 5. Conventional melting curves obtained by real-time PCR amplification targeting ND2 region of mtDNA using DNA extracts from honeybees. (**A**) A. dorsata (Tm = 69.2 ± 0.1) and A. mellifera (Tm = 72.4 ± 0.1). (**B**) A. cerana (Tm = 71.9 ± 0.2) and A. mellifera (Tm = 72.4 ± 0.1).

4. Discussion

Although previous attempts based on DNA barcoding of 16S rRNA and COI genes were helpful to inspect mislabeling [33], it was not functional to detect honey adulteration. In spite of the availability of species-specific primers to differentiate ACH from AMH [30], the primers developed by Kim et al. [30] were only applicable to honey originated from Korea but failed to differentiate ACH and AMH originated from China [26]. Since Soares et al. [31] only developed species-specific primers (AC1-F/AC1-R) to amplify

111 bp of tRNAlux-cox2 intergenic region of *A. cerana* mtDNA, the new *A. mellifera* species-specific primers were needed for the adulteration test. In addition, although AC1-F/AC1-R primers were useful in the discrimination of ACH from AMH, unlike ACF2/ACR2, they also amplified the non-specific band from ADH DNA extract, suggesting AC1-F/AC1-R was not enough to distinguish ACH from ADH (Supplementary Figure S1). In this study, we provided not only the first species-specific primer set to identify ADH, but also two new species-specific primer sets to identify ACH and AMH. Notably, our newly designed primers successfully amplified specific bands only from the targeted DNA sample and were able to discriminate both ACH and ADH from AMH, and vice versa, thus providing new DNA-based assay for testing entomological origin of honey (refer to Figure 4A,B).

Zhang et al. [26] developed a gDNA-based method for the identification of two different major honeys from domesticated honeybees in the market. Two sets of primers (C-F/C-R for *A. cerana* and M-F/M-R for *A. mellifera*) were designed to amplify the Major royal jelly protein 2 (MRJP2) gene, resulting in the different size of PCR product in the gel electrophoresis, making it useful to discriminate ACH from AMH. In addition, it is possible to identify the honey samples through real-time PCR-based Tm analysis. Although the predicted size of the PCR product was 212 bp for *A. cerana* and 393 bp for *A. mellifera*, but the length of amplicons for *A. mellifera* was 560 bp from the PCR, as the primers were designed based on complementary DNA (cDNA) without an intron. ACH and AMH samples were distinguishable using C-F/C-R and M-F/M-R primer sets; however, 560 bp tends to be long for accurate honey identification and adulteration test with relatively old honey samples, which can possibly have DNA degradation problems. Honey is a complex matrix, and its phenolic/H_2O_2 induced oxidative stress would lead the DNA that remained inside honey to be easily degraded as storage time increased [31,34–36]. Notably, Schnell et al. [37] reported the diminished rate of successful amplification of amplicon size larger than 380 bp and in the fragmented DNA, thus the amplification of short amplicons is preferable [38]. Thus our new species-specific primer sets with amplicon size ranging 224~377 bp (refer to Figures 2 and 3) would provide a better chance to successfully examine the old honey samples. Although the speed of degradation of DNA inside honey is not well understood and very difficult to predict accurately as different honey have different biochemical compositions, DNA degradation problem needs to be considered while examining the entomological origin of honey via DNA-based assay.

Real-time PCR-based identification of the entomological origin of honey was successfully developed previously to discriminate ACH and AMH using species-specific primers [26,31]. Refer to the Figure 5, the primer sets developed in the present study can be used to differentiate the entomological origin of three different types of honey. Although ADH and AMH can be simply differentiated using melting curve analysis, this method should be applied to differentiate ACH and AMH with caution due to the close Tm of the amplicons. Tm-based identification method is quick and accurate without the requirement of the gel electrophoresis step. Thus, it will provide a possible high-throughput analyses method for the identification of the origin of honey.

The cost of conducting analysis for one honey sample using the combination of two-round PCR and subsequent gel electrophoresis using the methodology described in current research is 6.4$ per honey sample (DNA extraction kit, PCR master mix, agarose powder, ladder, TBE buffer, staining dye and primer cost), however, these expenses for authentication analysis using Real-Time PCR technic is about 5.2$ (DNA extraction, master mix, and primer cost). The electricity and labor cost required to run the equipment have not been considered in our calculation. Although the cost of the authentication analysis per sample is slightly lower using Real-Time PCR, it is more expensive to establish such facilities in comparison to the conventional PCR method. Furthermore, according to our calculations, the duration of analysis using Real-Time PCR is slightly longer (~35 min) than conventional PCR.

5. Conclusions

Three species-specific primer sets targeting the *NADH dehydrogenase 2 (ND2)* region of mtDNA were designed and successfully applied to trace the entomological origin of honey produced by different honeybees, *A. cerana*, *A. dorsata* and *A. mellifera*. In addition, the *A. mellifera* specific primer set is applicable in honey fraud detection. The possibility of using melting curve analysis in discrimination of the origin of honey using the same primer sets is also confirmed. Our preliminary studies indicated the impossibility of providing species-specific primers with a smaller size of PCR product in the mitochondrial DNA (except the one provided by Soares et al. [7] for ACH). However, further studies targeting nuclear DNA are required. PCR-based method using species-specific primers provides a rapid and cost-effective method to screen the entomological origin of honey. Therefore, the development of new primer sets to identify honey produced by other species of honeybees will be valuable. On the other hand, more studies are needed to understand the pace of DNA degradation in honey and the applicability and limitations of using molecular methods in the authentication of older honey samples.

Supplementary Materials: The following supporting information can be downloaded at: https://www.mdpi.com/article/10.3390/foods11070928/s1, Table S1: The honey samples and their labels and localities used for analysis. Table S2: Mitochondrial sequences of honeybees (*A. cerana*, *A. dorsata* and *A. mellifera*) used to design species-specific primers. Figure S1: Preliminary specificity test of ACF/ACR (Suarez et al., 2018) using DNA extracts of honeybees [39–57].

Author Contributions: S.M.N.: Visualization, methodology, design primers, Sequence assembling and blast, investigation, writing the original draft. F.Y.: Visualization, writing the original draft, investigation, methodology. H.W.C.: Project administration, supervision, review and editing. C.J.: Supervision, funding acquisition, resources, review and editing. All authors have read and agreed to the published version of the manuscript.

Funding: This study was supported by the BSRP through the National Research Foundation of Korea (NRF), ministry of Education (grant number NRF-2018R1A6A1A03024862), and Rural Development Administration (RDA agenda PJ01574604 on honeybee pollination).

Institutional Review Board Statement: Not applicable.

Informed Consent Statement: Not applicable.

Data Availability Statement: All the sequences generated in this research, have been deposited in GenBank under accession numbers MW660861-MW660880.

Acknowledgments: We are grateful to Bajaree Chuttong, Chiang Mai University who provided us with honey and bee samples from Thailand. We also thank Ratna Thapa for facilitating SMA and CJ's expedition trip to Nepal. This study was supported by the BSRP through the National Research Foundation of Korea (NRF), ministry of Education (grant number NRF-2018R1A6A1A03024862).

Conflicts of Interest: S.M.N. and F.Y. are a research professor and Ph.D. student at ANU, respectively, and received a full-time salary for this work. C.J. has received research grants from NRF, and supervised the work. H.W.C. has served on advisory for the project. All declare no conflicts of interest on this paper.

References

1. Ajibola, A.; Chamunorwa, J.P.; Erlwanger, K.H. Dietary supplementation with natural honey promotes growth and health of male and female rats compared to cane syrup. *Sci. Res. Essays* **2013**, *8*, 543–553.
2. Soares, S.; Amaral, J.S.; Oliveira, M.B.; Mafra, I. A comprehensive review on the main honey authentication issues: Production and origin. *Compr. Rev. Food Sci. Food Saf.* **2017**, *16*, 1072–1100. [CrossRef] [PubMed]
3. Oldroyd, B.P.; Wongsiri, S. *Asian Honey Bees: Biology, Conservation and Human Interactions*; Harvard University Press: Cambridge, MA, USA, 2006.
4. Wongsiri, S.; Chanchao, C.; Lekprayoon, C.; Wattanasermkit, K.; Deowanish, S.; Leepitakrat, S. Honeybee diversity and management in the new millennium in Thailand. In Proceedings of the 7th International Conference on Tropical Bees, Chiang Mai, Thailand, 19–25 March 2000; pp. 9–14.

5. Corlett, R.T. Honeybees in natural ecosystems. In *Honeybees of Asia*; Hepburn, H.R., Radloff, S.E., Eds.; Springer-Verlag: Berlin, Germany, 2011; pp. 215–226.
6. Partap, U. The pollination role of honeybees. In *Honeybees of Asia*; Hepburn, H.R., Radloff, S.E., Eds.; Springer-Verlag: Berlin, Germany, 2011; pp. 227–255.
7. Hall, T.A. BioEdit: A User-Friendly Biological Sequence Alignment Editor and Analysis Program for Windows 95/98/NT. *Nucleic Acids Symp. Ser.* **1999**, *41*, 95–98.
8. Yong, P.L.; Othman, M.S.H. *Economic Value of Honey Bees, Peninsular Malaysia*; Forestry Department Peninsular Malaysia (FDPM): Kuala Lumpur, Malaysia, 2007.
9. Moniruzzaman, M.; Khalil, M.I.; Sulaiman, S.A.; Gan, S.H. Physicochemical and antioxidant properties of Malaysian honeys produced by *Apis cerana*, *Apis dorsata* and *Apis mellifera*. *BMC Complement. Altern. Med.* **2013**, *13*, 43. [CrossRef] [PubMed]
10. Ruttner, F. *Biogeography and Taxonomy of Honeybees*; Springer: Heidelberg/Berlin, Germany, 1988.
11. Jung, C.; Lee, M. Beekeeping in Korea: Past, present, and future challenges. In *Asian Beekeeping in the 21st Century*; Chantawannakul, P., Williams, G., Neumann, P., Eds.; Springer: Berlin/Heidelberg, Germany, 2016; pp. 175–197.
12. Lee, D.C.; Lee, S.Y.; Cha, S.H.; Choi, Y.S.; Rhee, H.I. Discrimination of native bee-honey and foreign bee-honey by SDS–PAGE. *Korean J. Food Sci. Technol.* **1998**, *30*, 1–5.
13. He, X.; Wang, W.; Qin, Q.; Zeng, Z.; Zhang, S.; Barron, A.B. Assessment of flight activity and homing ability in Asian and European honey bee species, *Apis cerana* and *Apis mellifera*, measured with radio frequency tags. *Apidologie* **2012**, *44*, 38–51. [CrossRef]
14. Jaafar, M.; Othman, M.; Yaacob, M.; Talip, B.; Ilyas, M.; Ngajikin, N.; Fauzi, N. A review on honey adulteration and the available detection approaches. *J. Integr. Eng.* **2020**, *12*, 125–131.
15. Bogdanov, S.; Ruoff, K.; Oddo, L.P. Physico-chemical methods for the characterisation of unifloral honeys: A review. *Apidologie* **2004**, *35* (Suppl. 1), S4–S17. [CrossRef]
16. Sahinler, N.; Sahinler, S.; Gul, A. Biochemical composition of honeys produced in Turkey. *J. Apic. Res.* **2004**, *43*, 53–56. [CrossRef]
17. Guler, A.; Bakan, A.; Nisbet, C.; Yavuz, O. Determination of important biochemical properties of honey to discriminate pure and adulterated honey with sucrose (*Saccharum officinarum* L.) syrup. *Food Chem.* **2007**, *105*, 1119–1125. [CrossRef]
18. Chen, L.; Xue, X.; Ye, Z.; Zhou, J.; Chen, F.; Zhao, J. Determination of Chinese honey adulterated with high fructose corn syrup by near infrared spectroscopy. *Food Chem.* **2011**, *128*, 1110–1114. [CrossRef]
19. Moore, J.C.; Spink, J.; Lipp, M. Development and application of a database of food ingredient fraud and economically motivated adulteration from 1980 to 2010. *J. Food Sci.* **2012**, *77*, R118–R126. [CrossRef]
20. Bottero, M.T.; Dalmasso, A. Animal species identification in food products: Evolution of biomolecular methods. *Vet. J.* **2011**, *190*, 34–38. [CrossRef] [PubMed]
21. Kumar, A.; Kumar, R.R.; Sharma, B.D.; Gokulakrishnan, P.; Mendiratta, S.K.; Sharma, D. Identification of species origin of meat and meat products on the DNA basis: A review. *Crit. Rev. Food. Sci. Nutr.* **2013**, *55*, 1340–1351. [CrossRef] [PubMed]
22. Amaral, J.; Meira, L.; Oliveira, M.; Mafra, I. Advances in authenticity testing for meat speciation. In *Advances in Food Authenticity Testing*; Downey, G., Ed.; Elsevier: Amsterdam, The Netherlands, 2016; pp. 369–414.
23. Willette, D.A.; Simmonds, S.E.; Cheng, S.H.; Esteves, S.; Kane, T.L.; Nuetzel, H.; Pilaud, N.; Rachmawati, R.; Barber, P.H. Using DNA barcoding to track seafood mislabeling in Los Angeles restaurants. *Conserv. Biol.* **2017**, *31*, 1076–1085. [CrossRef] [PubMed]
24. Won, S.; Lee, D.; Ko, S.H.; Kim, J.; Rhee, H. Honey major protein characterization and its application to adulteration detection. *Food Res. Int.* **2008**, *41*, 952–956. [CrossRef]
25. Zhang, Y.; Wang, S.; Chen, Y.; Wu, Y.; Tian, J.; Si, J.; Zhang, C.; Zheng, H.; Hu, F. Authentication of *Apis cerana* honey and *Apis mellifera* honey based on major royal jelly protein 2 gene. *Molecules* **2019**, *24*, 289. [CrossRef]
26. Zhang, Y.-Z.; Chen, Y.-F.; Wu, Y.-Q.; Si, J.-J.; Zhang, C.-P.; Zheng, H.-O.; Hu, F.-L. Discrimination of the entomological origin of honey according to the secretions of the bee (*Apis cerana* or *Apis mellifera*). *Food Res. Int.* **2019**, *116*, 362–369. [CrossRef]
27. Moritz, C.; Dowling, T.E.; Brown, W.M. Evolution of animal mitochondrial DNA: Relevance for population biology and systematics. *Annu. Rev. Ecol. Evol. Syst.* **1987**, *18*, 269–292. [CrossRef]
28. Song, S.; Pursell, Z.F.; Copeland, W.C.; Longley, M.J.; Kunkel, T.A.; Mathews, C.K. DNA precursor asymmetries in mammalian tissue mitochondria and possible contribution to mutagenesis through reduced replication fidelity. *Proc. Natl. Acad. Sci. USA* **2005**, *102*, 4990–4995. [CrossRef]
29. Zink, R.M.; Barrowclough, G.F. Mitochondrial DNA under siege in avian phylogeography. *Mol. Ecol.* **2008**, *17*, 2107–2121. [CrossRef] [PubMed]
30. Kim, C.; Lee, D.; Choi, S. Detection of Korean native honey and European honey by using duplex polymerase chain reaction and Immunochromatographic assay. *Food Sci. Anim. Resour.* **2017**, *37*, 599–605. [CrossRef] [PubMed]
31. Soares, S.; Grazina, L.; Mafra, I.; Costa, J.; Pinto, M.A.; Duc, H.P.; Oliviera, M.B.; Amaral, J. Novel diagnostic tools for Asian (*Apis cerana*) and European (*Apis mellifera*) honey authentication. *Food Res. Int.* **2018**, *105*, 686–693. [CrossRef] [PubMed]
32. R Core Team. *R: A Language and Environment for Statistical Computing*; R Foundation for Statistical Computing: Vienna, Austria, 2018; Available online: https://www.R-project.org/ (accessed on 20 December 2021).
33. Kek, S.P.; Chin, N.L.; Tan, S.W.; Yusof, Y.A.; Chua, L.S. Molecular identification of honey entomological origin based on bee mitochondrial 16S rRNA and COI gene sequences. *Food Control* **2017**, *78*, 150–159. [CrossRef]

34. Brudzynski, K.; Abubaker, K.; Miotto, D. Unraveling a mechanism of honey antibacterial action: Polyphenol/H2O2-induced oxidative effect on bacterial cell growth and on DNA degradation. *Food Chem.* **2012**, *133*, 329–336. [CrossRef] [PubMed]
35. Utzeri, V.J.; Ribani, A.; Fontanesi, L. Authentication of honey based on a DNA method to differentiate Apis mellifera subspecies: Application to Sicilian honey bee (*A. M. siciliana*) and iberian honey bee (*A. M. iberiensis*) honeys. *Food Control* **2018**, *91*, 294–301. [CrossRef]
36. Bovo, S.; Utzeri, V.J.; Ribani, A.; Cabbri, R.; Fontanesi, L. Shortgun sequencing of honey DNA can describe honey bee derived environmental signatures and the honey bee hologenome complexity. *Sci. Rep.* **2020**, *10*, 9279. [CrossRef]
37. Schnell, I.B.; Fraser, M.; Willerslev, E.; Gilbert, M.T.P. Characterisation of insect and plant origins using DNA extracted from small volumes of bee honey. *Arthropod Plant Interact.* **2010**, *4*, 107–116. [CrossRef]
38. Jain, S.A.; Jesus, F.T.; Marchioro, G.M.; Araujo, E.D. Extraction of DNA from honey and its amplification by PCR for botanical identification. *Food Sci. Technol.* **2013**, *33*, 753–756. [CrossRef]
39. Eimanifar, A.; Kimball, R.T.; Braun, E.L.; Ellis, J.D. The complete mitochondrial genome of the Cape honey bee Esch., Apis mellifera capensis (Insecta: Hymenoptera: Apidae). *Mitochondrial DNA Part B* **2016**, *1*, 817–819. [CrossRef]
40. Eimanifar, A.; Kimball, R.T.; Braun, E.L.; Fuchs, S.; Grünewald, B.; Ellis, J.D. The complete mitochondrial genome of an east African honey bee, Apis mellifera monticola Smith (Insecta: Hymenoptera: Apidae). *Mitochondrial DNA Part B* **2017**, *2*, 589–590. [CrossRef] [PubMed]
41. Eimanifar, A.; Kimball, R.T.; Braun, E.L.; Moustafa, D.M.; Haddad, N.; Fuchs, S.; Grunewald, B.; Ellis, J.D. The complete mitochondrial genome of the Egyptian honey bee, Apis mellifera lamarckii (Insecta: Hymenoptera: Apidae). *Mitochondrial DNA Part B* **2017**, *2*, 270–272. [CrossRef] [PubMed]
42. Eimanifar, A.; Kimball, R.T.; Braun, E.L.; Fuchs, S.; Grunewald, B.; Ellis, J.D. The complete mitochondrial genome of Apis mellifera meda (Insecta: Hymenoptera: Apidae). *Mitochondrial DNA Part B* **2017**, *2*, 268–269. [CrossRef] [PubMed]
43. Eimanifar, A.; Kimball, R.T.; Braun, E.L.; Ellis, J.D. Mitochondrial genome diversity and population structure of two western honeybee subspecies in the Republic of South Africa. *Sci. Rep.* **2018**, *8*, 1333. [CrossRef] [PubMed]
44. Fuller, Z.L.; Nino, E.L.; Patch, H.M.; Bedoya-Reina, O.C.; Baumgarten, T.; Muli, E.; Mumoki, F.; Ratan, A.; McGraw, J.; Frazier, M.; et al. Genome-wide analysis of signatures of selection in populations of African honey bees (Apis mellifera) using new web-based tools. *BMC Genom.* **2015**, *16*, 518. [CrossRef]
45. Gibson, J.D.; Hunt, G.J. The complete mitochondrial genome of the invasive Africanized honey bee, Apis mellifera scutellata (Insecta: Hymenoptera: Apidae). *Mitochondrial DNA Part A* **2016**, *27*, 561–562. [CrossRef]
46. Haddad, N.J. Mitochondrial genome of the Levant Region honey bee, Apis mellifera syriaca (Hymenoptera: Apidae). *Mitochondrial DNA Part A* **2016**, *27*, 4067–4068. [CrossRef]
47. Hu, P.; Lu, Z.X.; Haddad, N.; Noureddine, A.; Loucif-Ayad, W.; Wang, Y.Z.; Zhang, R.B.Z.A.L.; Guan, X.; Zhang, H.X.; Niu, H. Complete mitochondrial genome of the Algerian honey bee, Apis mellifera intermissa (Hymenoptera: Apidae). *Mitochondrial DNA Part A* **2016**, *27*, 1791–1792.
48. Ilyasov, R.A.; Park, J.; Takahashi, J.; Kwon, H.W. Phylogenetic uniqueness of honeybee Apis cerana from the Korean peninsula inferred from the mitochondrial, nuclear, and morphological data. *J. Apicul. Sci.* **2018**, *62*, 189–214. [CrossRef]
49. Nakagawaa, I.; Maedaa, M.; Chikanoa, M.; Okuyamaa, H.; Murrayb, R.; Takahashia, J. The complete mitochondrial genome of the yellow coloured honeybee Apis mellifera (Insecta: Hymenoptera: Apidae) of New Zealand. *Mitochondrial DNA Part B* **2018**, *3*, 66–67. [CrossRef]
50. Okuyama, H.; Hill, J.; Martin, S.J.; Takahashi, J. The complete mitochondrial genome of a Buckfast bee, Apis mellifera (Insecta: Hymenoptera: Apidae) in Northern Ireland. *Mitochondrial DNA Part B* **2018**, *3*, 338–339. [CrossRef] [PubMed]
51. Okuyama, H.; Tingek, S.; Takahashi, J. The complete mitochondrial genome of the cavity-nesting honeybee, Apis cerana (Insecta: Hymenoptera: Apidae) from Borneo. *Mitochondrial DNA Part B* **2017**, *2*, 475–476. [CrossRef] [PubMed]
52. Okuyama, H.; Jimi, R.; Wakamiya, T.; Takahashi, J. Complete mitochondrial genome of the honeybee Apis cerana native to two remote islands in Japan. *Conserv. Genet. Resour.* **2017**, *9*, 557–560. [CrossRef]
53. Takahashi, J.; Deowanish, S.; Okuyama, H. Analysis of the complete mitochondrial genome of the giant honeybee, Apis dorsata, (Hymenoptera: Apidae) in Thailand. *Conserv. Genet. Resour.* **2017**. [CrossRef]
54. Takahashi, J.; Wakamiya, T.; Kiyoshi, T.; Uchiyama, H.; Yajima, S.; Kimura, K.; Nomura, T. The complete mitochondrial genome of the Japanese honeybee, Apis cerana japonica (Insecta: Hymenoptera: Apidae). *Mitochondrial DNA Part B* **2016**, *1*, 156–157. [CrossRef] [PubMed]
55. Tan, H.W.; Liu, G.H.; Dong, X.; Lin, R.Q.; Song, H.Q.; Huang, S.Y.; Yuan, Z.G.; Zhao, X.Q. The complete mitochondrial genome of the Asiatic cavity nesting honeybee Apis cerana (Hymenoptera: Apidae). *PLoS ONE* **2011**, *6*, e23008. [CrossRef] [PubMed]
56. Wang, A.R.; Kim, J.S.; Kim, M.J.; Kim, H.K.; Choi, Y.S.; Kim, I. Comparative description of mitochondrial genomes of the honey bee Apis (Hymenoptera: Apidae): Four new genome sequences and Apis phylogeny using whole genomes and individual genes. *J. Apic. Res.* **2018**, *57*, 484–503. [CrossRef]
57. Yang, J.; Xu, J.; Wu, J.; Zhang, X.; He, S. The complete mitogenome of wild honeybee Apis dorsata (Hymenoptera: Apidae) from South-Western China. *Mitochondrial DNA Part B* **2019**, *4*, 231–232. [CrossRef]

Article

Determination of Floral Origin Markers of Latvian Honey by Using IRMS, UHPLC-HRMS, and ¹H-NMR

Kriss Davids Labsvards [1,2,*], Vita Rudovica [1], Rihards Kluga [1], Janis Rusko [1,2], Lauma Busa [1], Maris Bertins [1], Ineta Eglite [3], Jevgenija Naumenko [1], Marina Salajeva [1] and Arturs Viksna [1]

[1] Department of Chemistry, University of Latvia, Jelgavas Street 1, LV-1004 Riga, Latvia; vita.rudovica@lu.lv (V.R.); rihards.kluga@lu.lv (R.K.); janis.rusko@bior.lv (J.R.); lauma.busa@lu.lv (L.B.); maris.bertins@lu.lv (M.B.); jn18020@students.lu.lv (J.N.); ms18103@students.lu.lv (M.S.); arturs.viksna@lu.lv (A.V.)

[2] Institute of Food Safety, Animal Health and Environment "BIOR", Lejupes Street 3, LV-1076 Riga, Latvia

[3] Latvian Beekeeping Association, Rigas Street 22, LV-3004 Jelgava, Latvia; ineta.eglite@strops.lv

* Correspondence: kriss_davids.labsvards@lu.lv; Tel.: +371-26395784

Abstract: The economic significance of honey production is crucial; therefore, modern and efficient methods of authentication are needed. During the last decade, various data processing methods and a combination of several instrumental methods have been increasingly used in food analysis. In this study, the chemical composition of monofloral buckwheat (*Fagopyrum esculentum*), clover (*Trifolium repens*), heather (*Calluna vulgaris*), linden (*Tilia cordata*), rapeseed (*Brassica napus*), willow (*Salix cinerea*), and polyfloral honey samples of Latvian origin were investigated using several instrumental analysis methods. The data from light stable isotope ratio mass spectrometry (IRMS), ultra-high performance liquid chromatography coupled with high-resolution mass spectrometry (UHPLC-HRMS), and nuclear magnetic resonance (NMR) analysis methods were used in combination with multivariate analysis to characterize honey samples originating from Latvia. Results were processed using the principal component analysis (PCA) to study the potential possibilities of evaluating the differences between honey of different floral origins. The results indicate the possibility of strong differentiation of heather and buckwheat honeys, and minor differentiation of linden honey from polyfloral honey types. The main indicators include depleted $\delta^{15}N$ values for heather honey protein, elevated concentration levels of rutin for buckwheat honey, and qualitative presence of specific biomarkers within NMR for linden honey.

Keywords: honey; light stable isotope mass spectrometry; ultra-high performance liquid chromatography; high resolution mass spectrometry; nuclear magnetic resonance; principal component analysis; floral origins

1. Introduction

Due to its sweet taste and antibacterial properties, honey is in high demand in today's market. In 2018, approximately 2000 tons of honey from 103,000 beehives were produced in Latvia. In Europe more generally, the demand for honey is higher than local producers can produce, and therefore a large part is imported [1]. Honey is an expensive product when compared to other sweeteners. Counterfeit honey is considered to be honey that contains added amounts of other cheaper sweeteners. Directive 2014/63/EU of the European Parliament and the Council clearly defines what constitutes natural honey. To protect the interests of consumers and regulate the fair price of honey in today's market, methods of honey authenticity and quality indicators are constantly evolving. One or several modern instrumental methods are increasingly used with which quality characteristics are determined, as well as the authenticity of honey is assessed by applying chemometric methods [2].

Although IRMS has been used mainly to determine the presence of C_4 sugar additives, measurements of light stable isotopes have increasingly been used to determine the botanical or geographical origin of honey. $\delta^{13}C$ and $\delta^{15}N$ values for honey and proteins provide useful information in distinguishing samples of different floral types of honey (acacia, chestnut, citrus, eucalyptus, rhododendron, and polyfloral honey) [3].

Polyphenol compound concentrations are considered as potential useful variables for floral origins [4]. The polyphenol profile is a useful tool for geographical and floral origin assessment. A robust UHPLC-HRMS method for polyphenol quantification is often used [5–7]. Sugaring-out assisted liquid–liquid extraction (SULLE) sample preparation has been proven to be an optimal choice of honey studies using HRMS equipment [5].

Nuclear magnetic resonance (NMR) is increasingly used to evaluate the authenticity of honey. The information provided by the proton NMR spectrum, in combination with various chemometric methods, is used to distinguish between honey of different botanical origins. Depending on the task to be performed, chemometrics can be performed for the whole spectrum or only for a certain interval. In most cases, the region characteristic of aromatic compounds (9–6 ppm) or the region characteristic of aliphatic compounds (3–0.5 ppm) are used [8,9]. A study in Brazil successfully distinguished between citrus, eucalyptus, and wildflower honey, and some honey was found to be counterfeit [10]. A similar approach was used by Spiteri et al. for the assessment of geographical origin [11]. Samples of acacia honey from Eastern Europe and Italy were compared. Due to the different flora, characteristic floral markers were found in the samples [12].

Principal component analysis (PCA) in chemistry allows for the study of the properties of different datasets of chemical compounds. Determining which compounds have similar properties and which study objects form groups, one can also try to predict the properties of the study object or belonging to a group. Various instrumental analyses are practically effective for the analysis of principal components, wherein the spectral image is obtained, for example, the total ion chromatogram after the retention time, under different conditions [13]. Quantitative values of various honey compounds, isotope ratio values, etc., quality indicators can be used to analyze the principal components. Depending on the purpose of the study (counterfeits, origin of flowers, geographical regions, etc.), honey types are selected, chemical instrumental analyses are performed, and the results are used for the analysis of principal components to determine the formation of groups [10,14–16].

The main aim of this study was the use of different methodologies to classify the botanical origin of various types of monofloral Latvian honey to target the mislabeling of protected destination of origin (PDO) products. One of the goals was to gather the data on fresh samples collected directly from the beekeepers of Latvia instead of processed and commercially available honey. Further, we validated the true floral origin using melissopalynology analysis. Finally, we evaluated multiple criteria to classify individual monofloral variety honeys by using a combination of analytical methods (IRMS, UHPLC-HRMS and NMR) and statistical treatment of experimental data and PCA analysis.

2. Materials and Methods

2.1. Samples

A total of 78 different honey samples were collected directly from the beekeepers in the territory of Latvia, declared as of natural origin and of specific monofloral varieties. The true botanical origin of the samples was further examined by melissopalynology analysis [17] and later confirmed or deemed of lesser, polyfloral quality. The criteria of specific pollen for monofloral honey [18] was reached for 4 buckwheat (*Fagopyrum esculentum*) (>25%), 6 clover (*Trifolium repens*) (>45%), 3 heather (*Calluna vulgaris*) (>40%), 3 linden (*Tilia cordata*) (>17%), 4 rapeseed (*Brassica napus*) (>70%), and 3 willow (*Salix cinerea*) (>45%) honey samples. The other 55 honey samples were polyflorals and kept for honey analysis to make an assessment for the capability of potential floral origins indicators.

2.2. IRMS

2.2.1. Protein Extraction by Dialysis

Honey proteins were prepared according to the method described by Bilikova [19] with slight readjustments. A 15 g sample of honey was weighed, and 15 mL of deionized water was added. Semi-permeable SnakeSkin (10K MWCO) dialysis membrane was filled with a homogeneous clear honey solution. After dialysis, the purified protein solution was quantitatively transferred into a beaker and placed in the drying oven at 40 °C for about 48 h until the proteins were dried. Then, proteins were weighed and stored at 4 °C until IRMS analysis.

2.2.2. $\delta^{13}C$ and $\delta^{15}N$, and Total Carbon and Nitrogen Analysis

Continuous flow IRMS (Nu Horizon) coupled with an elemental analyzer (EuroEA3024) was used for the analysis. The complete combustion of the samples and the operation of the element analyzer were verified by performing stability tests on the equipment. Certified reference materials USGS-40 and USGS-41 were used as reference materials. The device conditions were prepared as described in previously published method [20].

2.3. UHPLC-HRMS

2.3.1. Chemicals

Analytical standards of 3,4-dihydroxybenzoic acid (>98.2%), acacetin (>98.7%), apigenin (>99%), caffeic acid (>98.5%), catechin (>99%), chlorogenic acid (>99%), chrysin (>99%), daidzein (>99%), galangin (>98.5%), gallic acid (>95.5%), genistein (>99%), (-)-epicatechin (>90.3%), folic acid (>91.2%), formononetin (>99%), isovitexin (>99%), luteolin (>99.9%), myricetin (>98%), o-coumaric acid (>99.7%), p-coumaric acid (>99.6%), p-hydroxybenzoic acid (>99.9%), pantothenic acid (>98.6%), phenylacetic acid (>99.7%), rhamnetin (>99%), rutin trihydrate (>94%), quercetin (>98%), sinapic acid (>96%), syringic acid (>98.5%), trans-ferulic acid (>99.8%), vanillic acid (>98.2%) were purchased from Extrasynthese (Lyon, France) or Sigma-Aldrich (St. Louis, MO, USA). The standard of (-)-cis, trans-abscisic acid (>99.9%) was purchased from Santa Cruz Biotechnology (Dallas, TX, USA), and kaempferol (>97%) was purchased from ChromaDex (Santa Ana, CA, USA). HPLC-MS grade acetonitrile (MeCN) and dimethyl sulfoxide (DMSO) were purchased from Merck (Darmstadt, Germany), while formic acid (FA), hydrochloric acid (HCl), and sodium chloride (NaCl) were purchased from Sigma-Aldrich.

2.3.2. SULLE Sample Preparation

Samples were prepared by the previously published SULLE method [5]. A total of 0.5 g of honey was added in an Eppendorf tube within 0.5 mL of 10% NaCl in 0.01M HCl (pH = 2). A total of 1 mL of MeCN was added to the mixture, and the tube was vortexed for another 1 min at 2000 rpm followed by 1 min centrifugation at 15,000 rpm. The upper organic phase was collected in a 2 mL crimp top chromatography vial. The procedure was repeated until the total collected organic phase amount of about 1.9 mL. The organic phase was dried under a gentle nitrogen flow at room temperature and reconstituted in 0.5 mL of water/MeCN mixture (98:2 v/v) with added 0.1% FA. Extracts were stored at 4 °C, in the dark, before the analysis.

2.3.3. UHPLC-HRMS Systems

Liquid chromatography analysis was performed using a Dionex UltiMate 3000 UHPLC system (Thermo Scientific, Oleten, Switzerland) equipped with a Kinetex PFP column (3.00 × 100 mm, 1.7 µm, 100 Å), obtained from Phenomenex (Torrance, CA, USA). LC system was coupled to a high-resolution mass spectrometer Q Exactive (Thermo Scientific, Bremen, Germany). LC parameters: 5 µL injection volume, 40 °C column temperature, 10 °C sample temperature, flow rate set to 0.450 mL·min^{-1}, diverter valve was switched to mass spectrometer at 1.3 min. The mobile phase A (0.1% formic acid in H_2O) and B (0.1% formic acid in MeCN) were used in gradient mode: 4 min preinjection equilibration held at 2% B; 0–3 min at 2–5% B; 3–9 min at 5–98% B; 9–13 hold at 98% B; 13–14 return to the initial 2% B.

Heated electrospray (HESI) interface was used in positive and negative ionization mode, and polarity switching method was used with the following parameters: ion source voltage in negative/positive ionization (2500 V/3500 V), 280 °C temperature for ion transfer tube, 450 °C evaporator temperature.

2.4. NMR

2.4.1. Sample Preparation

The method proposed by Schievano et al. was used and adjusted for available equipment to acquire ^{1}H-NMR spectra of honey [21]. A total of 200 ± 3 mg of honey was dissolved in 1.0 mL of D_2O buffer solution. The resulting solution was transferred to an NMR tube, and ^{1}H-NMR spectra were acquired. D_2O buffer solution was prepared by dissolving 1.02 g of KH_2PO_4 and 0.96 mg of NaN_3 in 20 mL of D_2O. The buffer solution pH was adjusted to 4.4 with 85% H_3PO_4.

2.4.2. ^{1}H-NMR Spectra Acquisition

NMR spectra were acquired with Bruker BioSpin GmbH, Rheinstetten, Germany, Fourier300 spectrometer (working frequency of 300 MHz for ^{1}H) equipped with a 5 mm DUL 13C-1H/D Z-gradient EasyProbe. ^{1}H-NMR spectra were acquired with noesypr1d pulse program using 125 ms mixing time and −40 dBW presaturation power level during recycle delay and mixing time, 2 s relaxation delay (D1), 6103 Hz spectral width, 64k points of time-domain (TD), and 8 dummy scans (DS). The acquisition time for one scan was 5.37 s. Constant receiver gain (rg = 3) was used.

2.4.3. ^{1}H-NMR Spectra Processing

Acquired ^{1}H-NMR spectra were processed with MestReNova software (version 14.1.1). FID was zero-filled to 128k points, and exponential apodization (0.3 Hz) was used. Manual phase correction and automatic baseline correction (Whittaker smoother) were performed. Chemical shifts were referenced to α-glucopyranose doublet (δ = 5.320 ppm). ^{1}H spectra were binned using signal integral sum 0.5–3.0, 6.0–9.0 ppm with a bin width of 0.01 ppm. The binned spectra were normalized to the total area.

2.5. Statistical Analysis

The data processing was performed using statistical software Minitab 17.1.0. One-way ANOVA analysis of variance was performed in order to assess the significant differences of the variable between monofloral and polyfloral honey samples. The confidence level (p = 0.05) was used for every ANOVA test. Tukey comparison procedure for assuming equal variances was used for every variable obtained from IRMS and UHPLC-HRMS methods while Fisher comparison was used once for total N assessment in honey proteins. Principal component analysis was performed for data reduction in order to find potential chemical compound biomarkers for floral origins. The correlation matrix was used for analysis. As a pre-step, the software performed standardization of variables, meaning a variable was rescaled to have a mean of zero and a standard deviation of one. Principal component scores and their correlation coefficients are stored in a Supplementary Excel file. The formation of monofloral group clusters or positions in the score plot was used to assess the potential of variable capability as a marker.

3. Results and Discussion

3.1. IRMS Analysis of Honey Proteins

C and N isotope ratio and total weight fraction of monofloral and polyfloral honey samples is presented in Table 1. The ANOVA one-way results show that there was no significant variance, with a confidence level of 95% between monofloral and polyfloral honey, by using $\delta^{13}C$ values ($p = 0.08$). Tukey test simultaneous differences of means for $\delta^{13}C$ are described in Supplementary Material (Figure S1).

Table 1. IRMS analyses results by floral origins of honey.

Floral Origins	$\delta^{13}C$, ‰		$\delta^{15}N$, ‰		Total C, %		Total N, %		
	Mean	SD	Mean	SD	Mean	SD	Mean	SD	N
Buckwheat	−28.7	0.7	6.8	1.5	48.1	1.4	9.6	0.5	4
Clover	−27.7	0.9	6.5	1.7	50	4	8	2	6
Heather	−28.13	0.10	−2.3	1.0	47.4	0.6	10.0	0.6	3
Linden	−26.7	0.2	5.8	0.7	50	2	7.0	0.7	3
Rapeseed	−27.5	0.5	4.9	1.1	53	4	6.4	0.8	4
Willow	−27.6	0.5	6.5	1.0	56	7	6	2	3
Polyfloral	−27.4	0.9	4	3	51	6	8	2	55

The carbon isotope ratio in honey proteins is directly influenced by carbon fixation in plants from which bees are gathering honey. Therefore, carbon isotope analysis is mainly used for C4 plant additive determination in honey. Nevertheless, the $\delta^{13}C$ values are often used for floral origins determination [3,22].

After extractions of sugars, the honey proteins showed $\delta^{13}C$ values in a range of −25.47‰ to −29.64‰, which is characteristic for C3 plants [23]. Moreover, in comparing polyfloral honey proteins ($\delta^{13}C = -27.4 \pm 0.9$‰) with other results, we found that $\delta^{13}C$ values are more depleted than of Mediterranean region honey proteins. $\delta^{13}C$ values are dependent on the amount of sun exposure to plants and air humidity; therefore, an increase of sunny days and less precipitations increases the $\delta^{13}C$ values [24].

The nitrogen isotope ratio for honey proteins reflects the nitrogen content of the soil where plants from which bees have gathered the nectar grow. $\delta^{15}N > 0.0$‰ values indicate that the nitrogen is biologically fixed in soil, and values near 0.0‰ show that the nitrogen is obtained from air. Results show clover honey proteins are enriched with heavy nitrogen isotope, although the plant is considered as gathering nitrogen via *Rhizobium* bacteria [25]. Exceptional honey proteins were extracted from heather honey, indicating depleted nitrogen ratio values and statistically different significance ($p = 0.001$) using ANOVA one-way Tukey tests (see Figure S2). In total, 11 out of 78 honey proteins showed negative $\delta^{15}N$ values. These samples of honey were heather monoflorals and polyflorals that had reported the presence of heather (*Calluna vulgaris*) pollen (see the Supplementary Excel file).

Total carbon and nitrogen in honey proteins were found to have no particular statistical difference using the ANOVA test. p-values were found for total carbon ($p = 0.5$) and total nitrogen ($p = 0.06$), although total nitrogen p-values were close to 0.05, which suggests that results could be capable for further floral origin discrimination investigation. Using the Fisher test, we found that there are differences in heather and buckwheat (increased total nitrogen) honey proteins between willow and rapeseed (decreased total nitrogen) (see Figure S3). Total nitrogen in proteins generally is ≈16% [26]. Obtained nitrogen mass fraction results suggest that after dialysis, pure protein is not obtained, but instead a mixture of protein and other molecularly large compounds that could not be separated via dialysis such as lipids and pollen [27,28].

3.2. UHPLC-HRMS Analysis

A total of 31 organic compounds (13 phenolic acids, 14 flavonoids, 2 vitamins, 2 plant hormones) were successfully quantified in polyfloral honey and various monoflorals. The biochanin A, biotin, and procyanidin A2 were found only in a few samples near the LOQ, and these compounds were omitted for future assessments.

In Figure 1, concentrations for the 27 most common found compounds in polyfloral honey of origins of Latvia are shown. The highest concentrations of phenolic acids were obtained for p-hydroxybenzoic acid (3923 ± 3522 µg/kg), abscisic acid (4174 ± 2238 µg/kg), p-coumaric acid (2685 ± 1271 µg/kg), and ferulic acid (1638 ± 572 µg/kg) while kaempferol (1432 ± 728 µg/kg) was the flavonoid and pantothenic acid B5 (986 ± 412 µg/kg) was vitamin with the highest average concentrations.

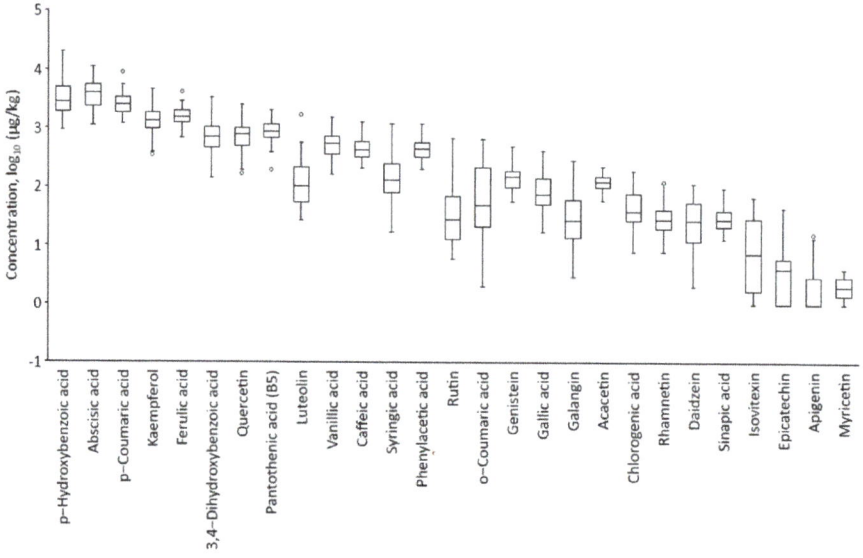

Figure 1. Boxplots of organic compound concentration (µg/kg) in polyfloral honey determined by UHPLC-HRMS; results converted in decimal logarithm scale.

Formononetin, chrysin, and folic acid were not shown by boxplot because these compounds were found over LOQ only 6 to 21 polyflorals, suggesting these compounds are characteristic of a specific floral origin. Apigenin was also omitted, although it was found in 43 polyfloral honey samples but slightly over LOQ, and the mean value was 2 ± 3 µg/kg.

The results come in good agreement with another study showing similar concentration levels of the same compounds, except apigenin, which was found in larger concentrations by Lo Dico et al. [7]. The one-way ANOVA tests of Tukey comparison were performed to honey groups of different floral origins. The six compounds showed statistically significant differences that could be used for monofloral honey samples or speciation of honey floral origins. Rutin interval plot and graphical summary of differences of mean are shown in Figure 2.

Rutin showed a statistically significant difference in buckwheat honey and in the other types of honey. In buckwheat honey, rutin showed a concentration of 572 ± 167 µg/kg, while polyfloral honey contained from <5 (LOQ) to 696 µg/kg with a mean of 53 µg/kg. Two polyfloral samples (P5 and P42) had notably higher concentrations of rutin, corresponding to 649 and 696 µg/kg, respectively, equivalent to high buckwheat pollen presence for polyflorals (17 and 24%, respectively). It was less found in linden and rapeseed honey and not found at all in heather honey. This comes in good agreement with melissopalynology results, as buckwheat (*Fagopyrum esculentum*) pollen was found in clover and willow monofloral honeys in a range of 0–6%. Other statistically significant differences within honey floral origins were found using vanillic acid, quercetin, p-hydroxybenzoic acid, p-coumaric acid, and pantothenic acid B5 (see Table 2). The monofloral clover and willow honey interfered to discriminate buckwheat honey from other types of floral origins using p-hydroxybenzoic acid and p-coumaric acid concentrations. Interference could be explained by buckwheat pollen presence in clover and willow monofloral honey. The quercetin concentrations showed a statistical difference between buckwheat and heather honey. While quercetin has no potential as a specific floral marker, it would be very helpful, since both share similar visual properties as dark-colored honeys [29,30]. Similarly, pantothenic acid and vanillic acid can be used for specific floral origin request determination, or could be a helpful indicator with a combination of other variables.

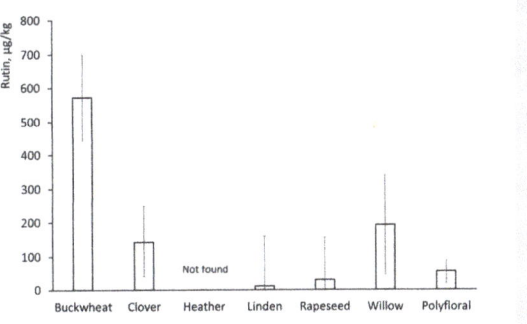
(a)

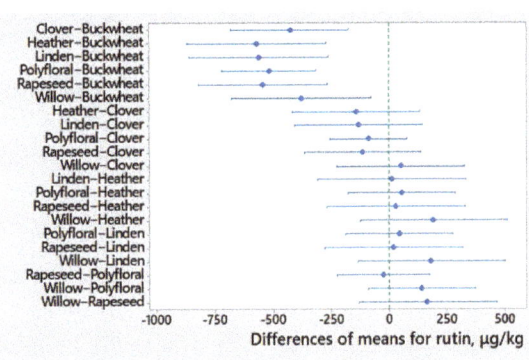
(b)

Figure 2. One-way ANOVA test of statistically significant difference between monofloral buckwheat, clover, heather, linden, rapeseed, willow, and polyfloral honey (**a**) using interval plot (µg/kg) as graphical summary with 95% confidence interval bars, and (**b**) using Tukey comparison of 95% confidence intervals.

Table 2. Comparison of chemical compound mass concentrations (µg/kg) quantified by UHPLC-HRMS that share statistically significant differences between groups of floral origins.

Floral Origins	Mean ± SD, µg/kg				
	p-Hydroxybenzoic Acid	p-Coumaric Acid	Pantothenic Acid (B5)	Quercetin	Vanillic Acid
Buckwheat	13,863 ± 4472 [A]	5561 ± 1159 [A]	910 ± 247 [AB]	1297 ± 511 [A]	602 ± 329 [AB]
Clover	7907 ± 4809 [AB]	3963 ± 991 [AB]	764 ± 193 [AB]	523 ± 204 [AB]	477 ± 164 [AB]
Heather	2984 ± 494 [B]	2519 ± 738 [B]	1513 ± 250 [A]	198 ± 86 [B]	190 ± 29 [B]
Linden	1423 ± 1004 [B]	2509 ± 161 [B]	558 ± 243 [B]	475 ± 390 [AB]	447 ± 337 [AB]
Rapeseed	1740 ± 248 [B]	2341 ± 499 [B]	577 ± 87 [B]	986 ± 167 [AB]	725 ± 28 [AB]
Willow	6753 ± 3252 [AB]	4550 ± 1529 [AB]	1017 ± 131 [AB]	726 ± 445 [AB]	1014 ± 619 [A]
Polyfloral	3923 ± 3522 [B]	2685 ± 1271 [B]	986 ± 412 [AB]	824 ± 419 [AB]	585 ± 288 [AB]

[AB]—results marked with a different superscript letter are significantly different using ANOVA one-way Tukey test ($p < 0.05$). Letter "A" indicates affiliation to a group with higher means and letter "B" indicates affiliation to a group with lower means, while "AB" shows affiliation for both groups.

Comparing honey of Polish origins, the authors of [31] found similar levels of p-coumaric acid and quercetin in heather honey. However, Latvian honey showed lower concentrations of chrysin, galangin, and apigenin but higher concentrations of luteolin than Polish honey. Rapeseed honey of Romanian origins share similar levels of chlorogenic acid and p-coumaric acid but increased of gallic acid, p-hydroxybenzoic acid, 3,4-dihydroxybenzoic acid, vanillic acid, caffeic acid, and myricetin [32]. In another study, p-hydroxybenzoic acid is mentioned as a commonly found compound in clover and heather honey. Moreover, p-coumaric and vanillic acid are reported as commonly found in heather honey, while our study shows that concentration levels were not different from Latvian polyfloral honey. Quercetin is usually found in clover honey [33]. Regardless of other studies, recent preliminary UHPLC-HRMS results of Latvian honey showed rutin as a suggestable indicator for buckwheat honey. However, increased rutin concentration levels for a few polyflorals containing notable buckwheat pollen percentage were also observed. This suggests a need for further investigation to determine a threshold level of rutin in order to distinguish buckwheat honey from polyfloral honeys.

3.3. Principal Component Analysis

All data for PC scores and loadings are available in the Supplementary Information in the form of excel spreadsheets. The carbon and nitrogen isotope ratio and total element percentage were used for PCA to determine the Latvian honey floral origins using a single IRMS method. The variables were standardized, and a correlation matrix was used since variables were expressed in different units of measurement. Eigenvalues were expressed in the scree plot (see Figure S4). PC1-PC3 described variability by 94.6%, and these components were used for evaluation. A total of 15 samples were considered as outliners using a Mahalanobis distance criteria and were withdrawn from PCA. The outliner samples were coded as follows: monofloral buckwheat (B1), clover (C1), rapeseed (R2), willow (W1), and polyfloral (P2, P5, P8, P18, P20, P23, P24, P26, P33, P42, P47) honey. In Figure 3a, the heather honey formed a cluster away from other honey samples because of PC1. After the investigation of loading coefficients (see Figure 3b), it appears that PC1 had a high positive correlation (r = 0.50) of nitrogen isotope ratio and total carbon in proteins (r = 0.53) but negative correlation of total nitrogen in proteins (r = −0.67). The monoflorals of heather honey were significantly different ($p = 0.001$) of depleted $\delta^{15}N$ values, while total carbon and total nitrogen showed no significant differences. Nevertheless, in comparing the means of the heather honey and other types of origins, we found that the mean of total carbon was the lowest, and total nitrogen was highest for heather honey. Polyflorals (P3, P6, P14, P55) that formed cluster with monofloral heather honey also contained heather pollen (31%, 10%, 38%, 4%). The honey sample P46 had 6% of heather pollen content and it was the only honey sample that had heather pollen more than 4%; moreover, it was not located in the cluster. Other polyfloral honey samples with heather pollen >4% (P2—24%; P8—35%; P18—22%) were classified as outliners and had similar $\delta^{15}N$ values but increased means of total carbon and decreased means of total nitrogen. This highlights the need to monitor the total carbon and nitrogen content in honey protein IRMS analysis when monofloral heather honey purity must be assessed.

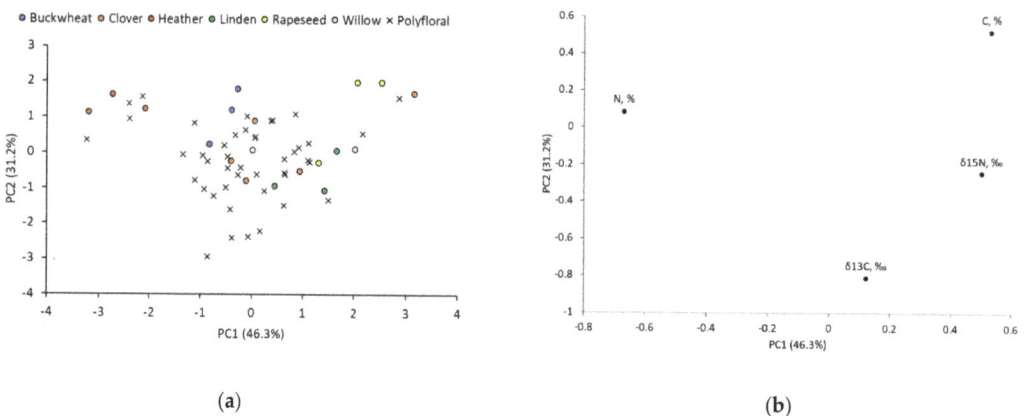

Figure 3. PCA of monofloral buckwheat, clover, heather, linden, rapeseed, willow, and polyfloral honey samples: (**a**) score plot between PC1 and PC2, and (**b**) loading plot of variables obtained by IRMS ($\delta^{13}C$, $\delta^{15}N$, total C and N in honey proteins).

The concentrations of phenolic acids, flavonoids, vitamins, and plant hormones in honey were used for PCA of UHPLC-HRMS assessment. The catechin, chrysin, folic acid, and formononetin were omitted for evaluation, and 27 compound concentrations (µg/kg) were standardized and a correlation matrix was constructed. After evaluation of the scree plot (see Figure S5), we used the PC1-PC3 for further analysis, since they cover the most variability of data (45.7%). A total of 11 samples were considered as outliners using a Mahalanobis distance criteria and were withdrawn from PCA. The outliner samples were coded as follows: monofloral clover (C2, C5) and polyfloral (P7, P8, P20, P26, P29, P37, P43, P46, P53) honey. The formation of an exceptional cluster such as with IRMS results were not observed. The PC2 has strong positive correlation of p-hydroxybenzoic acid (r = 0.39), rutin (r = 0.38), and p-coumaric acid (r = 0.34), and these compounds were previously discussed as potential buckwheat honey floral markers. The location of buckwheat honey in score plot (see Figure 4a) was outside of the majority of samples, but buckwheat honey was found to have higher PC2 scores than other honey. The polyflorals (P5, P42, P51, P54) with similar PC2 scores also had buckwheat pollen (17%, 24%, 16%, 4%). The PC3 was not selective for certain floral group but depended on the ratio of 3,4-dihydroxybenzoic acid (r = 0.43) and abscisic acid (r = −0.35) concentrations.

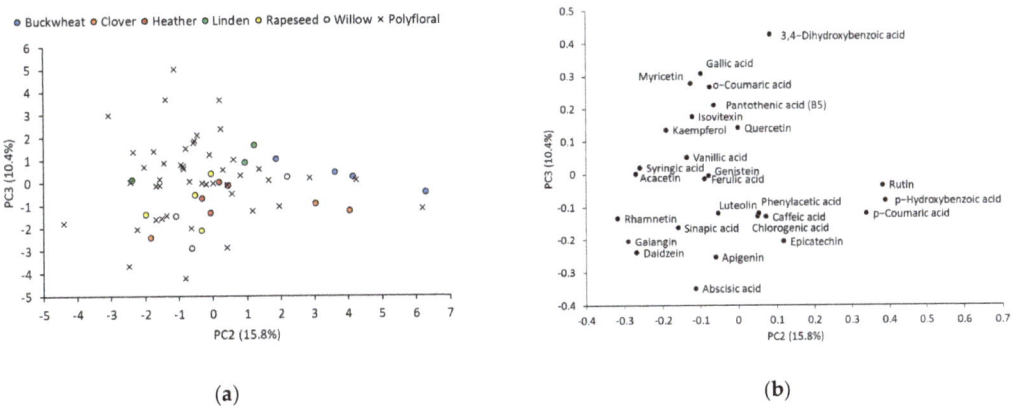

Figure 4. PCA of monofloral buckwheat, clover, heather, linden, rapeseed, willow, and polyfloral honey samples: (**a**) score plot between PC2 and PC3, and (**b**) loading plot of 27 organic compound concentrations obtained by UHPLC-HRMS.

Due to the presence of a wide range of chemical compounds found in honey, NMR is considered to be one of the most prominent methods for food analysis [34]. The complete identification of chemical compounds from ^{1}H-NMR spectra is a difficult task because of compound low concentrations and signal overlays. Nevertheless, the honey of similar floral origins share a similar ^{1}H-NMR spectra image, and therefore principal component analysis (PCA) was used to conduct an assessment of honey without full quality analysis. The ^{1}H-NMR spectra of honey samples were transformed into the spectral bins from 0.5 to 3 ppm (aliphatic region) and 6 to 9 ppm (aromatic region), with a bin width of 0.01 ppm before the principal component analysis. Carbohydrate region (3–6 ppm) was excluded for the PCA due to the presence of high intensity peaks that are sensitive to scaling method prior to PCA and strongly affect cluster forming in a PCA plot [35]. Furthermore, minor and specific carbohydrate ^{1}H NMR signals were not resolved using 300 MHz NMR spectrometer, which could improve PCA discrimination [21,36,37]. The obtained scree plot and PCA plots of Latvian honey samples are shown in Figure S6. The PC1-PC3 had exceptionally high contribution of data variability, explaining 29.2% of the variance from 552 variables. PCA plot with PC1 and PC3 of studied Latvian honeys of monofloral honeys (Figure 5a) could be described in several groups as follows: (1) buckwheat, clover, and willow honeys with mostly negative PC1; (2) linden honeys with positive PC1 and mostly positive PC3; and (3) heather honeys with positive PC1 and negative PC3. Rapeseed honeys showed cluster near PC1 and PC3 cross-point that indicated absence of specific compounds. Honey grouping could be explained by using PC1 and PC3 loading plots (see Figure 5b,c). In the case of buckwheat, clover, and willow honeys, ^{1}H-NMR spectral bins with δ = 6.87–6.82, 7.16–7.21, 1.67–1.74, and 0.90–1.02 ppm contributed to negative PC1 score. This can be explained with the presence of tyrosine (δ = 6.87–6.82 and 7.16–7.21 ppm), leucine (δ = 1.67–1.74 ppm), and isoleucine and valine (δ = 0.90–1.02 ppm).

These amino acids have been previously found in a higher level for buckwheat honey [38]. Surprisingly, in Latvian monofloral clover and willow honeys, these amino acids were found as well. For the linden monofloral honey, ^{1}H-NMR spectral bins with δ = 2.40–2.47 and 7.15–7.23 ppm contributed to positive PC1 score, and bins with δ = 6.14–6.18 and 7.15–7.23 ppm for positive PC3 score. Linden honey ^{1}H-NMR spectra-specific bins can be attributed to the cyclohexa-1,3-diene-1-carboxylic acid (CDCA) derivatives (δ = 6.14–6.18 and 7.15–7.23 ppm) that are specific markers of monofloral linden honey [38]. Lastly, the heather honey showed resolved cluster position in PCA plot that was mostly affected by ^{1}H-NMR spectral bins with δ = 7.28–7.32 and 2.37 ppm. These bins can be assigned to the previously found carboxylic acids, such as phenylacetic acid, 3-phenyllactic acid and benzoic acid (δ = 7.28–7.32 ppm), and pyruvic acid (δ = 2.37 ppm) [38,39]. Typical binned ^{1}H-NMR spectra of analyzed monofloral honeys with the assigned compounds are shown in Figure S7. It was shown that PCA in combination with ^{1}H-NMR showed clear separation of monofloral heather honey from other studied honeys. Unfortunately, monofloral honeys with negative PC1 could not be resolved in separate clusters, and other statistical methods should thus be used (e.g., OPLS-DA) [38,40].

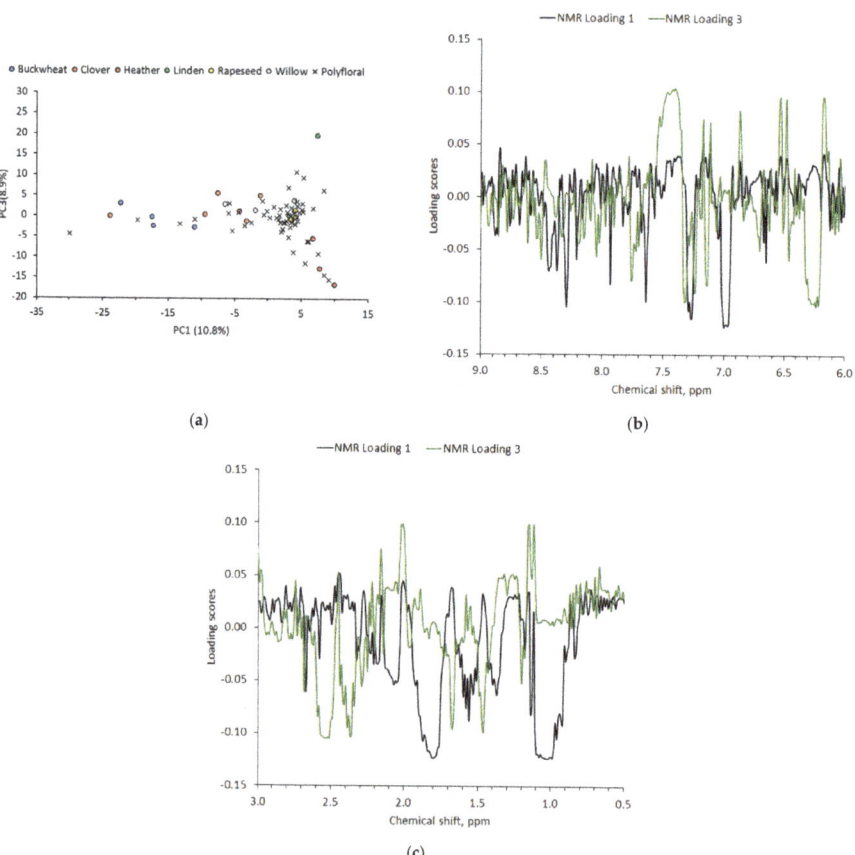

Figure 5. PCA of monofloral buckwheat, clover, heather, linden, rapeseed, willow, and polyfloral honey samples: (**a**) score plot between PC1 and PC3; loading plots of PC1 and PC3 for ^{1}H-NMR spectra with bin width 0.01 ppm for intervals (**b**) 9.0–6.0 ppm and (**c**) 3.0–0.5 ppm.

4. Conclusions

The chemical profile of monofloral buckwheat, clover, heather, linden, rapeseed, willow, and polyfloral honey samples of Latvian origins was assessed by IRMS, UHPLC-HRMS, and NMR methods in order to find suitable indicators that could be used for the classification of botanical origin of honey. The depletion in $\delta^{15}N$ values in honey proteins was suggested as indicator for heather honey ($\delta^{15}N = -2.3 \pm 1.0‰$). Moreover, the total N in proteins indicated potential distinctiveness between the pairs of willow and rapeseed honey, and buckwheat and heather honey. After the data treatment using PCA, the total nitrogen and total carbon in honey proteins were taken into account for recognition of heather honey origins. Out of 31 organic compounds quantified by UHPLC-HRMS, rutin showed a selective difference as a buckwheat honey indicator. p-Hydroxybenzoic acid, p-coumaric acid, pantothenic acid (B5), quercetin, and vanillic acid were found to have statistically different concentration levels within different monofloral honey types and could be used as specific indicators for monofloral honey purity. The polyphenol profile comes in good agreement with other studies, with the exception of the few compounds that were reported with higher concentrations in foreign country honey. The NMR qualitative analysis showed distinguishment among monofloral buckwheat, heather, and linden honey. Using NMR tyrosine, proline, alanine, and lactic acid, we found characteristic chemical shifts in buckwheat honey, with monosubstituted benzene derivatives and ethanol in heather honey and CDCA derivatives in linden honey. This study proves the validity of the combination of multiple analytical methods, statistical data treatment, and PCA to differentiate various natural monofloral honey classes, thus guaranteeing botanical authentication and the honey quality and origin.

Supplementary Materials: The following are available online at https://www.mdpi.com/article/10.3390/foods11010042/s1, Figures S1 and S2: Tukey simultaneous 95% confidence intervals differences of mean for $\delta^{13}C$ and $\delta^{15}N$. Figure S3: Fisher simultaneous 95% confidence interval differences of mean for total N (%). Figures S4–S6: PCA scree plots ($\delta 13C$, $\delta 15N$, total C and N in proteins; 27 organic compound quantified by UHPLC-HRMS; ^{1}H-NMR spectra intervals 9.0–6.0 and 3.0–0.5 ppm). Figure S7: ^{1}H-NMR spectra interval (0.5–3.0 and 4.5–8.5 ppm) overlayed comparison of monofloral buckwheat, clover, willow, linden, heather, and rapeseed honey. Supplementary workbook.

Author Contributions: Conceptualization and resources, K.D.L., J.R. and A.V.; methodology, validation, formal analysis, and investigation, K.D.L., M.B., J.N. (IRMS), J.R. (UHPLC-HRMS), R.K., M.S. (NMR), I.E. (Melissopalynology); writing—original draft preparation, K.D.L. and V.R.; visualization, K.D.L. and R.K.; writing—review and editing, V.R., L.B., J.R., R.K., M.B. and A.V.; supervision, K.D.L., V.R. and A.V.; project administration and funding acquisition, A.V. All authors have read and agreed to the published version of the manuscript.

Funding: This research was funded by the grant "Recognition of monofloral honey of Latvian origin using nuclear magnetic resonance, chromatographic, isotope ratio mass spectrometry and chemometric methods" grant no. LZP-2020/2-0200.

Data Availability Statement: Data are contained within the article or Supplementary Material.

Acknowledgments: "Strengthening of the capacity of doctoral studies at the University of Latvia within the framework of the new doctoral model", identification no. 8.2.2.0/20/I/006.

Conflicts of Interest: The authors declare no conflict of interest.

References

1. National Apiculture Programmes 2020–22. Available online: https://ec.europa.eu/info/sites/default/files/food-farming-fisheries/animals_and_animal_products/documents/honey-apiculture-programmes-overview-2020-2022.pdf (accessed on 11 November 2021).
2. Geana, E.I.; Ciucure, C.T. Establishing authenticity of honey via comprehensive Romanian honey analysis. *Food Chem.* **2020**, *306*, 125595. [CrossRef]
3. Bontempo, L.; Camin, F.; Ziller, L.; Perini, M.; Nicolini, G.; Larcher, R. Isotopic and elemental composition of selected types of Italian honey. *Meas. J. Int. Meas. Confed.* **2017**, *98*, 283–289. [CrossRef]

4. Puścion-Jakubik, A.; Borawska, M.H.; Socha, K. Modern methods for assessing the quality of Bee Honey and botanical origin identification. *Foods* **2020**, *9*, 1028. [CrossRef]
5. Rusko, J.; Vainovska, P.; Vilne, B.; Bartkevics, V. Phenolic profiles of raw mono- and polyfloral honeys from Latvia. *J. Food Compos. Anal.* **2021**, *98*, 103813. [CrossRef]
6. Kečkeš, S.; Gašić, U.; Veličković, T.Ć.; Milojković-Opsenica, D.; Natić, M.; Tešić, Ž. The determination of phenolic profiles of Serbian unifloral honeys using ultra-high-performance liquid chromatography/high resolution accurate mass spectrometry. *Food Chem.* **2013**, *138*, 32–40. [CrossRef]
7. Lo Dico, G.M.; Ulrici, A.; Pulvirenti, A.; Cammilleri, G.; Macaluso, A.; Vella, A.; Giaccone, V.; Lo Cascio, G.; Graci, S.; Scuto, M.; et al. Multivariate statistical analysis of the polyphenols content for the discrimination of honey produced in Sicily (Southern Italy). *J. Food Compos. Anal.* **2019**, *82*, 103225. [CrossRef]
8. Kuballa, T.; Brunner, T.S.; Thongpanchang, T.; Walch, S.G.; Lachenmeier, D.W. Application of NMR for authentication of honey, beer and spices. *Curr. Opin. Food Sci.* **2018**, *19*, 57–62. [CrossRef]
9. Consonni, R.; Bernareggi, F.; Cagliani, L.R. NMR-based metabolomic approach to differentiate organic and conventional Italian honey. *Food Control* **2019**, *98*, 133–140. [CrossRef]
10. Boffo, E.F.; Tavares, L.A.; Tobias, A.C.T.; Ferreira, M.M.C.; Ferreira, A.G. Identification of components of Brazilian honey by 1H NMR and classification of its botanical origin by chemometric methods. *LWT—Food Sci. Technol.* **2012**, *49*, 55–63. [CrossRef]
11. Spiteri, M.; Jamin, E.; Thomas, F.; Rebours, A.; Lees, M.; Rogers, K.M.; Rutledge, D.N. Fast and global authenticity screening of honey using 1H-NMR profiling. *Food Chem.* **2015**, *189*, 60–66. [CrossRef]
12. Schievano, E.; Stocchero, M.; Zuccato, V.; Conti, I.; Piana, L. NMR assessment of European acacia honey origin and composition of EU-blend based on geographical floral markers. *Food Chem.* **2019**, *288*, 96–101. [CrossRef]
13. Brereton, R.G. *Chemometrics: Data Analysis for the Laboratory and Chemical Plant*; John Wiley & Sons: Chichester, UK, 2003; pp. 185–220.
14. Vasić, V.; Đurđić, S.; Tosti, T.; Radoičić, A.; Lušić, D.; Milojković-Opsenica, D.; Tešić, Ž.; Trifković, J. Two aspects of honeydew honey authenticity: Application of advance analytical methods and chemometrics. *Food Chem.* **2020**, *305*, 125457. [CrossRef] [PubMed]
15. Wang, J.; Xue, X.; Du, X.; Cheng, N.; Chen, L.; Zhao, J.; Zheng, J.; Cao, W. Identification of Acacia Honey Adulteration with Rape Honey Using Liquid Chromatography–Electrochemical Detection and Chemometrics. *Food Anal. Methods* **2014**, *7*, 2003–2012. [CrossRef]
16. Geană, E.I.; Ciucure, C.T.; Costinel, D.; Ionete, R.E. Evaluation of honey in terms of quality and authenticity based on the general physicochemical pattern, major sugar composition and $\delta 13C$ signature. *Food Control* **2020**, *109*, 106919. [CrossRef]
17. Wen, Y.Q.; Zhang, J.; Li, Y.; Chen, L.; Zhao, W.; Zhou, J.; Jin, Y. Characterization of Chinese unifloral honeys based on proline and phenolic content as markers of botanical origin, using multivariate analysis. *Molecules* **2017**, *22*, 735. [CrossRef]
18. Requirements for Food Quality Schemes, Procedures for the Implementation, Operation, Monitoring, and Control Thereof. Annex 7. Available online: https://likumi.lv/ta/en/en/id/268347-requirements-for-food-quality-schemes-procedures-for-the-implementation-operation-monitoring-and-control-thereof (accessed on 11 November 2021).
19. Bilikova, K.; Krakova, T.K.; Yamaguchi, K.; Yamaguchi, Y. Major royal jelly proteins as markers of authenticity and quality of honey. *Arh. Hig. Rada Toksikol.* **2015**, *66*, 259–267. [CrossRef] [PubMed]
20. Shtangeeva, I.; Buša, L.; Viksna, A. Carbon and nitrogen stable isotope ratios of soils and grasses as indicators of soil characteristics and biological taxa. *Appl. Geochem.* **2019**, *104*, 19–24. [CrossRef]
21. Schievano, E.; Sbrizza, M.; Zuccato, V.; Piana, L.; Tessari, M. NMR carbohydrate profile in tracing acacia honey authenticity. *Food Chem.* **2020**, *309*, 125788. [CrossRef]
22. Chen, C.T.; Chen, B.Y.; Nai, Y.S.; Chang, Y.M.; Chen, K.H.; Chen, Y.W. Novel inspection of sugar residue and origin in honey based on the 13C/12C isotopic ratio and protein content. *J. Food Drug Anal.* **2019**, *27*, 175–183. [CrossRef]
23. Wu, L.; Du, B.; Vander Heyden, Y.; Chen, L.; Zhao, L.; Wang, M.; Xue, X. Recent advancements in detecting sugar-based adulterants in honey—A challenge. *TrAC—Trends Anal. Chem.* **2017**, *86*, 25–38. [CrossRef]
24. Schellenberg, A.; Chmielus, S.; Schlicht, C.; Camin, F.; Perini, M.; Bontempo, L.; Heinrich, K.; Kelly, S.D.; Rossmann, A.; Thomas, F.; et al. Multielement stable isotope ratios (H, C, N, S) of honey from different European regions. *Food Chem.* **2010**, *121*, 770–777. [CrossRef]
25. Fang, L.; He, X.; Zhang, X.; Yang, Y.; Liu, R.; Shi, S.; Shi, X.; Zhang, Y. A small amount of nitrogen transfer from white clover to citrus seedling via common arbuscular mycorrhizal networks. *Agronomy* **2021**, *11*, 32. [CrossRef]
26. Rouwenhorst, R.J.; Frank Jzn, J.; Scheffers, W.A.; van Dijken, J.P. Determination of protein concentration by total organic carbon analysis. *J. Biochem. Biophys. Methods* **1991**, *22*, 119–128. [CrossRef]
27. Chua, L.S.; Lee, J.Y.; Chan, G.F. Honey protein extraction and determination by mass spectrometry. *Anal. Bioanal. Chem.* **2013**, *405*, 3063–3074. [CrossRef]
28. Bocian, A.; Buczkowicz, J.; Jaromin, M.; Hus, K.K.; Legáth, J. An effective method of isolating honey proteins. *Molecules* **2019**, *24*, 2399. [CrossRef]
29. Karabagias, I.K.; Maia, M.; Karabagias, V.K.; Gatzias, I.; Badeka, A.V. Characterization of eucalyptus, chestnut and heather honeys from Portugal using multi-parameter analysis and chemo-calculus. *Foods* **2018**, *7*, 194. [CrossRef]

30. Bleha, R.; Shevtsova, T.V.; Živčáková, M.; Korbářová, A.; Ježková, M.; Saloň, I.; Brindza, J.; Synytsya, A. Spectroscopic discrimination of bee pollen by composition, color, and botanical origin. *Foods* **2021**, *10*, 1682. [CrossRef]
31. Halagarda, M.; Groth, S.; Popek, S.; Rohn, S.; Pedan, V. Antioxidant activity and phenolic profile of selected organic and conventional honeys from Poland. *Antioxidants* **2020**, *9*, 44. [CrossRef] [PubMed]
32. Pauliuc, D.; Dranca, F.; Oroian, M. Antioxidant activity, total phenolic content, individual phenolics and physicochemical parameters suitability for Romanian honey authentication. *Foods* **2020**, *9*, 306. [CrossRef]
33. Cianciosi, D.; Forbes-Hernández, T.Y.; Afrin, S.; Gasparrini, M.; Reboredo-Rodriguez, P.; Manna, P.P.; Zhang, J.; Lamas, L.B.; Flórez, S.M.; Toyos, P.A.; et al. Phenolic compounds in honey and their associated health benefits: A review. *Molecules* **2018**, *23*, 2322. [CrossRef]
34. Gerginova, D.; Simova, S.; Popova, M.; Stefova, M.; Stanoeva, J.P.; Bankova, V. NMR profiling of North Macedonian and Bulgarian honeys for detection of botanical and geographical origin. *Molecules* **2020**, *25*, 4687. [CrossRef] [PubMed]
35. Schievano, E.; Tonoli, M.; Rastrelli, F. NMR Quantification of Carbohydrates in Complex Mixtures. A Challenge on Honey. *Anal. Chem.* **2017**, *89*, 13405–13414. [CrossRef] [PubMed]
36. Consonni, R.; Cagliani, L.R.; Cogliati, C. Geographical discrimination of honeys by saccharides analysis. *Food Control* **2013**, *32*, 543–548. [CrossRef]
37. Consonni, R.; Cagliani, L.R.; Cogliati, C. NMR characterization of saccharides in italian honeys of different floral sources. *J. Agric. Food Chem.* **2012**, *60*, 4526–4534. [CrossRef]
38. Kortesniemi, M.; Slupsky, C.M.; Ollikka, T.; Kauko, L.; Spevacek, A.R.; Sjövall, O.; Yang, B.; Kallio, H. NMR profiling clarifies the characterization of Finnish honeys of different botanical origins. *Food Res. Int.* **2016**, *86*, 83–92. [CrossRef]
39. He, C.; Liu, Y.; Liu, H.; Zheng, X.; Shen, G.; Feng, J. Compositional identification and authentication of Chinese honeys by 1H NMR combined with multivariate analysis. *Food Res. Int.* **2020**, *130*, 108936. [CrossRef]
40. Zieliński, Ł.; Deja, S.; Jasicka-Misiak, I.; Kafarski, P. Chemometrics as a tool of origin determination of polish monofloral and multifloral honeys. *J. Agric. Food Chem.* **2014**, *62*, 2973–2981. [CrossRef]

Article

Evaluating the Presence and Contents of Phytochemicals in Honey Samples: Phenolic Compounds as Indicators to Identify Their Botanical Origin

Lua Vazquez [1], Daniel Armada [1], Maria Celeiro [1], Thierry Dagnac [2,*] and Maria Llompart [1,*]

1 CRETUS, Department of Analytical Chemistry, Nutrition and Food Science, Universidade de Santiago de Compostela, E-15782 Santiago de Compostela, Spain; lua.vazquez.ferreiro@usc.es (L.V.); daniel.armada.alvarez@usc.es (D.A.); maria.celeiro.montero@usc.es (M.C.)
2 Galician Agency for Food Quality-Agronomic and Agrarian Research Centre (AGACAL-CIAM), Unit of Food and Feed Safety and Organic Contaminants, Apartado 10, E-15080 A Coruña, Spain
* Correspondence: thierry.dagnac@xunta.gal (T.D.); maria.llompart@usc.es (M.L.)

Citation: Vazquez, L.; Armada, D.; Celeiro, M.; Dagnac, T.; Llompart, M. Evaluating the Presence and Contents of Phytochemicals in Honey Samples: Phenolic Compounds as Indicators to Identify Their Botanical Origin. *Foods* **2021**, *10*, 2616. https://doi.org/10.3390/foods10112616

Academic Editors: Olga Escuredo and M. Carmen Seijo

Received: 4 October 2021
Accepted: 25 October 2021
Published: 28 October 2021

Publisher's Note: MDPI stays neutral with regard to jurisdictional claims in published maps and institutional affiliations.

Copyright: © 2021 by the authors. Licensee MDPI, Basel, Switzerland. This article is an open access article distributed under the terms and conditions of the Creative Commons Attribution (CC BY) license (https://creativecommons.org/licenses/by/4.0/).

Abstract: Honey is a natural product well known for its beneficial properties. It contains phytochemicals, a wide class of nutraceuticals found in plants, including compounds with highly demonstrated antimicrobial and antioxidant capacities as phenolic compounds and flavonoids. The main goal of this work is the development of a miniaturized and environmentally friendly methodology to obtain the phenolic profile of Galician honeys (Northwest Spain) from different varieties such as honeydew, chestnut, eucalyptus, heather, blackberry and multi-floral. The total phenolic content (TPC) and antioxidant activity (AA) were also evaluated. As regards sample preparation, miniaturized vortex (VE) and ultrasound assisted extraction (UAE) employing aqueous-based solvents were performed. Individual quantification of 41 target phenolic compounds was carried out by liquid chromatography-tandem mass spectrometry (LC-MS/MS). Results revealed the presence of 25 phenolic compounds in the 91 analyzed samples, reaching concentrations up to 252 µg g^{-1}. Statistical tools such as analysis of variance (ANOVA) and principal component analysis (PCA) were employed to obtain models that allowed classifying the different honeys according to their botanical origin. Obtained results, based on TPC, AA and ∑phenolic compounds showed that significant differences appeared depending on the honey variety, being several of the identified phenol compounds being responsible of the main differentiation.

Keywords: honey; polyphenols; phenolic profile; total phenolic content; antioxidant activity; liquid chromatography; tandem mass spectrometry; principal components analysis

1. Introduction

Honey is a natural food product well known not only for its nutritional value, but also for its antimicrobial, antiviral, antifungal, anticancer, and antidiabetic properties, as several in vitro and in vivo studies have demonstrated [1]. From a compositional point of view, honey is a highly concentrated solution of complex mixture of sugars: fructose (38%), glucose (31%), water (17%), maltose (7%), as well as trisaccharides, other higher carbohydrates, sucrose, minerals, vitamins, and enzymes. Its composition depends strongly on the plant species from which the nectar or the honeydew was collected, and other factors, such as postharvest treatments, geographical, environmental or climate conditions [2,3]. Honey is among the top ten foods with the highest adulteration rate in the European Union, that implies a detrimental to its quality and consumers safety [4]. To protect this valuable food, a Codex standard for honey was adopted by the Codex Alimentarius Commission in 1981, being further revised in 1987, 2001 and 2019, to regulate its production and storage, establishing parameters to guarantee its quality [5]. In 2001, the European Council, following the Codex recommendations, established the Directive

2001/110/EC [6], amended 2014/63/EU [7] that laid down the production and trading parameters of honey within the member states of the EU. However, several countries issue national provisions, decisions, and guidelines defining their own physicochemical, organoleptic and microscopic characteristics, enhancing the difficulties of the applicability of harmonized regulations [8].

The identification of honey botanical origin is a valuable information to assure honey quality. In this way, the analysis of its phenolic composition has been employed as a tool for its classification and authentication [1,9,10]. Phenolic compounds are secondary metabolites of plants generally involved in their defense against ultraviolet radiation or pathogens and have been recognized as the main responsible for the antioxidant activity of honey [11–13]. The most abundant phenol- types in honey are flavonoids, especially flavones and flavanols, as well as phenolic acids derived from benzoic and cinnamic acids [2,14].

Several analytical procedures have been reported to determine honey physicochemical properties including colour, viscosity, pH, moisture, free acidity, electrical conductivity, sugars, HMF (hydroxymethylfurfural) content, formol index and insoluble solids [15–19], but due to the high number of existing honey varieties, more specific techniques are needed. The use of chromatography coupled to mass detectors (MS) to obtain a deep chemical characterization of this product is a very valuable option. However, the major drawback for honey analysis is sample preparation since it is a very complex matrix. To establish the honey aromatic profile, the combination of solid-phase microextraction (SPME) with gas chromatography-mass spectrometry (GC-MS) has been the main employed technique [20–22]. On the other hand, for the determination of more polar analytes, including phenolic compounds, traditional sample preparation involves the use of solid-liquid or liquid-liquid (SLE, LLE) before LC-MS or HPLC-UV analysis. However, these techniques are long time consuming, requiring large amounts of organic solvents and further clean-up steps before analysis. Microwave-assisted extraction (MAE) and ultrasound assisted extraction (UAE) have been also proposed as extraction techniques to determine phenolic compounds in honey, but their use was not satisfactory in the presence of thermosensitive flavonoids such as quercetin, kaempferol or myricetin, that are almost degraded as consequence of radiation. On the other hand, both extraction techniques seemed to be suitable for the extraction of phenolic acids [20].

Therefore, the goal of this work is the development of a miniaturized analytical methodology to obtain the phenolic profile of Galician honeys (Northwest Spain) from different varieties and nectar sources. A green, fast and low-cost sample preparation strategy based on vortex extraction (VE) followed by ultrasound assisted extraction (UAE) employing aqueous- based solvents was assessed. Individual quantification of 41 target phenolic compounds was carried out by liquid chromatography-tandem mass spectrometry (LC-MS/MS). Other indexes such as the total phenolic content (TPC) and antioxidant activity (AA) were also evaluated. Finally, advanced statistical tools such as analysis of variance (ANOVA) and principal component analysis (PCA) were employed to obtain models that allow classifying the different honeys according to their origins.

2. Materials and Methods

2.1. Chemicals, Reagents and Materials

The target phenolic compounds, their CAS numbers, molecular mass, log Kow, retention time and MS/MS transitions are summarized in Table 1. Methanol and ultrapure water, both MS grade, were supplied by Scharlab (Barcelona, Spain). Hydrochloric acid, formic acid, Folin–Ciocalteu's phenol reagent (2M), 2,2-diphenyil-1-picrylhydrazyl (DPPH), and 6-hydroxy-2,5,7,8-tetramethylchroman-2-carboxylic acid (Trolox®) were purchased from Sigma–Aldrich (Darmstadt, Germany). Sodium carbonate was supplied by Panreac (Barcelona, Spain).

Table 1. Target phenolic compounds: CAS number, molecular mass (Mm), log Kow, retention time, ionization mode and MS/MS transitions.

Phenolic Compounds	CAS	Mm (g mol^{-1})	log Kow	Retention Time (min)	Ionization Mode [1]	MS/MS Transitions [2]
gallic acid	149-91-7	170.1	0.70	2.61	−	169.02 → 125.04 (17) 169.02 → 153.1 (15)
phloroglucinic acid	71989-93-0	188.1	1.28	4.22	+	168.98 → 150.99 (17) 168.98 → 83.02 (23) 168.98 → 107.02 (22)
β-resorcylic acid [3]	89-86-1	154.1	1.63	5.00	+	153.00 → 109.05 (16) 153.00 → 65.09 (19) 153.00 → 67.07 (23)
protocatechuic acid [3]	99-50-3	154.1	0.86	5.00	+	152.98 → 109.04 (17) 152.98 → 91.04 (28) 152.98 → 108.03 (26)
caftaric acid	67879-58-7	312.2	0.21	4.78	−	310.96 → 178.97 (17) 310.96 → 148.96 (14)
protocatechualdehyde	139-85-5	138.1	1.09	5.05	+	137.07 → 136.11 (21) 137.07 → 91.09 (24) 137.07 → 92.13 (25)
procyanidin B1	20315-25-7	578.5		5.07	−	577.03 → 407.06 (26) 577.03 → 288.93 (25) 577.03 → 424.97 (26)
p-hydroxybenzoic acid	99-96-7	138.1	1.58	5.35	−	137.00 → 93.00 (17) 137.00 → 65.00 (27)
gentisic acid	490-79-9	117.1	1.74	5.38	+	152.96 → 108.00 (24) 152.96 → 81.02 (21) 152.96 → 109.01 (16)
catechin	18829-70-4	290.3	0.51	5.50	+	289.00 → 245.02 (17) 289.00 → 203.11 (22)
3-hydroxyphenylacetic acid	621-37-4	152.2	0.85	5.70	−	151.00 → 65.00 (20) 151.00 → 79.00 (20)
procyanidin B2	29106-49-8	578.5	2.29	5.96	−	577.03 → 407.06 (26) 577.03 → 288.93 (25) 577.03 → 424.97 (26)
gentisaldehyde	1194-98-5	138.1	1.53	6.11	+	136.99 → 108.02 (21) 136.99 → 81.08 (18) 136.99 → 109.04 (14)
chlorogenic acid	327-97-9	354.3	1.01	6.12	+	353.00 → 191.07 (22) 353.00 → 85.09 (43) 353.00 → 93.07 (45)
3-hydroxybenzaldehyde	100-83-4	122.1	1.29	6.22	+	121.02 → 93.05 (20) 121.02 → 92.05 (23) 121.02 → 120.04 (19)
4-hydroxybenzaldehyde	123-08-0	122.1	1.36	6.25	+	122.97 → 95.05 (13) 122.97 → 51.10 (36) 122.97 → 77.05 (20)
vanillic acid	121-34-6	168.2	1.43	6.29	−	167.00 → 108.00 (27) 167.00 → 152.00 (18)
γ-resorcylic acid	303-07-1	154.1	2.20	6.33	+	153.00 → 109.05 (17) 153.00 → 65.09 (21) 153.00 → 135.02 (16)
α-resorcylic acid	99-10-5	154.1	0.86	6.33	+	152.97 → 109.01 (15) 152.97 → 65.06 (16) 152.97 → 67.05 (20)
veratric acid	93-07-2	182.2	1.61	6.45	+	182.96 → 137.08 (6) 182.96 → 106.99 (22)

Table 1. Cont.

Phenolic Compounds	CAS	Mm (g mol^{-1})	log Kow	Retention Time (min)	Ionization Mode [1]	MS/MS Transitions [2]
caffeic acid	331-39-5	180.2	1.15	6.50	−	178.98 → 135.03 (19) 178.98 → 134.01 (28)
epicatechin	35323-91-2	290.3	0.51	6.56	+	289.00 → 245.02 (17) 289.00 → 203.11 (22)
epigallocatechin gallate	989-51-5	458.4	2.56	6.79	+	457.15 → 169.05 (21) 457.15 → 125.09 (42) 457.15 → 305.09 (21)
gallocatechin gallate	84650-60-2	458.4	2.56	7.29	+	457.15 → 169.05 (21) 457.15 → 125.09 (42) 457.15 → 305.09 (21)
procyanidin A2	41743-41-3	576.5	2.52	7.32	−	577.09 → 287.00 (32) 577.09 → 136.98 (62) 577.09 → 425.08 (13)
umbelliferone	93-35-6	162.1	1.58	7.80	+	162.99 → 107.04 (22) 162.99 → 77.05 (34) 162.99 → 91.05 (20)
p-coumaric acid	501-98-4	164.2	1.79	7.89	+	163.02 → 119.07 (18) 163.02 → 93.07 (37) 163.02 → 117.05 (38)
catechin gallate	130405-40-2	442.3	2.62	8.01	+	441.13 → 289.13 (20) 441.13 → 125.08 (42) 441.13 → 169.05 (24)
trans-ferulic acid	537-98-4	194.2	1.51	8.33	−	192.80 → 177.90 (12) 192.80 → 133.90 (16)
veratraldehyde	120-14-9	166.2	1.22	8.92	+	167.01 → 139.05 (13) 167.01 → 108.05 (21) 167.01 → 124.03 (18)
4-anisaldehyde	123-11-5	136.1	1.76	10.03	+	136.97 → 109.05 (12) 136.97 → 77.05 (23) 136.97 → 94.04 (18)
miquelianin	22688-79-5	478.4	0.20	10.32	+	479.09 → 461.50 (14) 479.09 → 302.96 (18)
rutin	153-18-4	610.5	0.15	10.35	−	609.18 → 270.92 (96) 609.18 → 178.87 (44) 609.18 → 300.01 (37)
isoquercitrin	482-35-9	463.4	0.76	10.43	+	465.07 → 256.90 (41) 465.07 → 302.97 (14)
myricetin	529-44-2	318.2	1.42	11.43	+	319.00 → 153.02 (31) 319.00 → 217.06 (31) 319.00 → 245.06 (27)
3,4,5-trimethoxycinnamic acid	90-50-6	238.2	1.58	11.59	+	239.03 → 221.04 (11) 239.03 → 162.99 (27) 239.03 → 190.01 (19)
3,5-dimethoxybenzaldehyde	7311-34-4	166.2	1.87	11.81	+	167.15 → 124.03 (17) 167.15 → 77.05 (26)
quercetin	117-39-5	302.2	1.48	12.10	+	303.09 → 229.10 (28) 303.09 → 153.04 (33)
kaempferol	520-18-3	286.2	1.96	12.57	−	285.07 → 184.91 (30) 285.07 → 239.12 (35)
apigenin	520-36-5	270.2	3.02	12.63	−	269.09 → 117.12 (37) 269.09 → 149.12 (26) 269.09 → 151.06 (26)
chrysin	480-40-0	254.2	3.52	13.24	+	253.13 → 143.18 (30) 253.13 → 63.20 (34) 253.13 → 145.16 (31)

[1] "−" and "+" indicate negative and positive ionization modes, respectively. [2] Underlined MS/MS transition used for quantification purpose. [3] Isomers: 2,4/3,4-dihydroxybenzoic acid.

Phenolic individual standard stock solutions (500–1000 µg mL^{-1}) were prepared in methanol. Further dilutions and mixtures were prepared in acidified water (0.1% formic acid)/methanol (80:20, v/v) (AW/MeOH). All solutions were stored at $-20\ °C$ and protected from light. All chemicals and reagents were of analytical grade.

A vortex stirrer by Velp Scientifica (Usmate, Italy) and an ultrasound bath (50 kHz) from JP Selecta (Barcelona, Spain) were employed to perform the extractions.

2.2. Honey Samples

Ninety-one honey samples from Galicia (Northwest Spain) were kindly supplied by the protected geographical indication (P.G.I.) Mel de Galicia. Samples were received in glass jars sealed with aluminum caps. They were stored in the original containers at controlled temperature (15 °C) and kept away from light until their analysis.

The methodology used for the study of the botanical characteristics was based on the determination of the pollen contained in the honey by centrifugation. In total, 52% of the honey samples contained between 2000 and 10,000 grains of pollen per gram of honey, according to the classes of Maurizio; 39% of these samples contained between 10,000 and 50,000 grains of pollen per gram of honey [19]. The pollen spectrum of the samples consisted of 82 different pollen types, with 45% of them likely to be labelled as monofloral, while the remaining 55% were considered multi-floral, in which was included 16% whose majority origin was honeydew (HD). As regards monofloral honeys, the chestnut (CN, 34%), the blackberry (BL, 27.3%), the eucalyptus (EU, 25%) and, to a lesser extent, the heather (HE, 13.7%) stand out.

It should be noted that, as for the main proportion of the honey produced in Galicia, the main types were *Castanea, Eucalyptus, Erica, Rubus* and *Cytisus*, all of them in the dominant category or as companion in the pollen spectrum of honey.

2.3. VE-UAE Procedure

Under the optimal experimental conditions (see Section 3.2), 0.1 g of honey were weighted in a 1.8 mL glass vial and 1 mL of acidified water (0.1% formic acid)/methanol (80:20, v/v) (AW/MeOH) was added. The vial was sealed with an aluminum cap furnished with PTFE-faced septa and the solution was stirred by vortex for 1 min. Afterwards, the vial was immersed in an ultrasound bath for 1 min (20 °C, 50 KHz). The obtained extract was filtered through 0.22 µm polytetrafluoroethylene (PTFE) filters and directly injected in the LC-MS/MS system for phenols analysis (see Section 2.6). The experimental procedure is summarized in Figure 1.

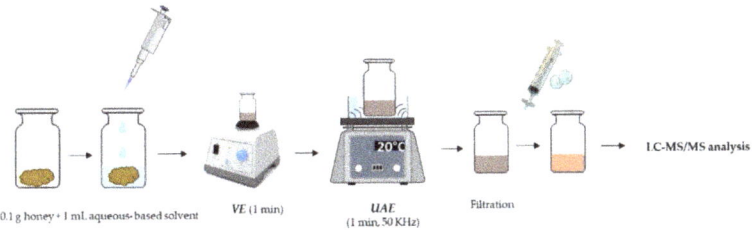

Figure 1. Schematic representation of the VE-UAE experimental procedure.

2.4. Determination of TPC

The total phenolic content (TPC) of honey samples was determined according to the Folin–Ciocalteu (FC) colorimetric method described by Singleton and Rosssi [23]. Honey sample preparation was performed employing a modified method of Pauliuc et al. [16]. Briefly, 0.5 g of honey sample were diluted in 5 mL of methanol/water (40:60, v/v, pH = 2, HCl) and magnetically stirred for 15 min. Afterwards, 1.3 mL of this solution was diluted (1:10, v/v) in water up to a final volume of 13 mL. Then, an aliquot of 5 mL was placed on a Falcon tube and 100 µL of Folin–Ciocalteu's phenol reagent and 1 mL of Na$_2$CO$_3$ solution

(20%, w/v) were added. The Falcon tubes were kept away from light for 30 min. Afterwards, the absorbance was measured at 760 nm in a UV-Vis spectrophotometer Shimadzu UVmini-1240 (Kyoto, Japan). The TPC was quantified employing a calibration curve prepared with gallic acid standards solutions ranging between 1–20 mg L^{-1} (R^2 = 0.9990) and expressed as mg of gallic acid equivalent (GAE) per 100 g of honey (mg GAE 100 g^{-1}).

2.5. Determination of AA

The antioxidant activity (AA) was determined by a modified method of Brand–Williams et al. [24]. Briefly, 200 µL of the honey solution (0.5 g of honey diluted in 5 mL of methanol/water, 40:60, v/v, pH = 2, HCl) were introduced in a Falcon tube and 3.9 mL of the DPPH reagent solution (0.1 mM in methanol) were added. After 30 min in the absence of light, the absorbance was measured at 515 nm. The AA was quantified employing a calibration curve prepared with Trolox® (0.1–0.9 mmol TRE g^{-1}, R^2 = 0.9970). The AA were expressed as micromoles of Trolox® equivalents (TRE) per 100 g of honey (µmol TRE 100 g^{-1}).

2.6. LC-MS/MS Analysis

The optimal instrumental conditions for the detection of the target phenols were adapted from Celeiro et al. [25]. LC-MS/MS analysis was performed employing a Thermo Scientific (San José, CA, USA) instrument based on a TSQ Quantum Ultra^{TM} triple quadrupole mass spectrometer equipped with a HESI-II (heated electrospray ionization) source and an Accela Open autosampler with a 20 µL loop. The chromatographic separation was achieved on a Kinetex C18 column (2.6 µm, 100 × 2.1 mm) with a guard column (SecurityGuard^{TM} ULTRA Holder) obtained from Phenomenex (Torrance, CA, USA). The injection volume was 10 µL and the column temperature was set at 50 °C. The mobile phase consisted of water (A) and methanol (B), both containing 0.1% formic acid. The eluted program started with 5% of B (held 5 min), it was up to 90% of B over 11 min (held 3 min). Then, initial conditions were reached in 5 min. The mobile phase flow rate was 200 µL min^{-1}. The total run time for each injection was 20 min. The mass spectrometer and the HESI-II source were working simultaneously in the positive and negative mode (see ionization mode for each target compound in Table 1). Selected reaction monitoring (SRM) acquisition mode was implemented monitoring 2 or 3 transitions per compound (see Table 1), for an unequivocal identification and quantification of the target compounds. The system was operated by Xcalibur 2.2 and Trace Finder^{TM} 3.2.

2.7. Statistical Analysis

Analysis of variance (ANOVA) and principal component analysis (PCA) were performed employing Statgraphics Centurion XVIII software package (Manugistics, Rockville, MD, USA).

3. Results

3.1. Selection of the Solvent

The 41 target phenols present a high polarity range, as can be seen in Table 1, with log K_{OW} values ranging between 0.2 and 3.5. Therefore, their chromatographic separation and response are expected to be highly dependent on the dilution solvent. Different aqueous-based solvents were tested since one of the objectives of this work is the development of a green methodology, reduced usage of toxic solvents.

Experiments were performed employing standard solutions containing the 41 target phenols at 200 µg L^{-1} prepared in: methanol (MeOH), acidified water with 0.1% formic acid (AW) and acidified water (0.1% formic acid)/methanol (80:20, v/v) (AW/MeOH). Results for some target phenols from high polar to low polar ones are shown in Figure 2. As can be seen, both aqueous based- solvents provided the highest chromatographic response for most compounds, especially for the highest polar ones, such as cinnamic- and benzoic- acids derivatives (gallic-, caftaric-, gentisic-, chlorogenic- acid, etc.). In contrast, the use of methanol to prepare the standard solutions for these compounds, resulted in chromatographic responses up to three times lower than those obtained with the aqueous-based solvents. Regarding medium polarity compounds, lower differences were observed for some compounds (α-resorcylic acid, umbelliferone, veratraldehyde, etc.) between the responses for the three tested solvents, whereas others achieved worse response for MeOH (epicatechin, p-coumaric acid, etc.). On the other hand, higher responses were obtained with AW/MeOH and MeOH for the low polar compounds, such as the flavonols quercetin and kaempferol and the flavones apigenin and chrysin.

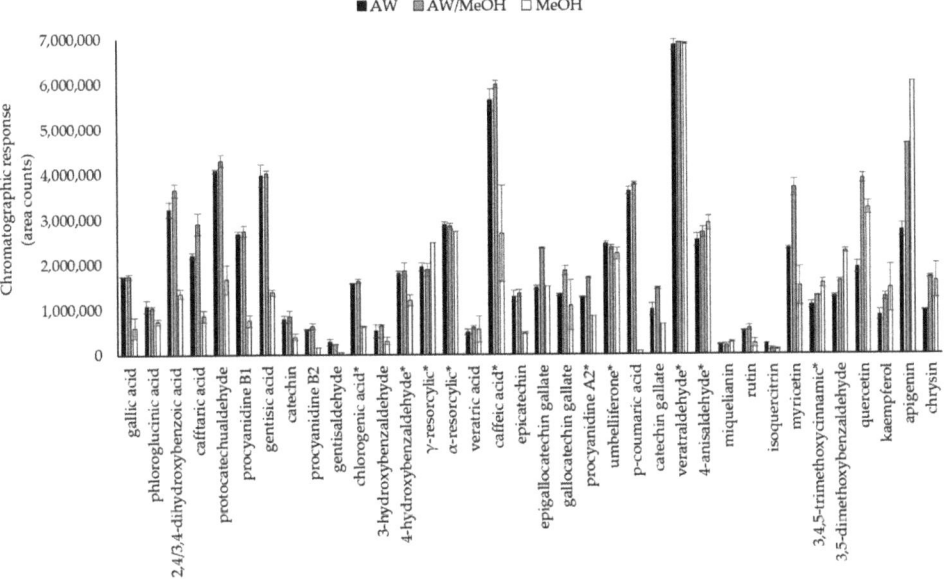

Figure 2. Chromatographic response for some phenols standard solutions (200 µg L^{-1}) prepared in AW: acidified water (0.1% formic acid); AW/MeOH: acidified water (0.1% formic)/methanol (80:20, v/v); MeOH: methanol. * Divided/10.

As it is well known, the solvent not only affects the chromatographic response (abundance) of the analytes, but also highly affects the retention efficiency and thus, the chromatographic peak shape. Figure 3 shows the comparison between the chromatographic peaks for protocatechualdehyde (Figure 3a) and chlorogenic acid (Figure 3b) (200 µg L^{-1}) prepared in AW, AW/MeOH and MeOH.

As can be seen, the standard prepared in methanol presented the worst peak shape, whereas standard prepared in AW and AW/MeOH showed in both cases satisfactory peak resolution. This behaviour was similar for most compounds, especially for those eluting first. Therefore, in view of these results, both aqueous based-solvent solutions were selected for further experiments.

3.2. VE-UAE Optimization

As previously commented, honey is a viscous and complex matrix, which makes not easy work with. Therefore, the selection of the most suitable extraction solvent is crucial to obtain the highest extraction efficiency. In this case, the extraction solvent and sample amount were optimized to obtain not only the highest extraction efficiency, but also to assess the possibility of miniaturizing the sample preparation procedure, fulfilling with the green chemistry principles.

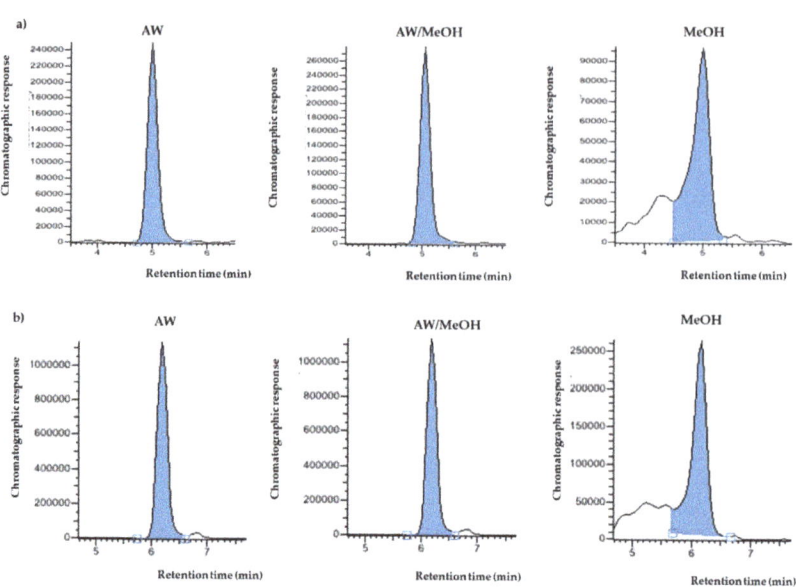

Figure 3. Peak shape comparison standard solutions (200 µg L^{-1}) prepared in AW, AW/MeOH and MeOH for: (**a**) protocatechualdehyde; (**b**) chlorogenic acid.

3.2.1. Extraction Solvent

Both aqueous-based solvents pre-selected in the preliminary studies, AW and AW/MeOH (see Section 3.1), were employed to prepare the honey samples. In this case, 0.1 g of honey in 1 mL of solvent were employed. Results for some detected compounds in two different honey samples from different origin, honeydew (HD) and multi-floral (MF) are represented in Figure 4a. In general, responses were similar for both aqueous- based solvents, excluding kaempferol, apigenin and chrysin. For these compounds, higher chromatographic responses, up to two times, were obtained in the two honey varieties, when AW/MeOH was employed as extractant. These results are in concordance with those previously obtained for the standard solutions of these flavones derivatives. For this reason, the solution AW/MeOH was selected.

3.2.2. Sample Amount

Until now, in most honey studies the employed amount of sample usually involves the use of several grams of honey [16,20,26]. Since one of the objectives of the work is to obtain a miniaturized methodology with a low sample, reagents and solvents consumption, two different sample sizes were evaluated: 0.1 g and 0.5 g diluted 1:10 w/v. Results are depicted in Figure 4b, for two honey samples varieties, HD and MF. As can be seen, in all cases responses were similar employing 0.1 g and 0.5 g, concluding that the use of only 0.1 g of honey were representative and homogeneous. Therefore, 0.1 g of honey sample and 1 mL of AW/MeOH were selected, allowing a miniaturization of the extraction VE-UAE procedure.

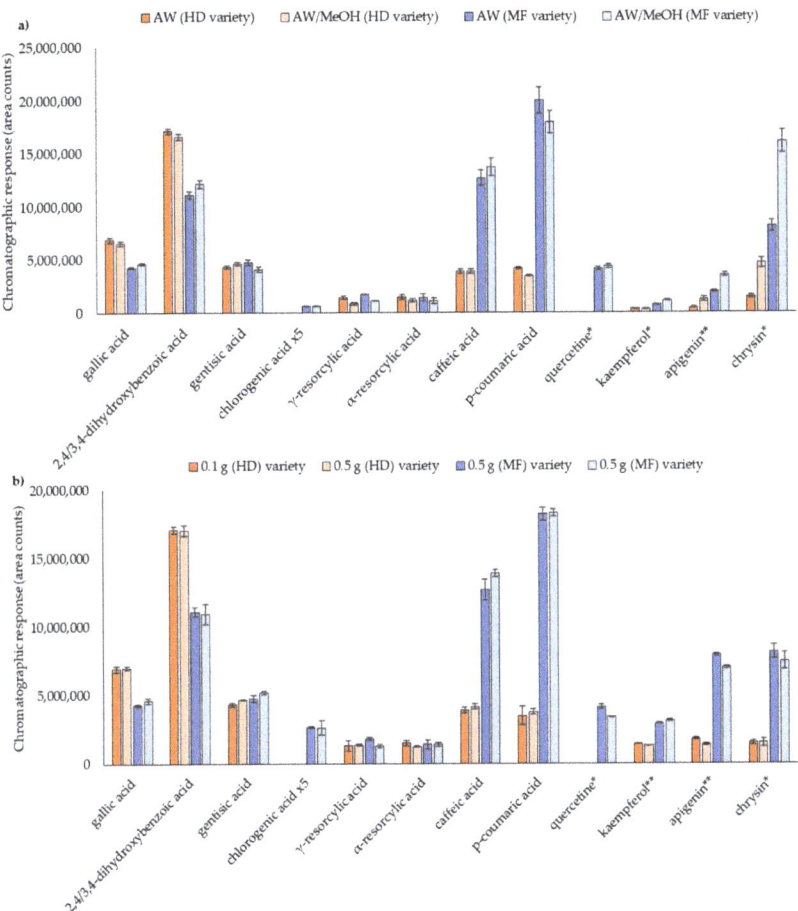

Figure 4. Optimization of the (**a**) extraction solvent; (**b**) sample size for two honey varieties (HD10 and MF32). * Response multiplied ×5; ** multiplied ×20.

3.3. VE-UAE-LC-MS/MS Performance

Under the optimal experimental conditions that involve the use of only 0.1 g of honey sample diluted in 1 mL of AW/MeOH, the whole VE-UAE-LC-MS/MS method was validated in terms of linearity, accuracy and precision. Limits of detection (LODs) were also calculated. Results are summarized in Table 2. Calibration curves were prepared in AW/MeOH containing the 41 target phenolic compounds, covering a concentration range for most compounds from 5 to 10,000 ng L^{-1}, with 11 concentration levels (5, 10, 20, 50, 100, 200, 500, 1000, 2000, 5000 and 10,000 ng L^{-1}) and three replicates per level. The method showed a good linearity, with coefficients of determination (R^2) higher than 0.99. Instrumental precision was evaluated within a day (n = 4) and amongst days (n = 5) for all the calibration concentration levels. Relative standard deviation (RSD) values for 200 µg L^{-1} are shown in Table 2, with mean values about 10%. To assess the accuracy of the proposed method, recovery studies were carried out employing a multi-floral honey sample (MF28). It is worth noting that only a few methods demonstrated accuracy for such a high number of phenolic compounds in honey samples and most of them imply the use of artificial matrices [20] or further experimental steps, mainly based on solid-phase extraction (SPE) to remove matrix components such as sugars [27]. The study was performed spiking the honey sample with the 41 target phenolic compounds at 2 µg g^{-1}. Results depicted in Table 2 show that recovery values ranged between 70 and 100% for most compounds with RSD values lower than 8%. Recovery percentages obtained in other studies that apply UAE were higher for some phenolic acids (gallic acid, p-coumaric acid) while higher values were obtained in the present study for other compounds, such as myricetin and kaempferol. The degradation of some flavonoids during the extraction procedure assisted with irradiation was demonstrated [20]. Those undesirable effects were not observed in the present study since UAE was only applied for 1 min.

Table 2. VE-UAE_LC-MS/MS performance: Linearity, precision, recoveries and LODs.

Phenolic Compounds	Linearity	Precision (RSD, %)		Recovery (%)	LOD (ng g^{-1})
	R^2	Intra-Day (n = 4)	Inter-Day (n = 5)		
gallic acid	0.9985	5.3	3.5	58.3 ± 9.0	39
phloroglucinic acid	0.9996	10	9.1	84.3 ± 4.7	137
β-resorcylic acid [1]	0.9954	11	7.2	78 ± 11	54
protocatechuic acid [1]	0.9937	3.4	4.5		
caftaric acid	0.9973	2.0	9.9	101 ± 5	22
protocatechualdehyde	0.9906	8.6	14	75.3 ± 5.2	16
procyanidine B1	0.9993	18	16	78 ± 11	17
gentisicacid	0.9996	14	16	90.5 ± 7.3	20
catechin	0.9917	6.5	15	61 ± 12	46
procyanidine B2	0.9926	3.9	18	59.0 ± 5.1	18
gentisaldehyde	0.9993	5.2	15	69.2 ± 1.0	91
chlorogenic acid	0.9992	4.3	2.6	59.2 ± 2.7	7.1
3-hydroxybenzaldehyde	0.9986	8.2	20	71.7 ± 4.4	17
4-hydroxybenzaldehyde	0.9978	13	15	113 ± 2	35
γ-resorcylic acid	0.9949	1.2	4.7	141 ± 2	14
α-resorcylic acid	0.9903	7.7	5.3	133 ± 7	12
veratric acid	0.9988	16	14	80.5 ± 4.6	40
caffeic acid	0.9962	4.3	5.6	76.1 ± 0.8	8.8
epicatechin	0.9946	11	9.7	54.7 ± 0.6	6.9
epigallocatechin gallate	0.9940	18	16	76.1 ± 8.0	121
gallocatechin gallate	0.9990	17	3.9	45.1 ± 8.3	79
procyanidine A2	0.9975	1.6	6.4	83.8 ± 4.2	12
umbelliferone	0.9928	8.9	12	106 ± 1	8.0
p-coumaric acid	0.9980	2.6	6.8	66.6 ± 1.7	7.9
catechin gallate	0.9988	7.9	11	71.7 ± 0.4	16
veratraldehyde	0.9999	4.2	4.1	77.3 ± 1.0	10
4-anisaldehyde	0.9953	5.2	9.1	88.9 ± 6.2	30
miquelianin	0.9928	13	6.1	115	192
rutin	0.9928	8.3	13	95.3 ± 6.7	29
isoquercitrin	0.9915	15	2.1	-	163
myricetin	0.9982	10	18	103 ± 19	216

Table 2. Cont.

Phenolic Compounds	Linearity R²	Precision (RSD, %) Intra-Day (n = 4)	Precision (RSD, %) Inter-Day (n = 5)	Recovery (%)	LOD (ng g^{-1})
3,4,5-trimethoxycinnamic acid	0.9944	12	12	86.9 ± 5.3	21
3,5-dimethoxybenzaldehyde	0.9985	15	13	76.5 ± 9.5	156
quercetin	0.9992	10	19	-	39
kaempferol	0.9906	5.9	9.7	61.5 ± 5.9	45
apigenin	0.9168	7.5	5.4	53.8 ± 2.8	7.0
chrysin	0.9948	10	6.6	70.3 ± 13	55
trans-ferulic acid	0.9972	7.5	4.2	105 ± 2.2	41
vanillic acid	0.9980	2.1	5.5	87.0 ± 5.7	24
p-hydroxybenzoic acid	0.9922	9.8	4.9	n.c.	7.0
3-hydroxyphenylacetic acid	0.9977	7.1	4.4	n.c.	60

[1] Sum of both isomers: 2,4/3,4- dihydroxybenzoic acid; n.c. Not calculated since the concentration in the sample was higher than the spike level (see sample MF28 in Table S2c).

Limits of detection (LODs) were calculated as the compound concentration giving a signal-to-noise ratio of three (S/N = 3) employing the honey sample spiked with the target compounds. Results depicted in Table 2 show that they were at the ng g^{-1} level for all target phenolic compounds.

3.4. Analysis of Real Honey Samples

3.4.1. TPC and AA

TPC and AA results for the 91 analyzed samples are summarized in Tables S1 and S2 for samples collected in 2018 and 2019, respectively. The ranges, mean and median concentrations for TPC, AA and individual phenolic compounds are shown in Table 3.

Results for TPC were similar in the two evaluated seasons. As shown in Table 3, TPC values ranged between 48–203 mg GAE 100g^{-1}. As expected, a relationship seems to exist between the total concentration of target phenolic compounds and the TPC values, since the highest TPC was found in the heather sample HE1 that shows the highest sum of phenolic compounds, 252 µg g^{-1}. On the other hand, most of EU and BL honey samples achieved low TPC.

The AA index ranged between 15–1017 µmol TRE 100 g^{-1} for the two seasons. Samples of honeydew achieved the highest antioxidant activity, reaching 1006 and 1017 µmol TRE 100 g^{-1} in sample HD4 and HD11, respectively. Results are in concordance, since most honey samples with high TPC values achieved high AA, as well. In the same way, the honey sample (EU11) with the lowest TPC (48 mg GAE 100 g^{-1}) also reached the minimum AA concentration (15 µmol TRE 100 g^{-1}).

Results of TPC and AA obtained in the Galician honeys were in consonance with those reported in other honeys from the same and different origin [19,28]. Thus, both indexes do not allow differentiating Galician honeys from other honeys, although they might allow to distinguish between the different honey varieties (see also Section 3.5).

Table 3. Range, mean concentration and median of phenolic compounds (μg g^{-1}), TPC (mg GAE 100g^{-1}) and AA (μmol TRE 100g^{-1}) for eucalyptus (EU), blackberry (BL), chestnut (CN) and chestnut/honeydew (CN/HD) honeys, honeydew (HD), heather (HE) and multi-floral (MF) honeys.

Phenolic Compounds	EU (N = 12)				BL (N = 8)				CN (N = 12)				CN/HD (N = 7)				HD (N = 12)				HE (N = 4)				MF (N = 36)			
	N	Range	Mean	Median	N	Range	Mean	Median	N	Range	Mean	Median	N	Range	Mean	Median	N	Range	Mean	Median	N	Range	Mean	Median	N	Range	Mean	Median
gallic acid	7	0.13–0.47	0.26	0.22	8	0.29–2.1	0.97	0.56	9	0.18–3.3	0.92	0.46	7	1.6–6.8	4.3	5.2	12	1.6–9.8	5.5	6.4	2	0.20–1.1	0.65	0.65	28	0.15–4.5	0.73	0.30
phloroglucinic acid	0				2	0.21–0.34	0.28	0.28	2	0.13–0.48	0.12	0.12	4	0.29–0.87	0.47	0.36	2	0.3–1.4	0.86	0.86	1	0.38	0.38	0.38	3	0.18–0.32	0.27	0.32
β-resorcylic acid/protocatechuic acid	12	0.22–0.55	0.40	0.41	8	0.42–4.6	1.9	1.5	12	0.11–0.13	1.5	0.74	7	2.5–8.0	5.5	6.1	12	3.3–10	7.0	6.8	4	0.36–2.5	0.95	0.49	36	0.26–11	1.5	0.72
caftaric acid	2	0.06–0.10	0.08	0.08	1	0.05	0.05	0.05	0				0				0				0				1	0.08	0.08	0.08
protocatechualdehyde	8	0.05–0.27	0.12	0.09	5	0.07–0.19	0.14	0.15	9	0.09–0.25	0.17	0.19	1	0.05	0.05	0.05	0				3	0.16–0.23	0.20	0.22	29	0.05–0.43	0.15	0.12
gentisic acid	11	0.07–0.44	0.17	0.11	8	0.10–1.3	0.69	0.70	12	0.16–2.0	0.90	0.80	7	1.7–3.0	2.4	2.6	12	1.6–3.8	2.7	2.7	4	0.12–0.98	0.36	0.18	32	0.06–5.2	1.2	0.87
gentisaldehyde	3	0.31–0.54	0.45	0.49	0				4	0.18–0.43	0.27	0.23	0				0				1	0.32	0.32	0.32	2	0.22–0.27	0.25	0.25
chlorogenic acid	5	0.05–0.10	0.09	0.10	6	0.05–0.09	0.07	0.07	10	0.05–0.18	0.08	0.07	2	0.06–0.15	0.11	0.11	3	0.06–0.10	0.08	0.08	3	0.05–0.12	0.09	0.11	25	0.05–0.19	0.10	0.08
3-hydroxybenzaldehyde	4	0.13–0.87	0.59	0.68	2	0.10–0.22	0.16	0.16	1	0.30	0.30	0.30	1	0.10	0.10	0.10	1	0.10	0.10	0.10	2	0.73–1.2	0.97	0.97	4	0.12–0.97	0.44	0.33
4-hydroxybenzaldehyde	4	1.2–1.3	1.3	1.3	1	0.28	0.28	0.28	4	0.33–1.5	0.71	0.51	0				0				3	1.2–1.8	1.5	1.7	9	0.10–2.0	0.82	0.32
γ-resorcylic acid	3	0.05–0.07	0.06	0.06	6	0.05–0.09	0.07	0.07	10	0.05–0.12	0.09	0.10	6	0.06–0.10	0.09	0.10	11	0.05–0.12	0.08	0.07	1	0.08	0.08	0.08	18	0.05–0.31	0.10	0.10
α-resorcylic acid	2	0.06–0.07	0.07	0.07	6	0.05–0.08	0.06	0.06	9	0.06–0.11	0.09	0.09	6	0.05–0.10	0.08	0.09	7	0.06–0.13	0.09	0.08	1	0.07	0.07	0.07	23	0.06–0.28	0.09	0.08
veratric acid	0				2	0.2–1.9	1.6	1.6	1	1.6	1.6	1.6	0				1	1.8	1.8	1.8	0				3	0.89–1.1	1.1	1.1
caffeic acid	12	0.07–0.37	0.20	0.17	8	0.35–0.81	0.55	0.57	12	0.06–0.85	0.38	0.30	7	0.22–0.62	0.37	0.33	12	0.07–0.57	0.28	0.30	4	0.12–0.42	0.22	0.17	36	0.05–0.64	0.27	0.26
p-coumaric acid	12	0.48–4.2	1.3	0.85	8	0.44–4.1	2.2	2.1	12	0.5–11	3.3	2.0	7	1.4–6.7	3.3	1.7	12	0.27–3.5	1.7	1.5	4	0.05–2.6	0.87	0.41	36	0.14–12	4.02	2.25
veratraldehyde	7	0.05–0.12	0.07	0.07	1	0.09	0.09	0.09	0				0				2	0.06–0.07	0.07	0.07	2	0.06–0.10	0.08	0.08	12	0.05–0.45	0.14	0.06
4-anisaldehyde	1	0.28	0.28	0.28	0				3	0.13–0.41	0.26	0.24	0				0				3	0.14–0.77	0.50	0.59	10	0.12–0.72	0.33	0.22
quercetin	7	0.25–0.80	0.51	0.51	7	0.25–1.4	0.83	1.02	10	0.22–1.3	0.46	0.38	5	0.39–0.62	0.46	0.42	11	0.16–0.63	0.35	0.31	1	0.79	0.79	0.79	28	0.29–1.1	0.56	0.51
kaempferol	7	0.23–0.44	0.31	0.25	6	0.25–0.49	0.40	0.45	12	0.27–0.60	0.44	0.44	7	0.24–0.41	0.34	0.35	7	0.22–0.38	0.28	0.26	2	0.25–0.31	0.28	0.28	15	0.22–0.71	0.34	0.33
apigenin	10	0.06–0.39	0.15	0.12	8	0.13–0.31	0.22	0.23	12	0.07–0.31	0.15	0.13	7	0.09–0.21	0.14	0.10	11	0.06–0.20	0.13	0.15	4	0.08–0.19	0.13	0.13	32	0.05–0.43	0.15	0.13
chrysin	12	0.50–3.4	1.9	1.9	8	2.2–5.9	3.8	3.4	12	0.53–4.5	2.8	2.9	7	1.7–4.2	2.6	2.2	12	0.94–4.4	2.6	2.5	4	1.8–3.1	2.3	2.1	36	0.51–5.9	2.63	2.40
trans-ferulic acid	11	0.48–0.91	0.69	0.69	8	0.62–1.7	1.1	1.2	8	0.36–2.0	0.93	0.90	7	0.42–1.0	0.72	0.74	10	0.27–0.96	0.6	0.7	4	0.15–1.1	0.48	0.34	23	0.19–1.5	0.64	0.64
p-hydroxybenzoic acid	12	0.77–2.3	1.4	1.3	8	0.68–2.3	1.4	1.4	12	1.2–3.4	2.0	1.8	7	1.1–2.5	1.8	1.7	12	0.83–2.9	1.6	1.6	4	1.4–4.0	2.7	2.6	36	0.77–5.1	2.3	2.1
3-hydroxyphenylacetic acid	5	12–123	66	61	3	3.7–63	32	29	3	2.8–12	7.5	8.0	2	1.7–9.5	5.6	5.6	7	0.41–11	4.2	2.0	3	54–242	140	125	10	1.3–138	24	9.9
vanillic acid	0				1	0.27	0.27	0.27	2	0.07–0.20	0.14	0.14	2	0.15–0.40	0.28	0.28	8	0.14–6.2	1.1	0.30	2	0.10–0.12	0.11	0.11	6	0.07–0.58	0.31	0.28
Σ [phenolic compounds]	12	3.4–130	35	11	8	11–71	27	16	12	7.1–31	16	15	7	21–30	25	23	12	18–35	26	26	4	15–252	116	99	36	6.5–144	21	15
TPC (mg GAE 100g^{-1})	12	48–139	85	66	8	48–111	85	96	12	64–192	118	119	7	119–192	148	147	12	95–197	152	151	4	102–203	163	173	36	55–193	116	116
AA (μmol TRE 100g^{-1})	12	15–846	239	176	8	111–428	232	211	12	140–540	265	210	7	228–854	592.43	611	12	420–1017	756	852	4	138–392	272	279	36	56–837	252	236

60

3.4.2. Individual Phenolic Content

Individual target phenolic compounds concentrations, as well as the sum of them for the 91 analyzed samples are summarized in Tables S1 and S2, and concentration ranges for the analyzed varieties are shown in Table 3.

Among the 41 target phenolic compounds, 22 were found in the samples of the 2018 season whereas 25 were detected in samples of 2019. The highest concentration of individual phenolic compounds was found in the heather variety (HE), with total phenolic compounds concentrations reaching 252 µg g^{-1}, especially owing to the high content of 3-hydroxyphenylacetic acid (242 µg g^{-1} in sample HE1). The sum of phenolic compounds was highly influenced by the concentration of this compound since it was found in 33 of the 91 analyzed samples at a mean value of 35 µg g^{-1} and in the range 0.41–242 µg g^{-1}. It is worth noting that these high 3-phenoxyphenylacetic acid contents do not confer high antioxidant activities to the HE honey, compared with those containing honeydew.

Regarding those samples that were not highly affected by 3-phenoxyphenylacetic acid, honeydew honeys (HD) contained high concentration of the sum of phenolic compounds, with 35 and 25 µg g^{-1} in 2018 and 2019 seasons, respectively. In the same way, the mixture chestnut/honeydew variety (CN/HD) reached concentrations up to 30 µg g^{-1} for the sum of the target phenolic compounds. On the other hand, the lowest concentration for the sum of phenolic compound was detected in a eucalyptus honey (EU11) with 3.4 µg g^{-1}. In general, results were similar in both seasons and the values were in concordance with the TPC and AA.

The most abundant phenolic compound, detected in all the analyzed honey samples, was p-hydroxybenzoic acid, in a concentration range from 0.68 to 5.1 µg g^{-1}. Other benzoic- and hydroxycinnamic- derivates acids, such as gallic or protocatechuicacid, were found at high concentrations, up to 10 µg g^{-1} in honeydew (HD), 8.0 µg g^{-1} in chestnut/honeydew (CN/HD) and 4.6 µg g^{-1} in blackberry (BL) varieties. It is important to note that 7 of the 14 chestnut samples collected during the 2019 campaign contained honeydew, which could contribute to the concentration increase. In HD samples, gentisic acid (present in 86 honeys) reached the highest mean concentration of 2.7 µg g^{-1}. Additionally, p-coumaric acid, which was found in all samples, achieved concentrations up to 12 µg g^{-1} in multi-floral (MF) honeys. P-coumaric acid content fluctuations could be observed within the same variety; however, the highest concentrations and fluctuations occurred in multi-floral (MF) honeys, with values ranging from 0.1 to 12 µg g^{-1}.

Veratric acid was detected only in six honey samples of several types (BL, CN, HD and MF) at concentrations around 1.5 µg g^{-1}.

As regards aldehydes, 3-hydroxybenzaldehyde and 4-hydroxybenzaldehyde appeared in several samples of different honey varieties (EU, BL HE, CN and MF) at concentrations of up to 1.2 and 2.0 µg g^{-1}, respectively. It is important to note the absence of these two aldehydes in the honeydew samples as well as in the chestnut honeys with honeydew (CN/HD), except in sample HD10 in which 3-hydroxybenzaldehyde appeared at a concentration of 0.10 µg g^{-1}.

Concerning other families, 2 of the 3 targeted flavonols were found in the samples: quercetin in 69 samples and kaempferol in 40 honey samples, whereas myricetin was not detected in any sample. Additionally, the flavones chrysin and apigenin were found in 91 and 84 honey samples, respectively. They were present at concentrations up to 5.9 µg g^{-1}, for chrysin in BL8, and 0.43 µg g^{-1}, for apigenin in MF5. In contrast, flavanol compounds (catechin, epicatechin, gallocatechin gallate) were not detected in the analyzed samples.

3.5. Chemometric Study

3.5.1. Analysis of Variance (ANOVA)

One way ANOVA was performed to assess statistical differences between the botanical origin of honeys based on their bioactive properties (TPC and AA) and their phenolic profile/composition at a 95% of confidence level. One way ANOVA was selected instead of two ways because the harvest year was not statistically significant (data not shown).

The mean values and box-and-whiskers plots obtained from TPC and AA values are depicted in Figure 5a,b, respectively.

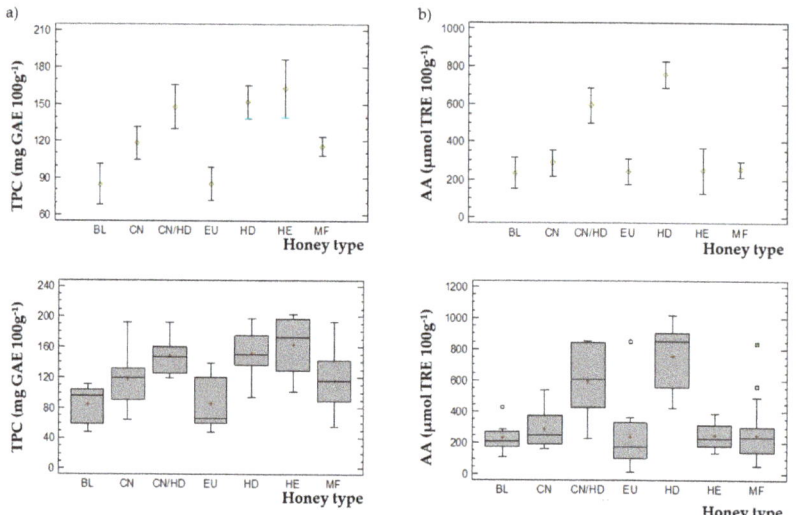

Figure 5. Mean values and box-and-whiskers plots for the different honey varieties based upon (a) TPC; (b) AA.

Concerning TPC, three different homogeneous groups could be determined and be easily visualized in Figure 5a. One group was formed by BL and EU honey varieties, the second one was formed by CN and MF, and the last group was formed by CN/HD, HD and HE honeys. These results were confirmed with those obtained in a multiple range least significant difference (LSD) test (data not shown).

Regarding AA (Figure 5b), three statistically different groups were obtained. The first one formed by HD, the second by the mixture CN/HD and the rest of honey varieties (BL, CN, EU, HE and MF) constituted the third one.

Considering the total concentration of phenolic compounds found in the analyzed samples (see Figure 6), only two groups could be differentiated. This was in concordance with the multiple range test: one formed by HE honey variety, and another group formed by the other varieties (BL, CN, CN/HD, EU, HD and MF). These results demonstrate the high influence of the concentration of 3-hydroxyphenylacetic acid in the sum of phenolic compounds, since it was detected at concentrations over to 200 µg g^{-1} in the HE samples, as already mentioned in Section 3.4.2.

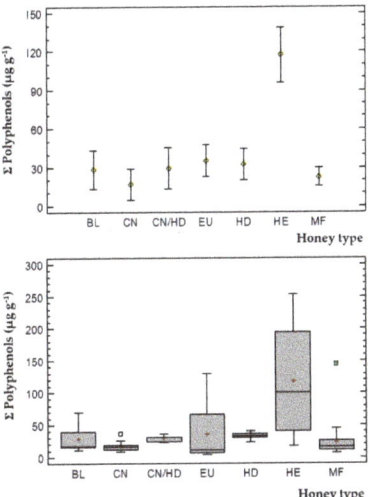

Figure 6. Mean values and box-and-whiskers plots for the different honey varieties based upon the sum of target phenolic compounds concentration.

3.5.2. Principal Components Analysis (PCA)

Honey classification based upon the presence of different target phenolic compounds was one of the objectives of this study. For this reason, a principal components analysis (PCA) was employed by means of a data matrix including the 91 analyzed samples and 25 variables given by the responses of the 25 phenolic compounds detected after LC-MS/MS analysis.

The phenolic compounds responses were auto-standardized by the Statgraphics software. Only principal components with the largest eigenvalues and greater than one were retained (Kaiser criterion). Six principal components (PC) were then retained and were enough to explain about 70% of variance (data not shown). As an example, the PC1 and PC2 and the PC1 and PC3 scatter plots for the 91 samples of different honey varieties are depicted in Figure 7a,b, respectively. A plot of component weights for PC1 and PC2 is also depicted in Figure 7c.

PC1 was mainly positively influenced by acids (gallic acid, β-resorcylic acid, protocatechuic acid and gentisic acid), and negatively by aldehydes (protocatechualdehyde, 3-hydroxybenzaldehyde, 4-hydroxybenzaldehyde and 4-anisaldehyde) (Figure 7c).

In contrast, PC2 was highly positively affected by phenolic acids (caffeic acid, *trans*-ferulic acid and p-coumaric acid), flavones (apigenin and chrysin), flavonols (quercetin and kaempferol) (Figure 7c).

As can be seen in Figure 7a,b, three different groups can be distinguished. Honeydew (HD) honeys as well as chestnut with honeydew (CN/HD) can be classified as one group positively highly influenced by PC1. On the other hand, three samples of heather honey (HE), negatively affected by PC1, can be gathered, whereas BL honeys are clearly differentiated according to PC2. Besides, a group including some of the EU honey samples can be differentiated.

Chestnut honeys are not clearly differentiated by any of the PC, although four of them (CN6, CN8, CN9 and CN10) show a simultaneous high concentration of quercetin, chrysin and *trans*-ferulic acid, as expressed in PC2 (Figure 7a). Blackberry honeys also contain high proportions of these three compounds plus kaempferol.

The plot of component weights depicted in Figure 7c indicates which compounds are dominant for each type of honey. For honeydew honey, gallic acid is the main chemical marker along with β-resorcylic acid and protocatechuic acid.

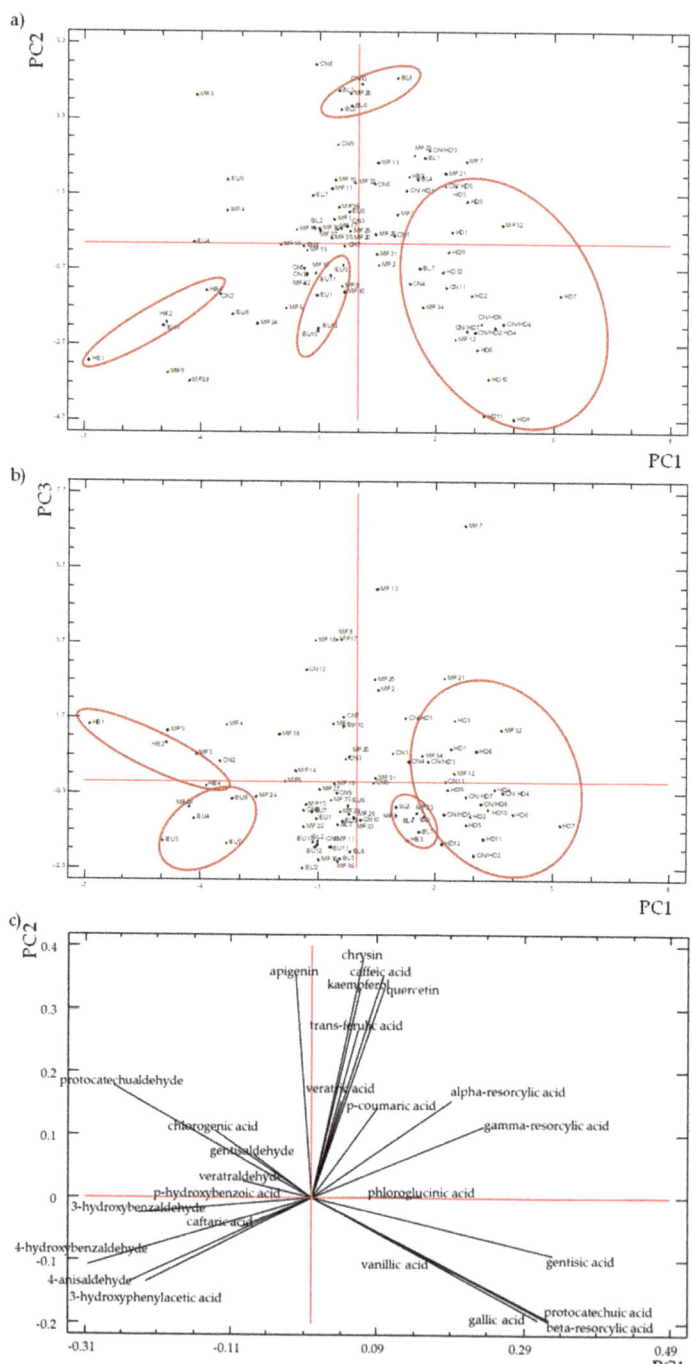

Figure 7. PCA analysis. Scatter plot of (**a**) PC1 and PC2; (**b**) PC1 and PC3; (**c**) plot of component weights for PC1 and PC2 of the 91 analyzed samples. HE: heather, HD: honeydew, CN: chestnut, CN/HD: chestnut/honeydew, EU: eucalyptus, MF: multi-floral, BL: blackberry.

In the case of heather honeys, 4-anisaldehyde, 3-hydroxyphenylacetic acid and 4-hydroxybenzaldehyde appear as main markers.

Most of the 32 multi-floral honeys are located at the centre of the PCA-2D component plots, confirming that with such a mixture of nectars coming from multiple plant species, no specific group and no specific origin can be identified.

Nevertheless, these results also show that PCA is a suitable approach to identify groups of honey from different botanical origins.

4. Conclusions

91 Galician honeys obtained from different botanical origins and nectar sources were analyzed to assess their similarities, differences and correlations in terms of phenolic profiles. A miniaturized, fast and environmentally friendly methodology based on VE-UAE-LC-MS/MS was successfully developed. Results revealed the presence of 25 out of the 41 target phenolic compounds in the 91 analyzed samples. TPC and AA were also evaluated, showing mean values around 121 mg GAE $100g^{-1}$ of honey and 340 µmol TRE $100g^{-1}$, respectively. ANOVA and PCA results based on TPC, AA and $\sum$phenolic compounds concentrations, revealed significant differences depending on the honey variety, demonstrating that phenolic compounds can be used as indicators to identify their floral origin. This study proves that the combination of chromatographic analysis with mass spectrometry detection and PCA are suitable tools to investigate the botanical authentication of honey and to guarantee its quality and origin.

Supplementary Materials: The following are available online at https://www.mdpi.com/article/10.3390/foods10112616/s1, Table S1: concentration (µg g^{-1}) of target phenolic compounds, TPC (mg GAE $100g^{-1}$) and AA (µmol TRE $100g^{-1}$) for: (a) EU, BL, CN and HD; (b) HE and MF honeys in the 2018 season. Table S2: concentration (µg g^{-1}) of target phenolic compounds, TPC (mg GAE $100g^{-1}$) and AA (µmol TRE $100g^{-1}$) for: (a) EU and BL; (b) CN, CN/HD and HD; (c) HE and MF honeys in the 2019 season.

Author Contributions: Conceptualization, M.L. and T.D.; methodology, L.V. and D.A.; validation, L.V. and D.A.; formal analysis, L.V. and D.A.; investigation, L.V. and D.A.; data curation, L.V., D.A. and M.L.; writing—original draft preparation, L.V., D.A. and M.C.; writing—review and editing, M.C., M.L. and T.D.; supervision, M.L.; project administration, M.L. and T.D.; funding acquisition, M.L. and T.D. All authors have read and agreed to the published version of the manuscript.

Funding: This research was funded by project GO FEADER 2018/054B (Xunta de Galicia). The authors belong to the National Network for the Innovation in miniaturized sample preparation techniques, RED2018-102522-T (Ministry of Science, Innovation and Universities, Spain), and to the Galician Competitive Research Groups IN607B 2019/13 and ED431 2020/06 (Xunta de Galicia). This study is based upon work from the Sample Preparation Study Group and Network, supported by the Division of Analytical Chemistry of the European Chemical Society. All these programmes are co-funded by FEDER (EU).

Data Availability Statement: Data are available within the present article and Supplementary Materials.

Acknowledgments: The authors wish to acknowledge the Regulator Council of Galician Honey (IXP, Mel de Galicia) for collecting and supplying the samples.

Conflicts of Interest: The authors declare no conflict of interest.

References

1. Cianciosi, D.; Forbes-Hernández, T.Y.; Afrin, S.; Gasparrini, M.; Reboredo-Rodriguez, P.; Manna, P.P.; Zhang, J.; Lamas, L.B.; Flórez, S.M.; Toyos, P.A.; et al. Phenolic Compounds in Honey and Their Associated Health Benefits: A Review. *Molecules* **2018**, *23*, 2322. [CrossRef] [PubMed]
2. Pyrzynska, K.; Biesaga, M. Analysis of phenolic acids and flavonoids in honey. *TrAC Trends Anal. Chem.* **2009**, *28*, 893–902. [CrossRef]
3. Jerković, I.; Tuberso, C.I.; Gugic, M.; Bubalo, D. Composition of Sulla (Hedysarum coronarium L.) Honey Solvent Extractives Determined by GC/MS: Norisoprenoids and Other Volatile Organic Compounds. *Molecules* **2010**, *15*, 6375–6385. [CrossRef]

4. Fakhlaei, R.; Selamat, J.; Khatib, A.; Razis, A.F.A.; Sukor, R.; Ahmad, S.; Babadi, A.A. The Toxic Impact of Honey Adulteration: A Review. *Foods* **2020**, *9*, 1538. [CrossRef]
5. Codex Alimentarius Commission. *Codex Alimentarius Commission Standards*. Ref. Nr. CL 1993/14-SH. Available online: https://www.fao.org/fao-who-codexalimentarius/sh-proxy/en/?lnk=1&url=https%253A%252F%252Fworkspace.fao.org%252Fsites%252Fcodex%252FStandards%252FCXS%2B12-1981%252FCXS_012e.pdf (accessed on 25 October 2021).
6. European Commission. Directive 2001/110/EC of 20 December relating honey. *Off. J. EC* **2001**, *L10*, 47–52.
7. Directive 2014/63/EU of the European parliament and of the council of 15 May 2014 amending council directive 2001/110/EC relating to honey. *Off. J. EU* **2014**, *L164*, 1–5.
8. Thrasyvoulou, A.; Tananaki, C.; Goras, G.; Karazafiris, E.; Dimou, M.; Liolios, V.; Kanelis, D.; Gounari, S. Legislation of honey criteria and standards. *J. Apic. Res.* **2018**, *57*, 88–96. [CrossRef]
9. Kečkeš, S.; Gašić, U.; Velickovic, T.C.; Milojković-Opsenica, D.; Natić, M.; Tešic, Z. The determination of phenolic profiles of Serbian unifloral honeys using ultra-high-performance liquid chromatography/high resolution accurate mass spectrometry. *Food Chem.* **2013**, *138*, 32–40. [CrossRef] [PubMed]
10. Gašić, U.; Milojković-Opsenica, D.M.; Tešić, Ž.L. Polyphenols as Possible Markers of Botanical Origin of Honey. *J. AOAC Int.* **2017**, *100*, 852–861. [CrossRef]
11. Jaganathan, S.K.; Mandal, M. Antiproliferative Effects of Honey and of Its Polyphenols: A Review. *J. Biomed. Biotechnol.* **2009**, *2009*, 830616. [CrossRef]
12. Khalil, M.L.; Sulaiman, S.A. The potential role of honey and its polyphenols in preventing heart disease: A review. *Afr. J. Tradit. Complement. Altern. Med.* **2010**, *7*, 315–321. [CrossRef] [PubMed]
13. Hossen, M.S.; Ali, M.Y.; Jahurul, M.H.A.; Abdel-Daim, M.M.; Gan, S.H.; Khalil, M.I. Beneficial roles of honey polyphenols against some human degenerative diseases: A review. *Pharm. Reports.* **2017**, *69*, 1194–1205. [CrossRef] [PubMed]
14. Ferreres, F.; Tomás-Barberán, F.A.; García-Vignera, C.; Tomás-Lorente, F. Flavonoids in honey of different geographical origin. *Eur. Food Res. Technol.* **1993**, *196*, 38–44. [CrossRef]
15. Pascual-Maté, A.; Osés, S.M.; Fernández-Muiño, M.A.; Sancho, M.T. Methods of analysis of honey. *J. Apic. Res.* **2018**, *57*, 38–74. [CrossRef]
16. Pauliuc, D.; Dranca, F.; Oroian, M. Antioxidant Activity, Total Phenolic Content, Individual Phenolics and Physicochemical Parameters Suitability for Romanian Honey Authentication. *Foods* **2020**, *9*, 306. [CrossRef]
17. Escuredo, O.; Rodríguez-Flores, M.S.; Rojo-Martínez, S.; Seijo, M.C. Contribution to the Chromatic Characterization of Unifloral Honeys from Galicia (NW Spain). *Foods* **2019**, *8*, 233. [CrossRef]
18. Seijo, M.C.; Escuredo, O.; Rodríguez-Flores, M.S. Physicochemical Properties and Pollen Profile of Oak Honeydew and Evergreen Oak Honeydew Honeys from Spain: A Comparative Study. *Foods* **2019**, *8*, 126. [CrossRef]
19. Escuredo, O.; Rodríguez-Flores, M.S.; Meno, L.; Seijo, M.C. Prediction of Physicochemical Properties in Honeys with Portable Near-Infrared (microNIR) Spectroscopy Combined with Multivariate Data Processing. *Foods* **2021**, *10*, 317. [CrossRef]
20. Biesaga, M.; Pyrzyńska, K. Stability of bioactive polyphenols from honey during different extraction methods. *Food Chem.* **2013**, *136*, 46–54. [CrossRef]
21. Cuevas-Glory, L.F.; Pino, J.A.; Santiago, L.S.; Sauri-Duch, E. A review of volatile analytical methods for determining the botanical origin of honey. *Food Chem.* **2007**, *103*, 1032–1043. [CrossRef]
22. Vazquez, L.; Celeiro, M.; Sergazina, M.; Dagnac, T.; Llompart, M. Optimization of a miniaturized solid-phase microextraction method followed by gas chromatography mass spectrometry for the determination of twenty four volatile and semivolatile compounds in honey from Galicia (NW Spain) and foreign countries. *Sustain. Chem. Pharm.* **2021**, *21*, 100451. [CrossRef]
23. Singleton, V.L.; Rossi, J.A. Colorimetry of total phenolics with phosphomolybdic-phosphotungstic acid reagents. *Am. J. Enol. Vit.* **1965**, *16*, 144–158.
24. Brand-Williams, W.; Cuvelier, M.E.; Berset, C. Use of a free radical method to evaluate antioxidant activity. *LWT Food Sci. Technol.* **1995**, *28*, 25–30. [CrossRef]
25. Celeiro, M.; Lamas, J.P.; Arcas, R.; Lores, M. Antioxidants Profiling of By-Products from Eucalyptus Greenboards Manufacture. *Antioxidants* **2019**, *8*, 263. [CrossRef] [PubMed]
26. Biesaga, M.; Pyrzynska, K. Liquid chromatography/tandem mass spectrometry studies of the phenolic compounds in honey. *J. Chromatogr. A* **2009**, *1216*, 6620–6626. [CrossRef]
27. Pascual-Maté, A.; Osés, S.M.; Fernández-Muiño, M.A.; Sancho, M.T. Analysis of Polyphenols in Honey: Extraction, Separation and Quantification Procedures. *Sep. Purif. Rev.* **2018**, *47*, 142–158. [CrossRef]
28. Attanzio, A.; Tesoriere, L.; Allegra, M.; Livrea, M.A. Monofloral honeys by Sicilian black honeybee (*Apis mellifera* ssp. *sicula*) have high reducing power and antioxidant capacity. *Heliyon* **2016**, *2*, e00193. [CrossRef] [PubMed]

Article

Unifloral Autumn Heather Honey from Indigenous Greek *Erica manipuliflora* Salisb.: SPME/GC-MS Characterization of the Volatile Fraction and Optimization of the Isolation Parameters

Marinos Xagoraris [1], Foteini Chrysoulaki [1], Panagiota-Kyriaki Revelou [1], Eleftherios Alissandrakis [2,3], Petros A. Tarantilis [1] and Christos S. Pappas [1,*]

[1] Laboratory of Chemistry, Department of Food Science and Human Nutrition, Agricultural University of Athens, 75 Iera Odos, 11855 Athens, Greece; mxagor@aua.gr (M.X.); fay_chrysoulaki@aua.gr (F.C.); p.revelou@aua.gr (P.-K.R.); ptara@aua.gr (P.A.T.)

[2] Laboratory of Quality and Safety of Agricultural Products, Landscape and Environment, Department of Agriculture, Hellenic Mediterranean University, Stavromenos, 71410 Heraklion, Greece; ealiss@hmu.gr

[3] Institute of Agri-Food and Life Sciences Agro-Health, Hellenic Mediterranean University Research Center, Stavromenos, 71410 Heraklion, Greece

* Correspondence: chrispap@aua.gr; Tel.: +30-2105294262

Citation: Xagoraris, M.; Chrysoulaki, F.; Revelou, P.-K.; Alissandrakis, E.; Tarantilis, P.A.; Pappas, C.S. Unifloral Autumn Heather Honey from Indigenous Greek *Erica manipuliflora* Salisb.: SPME/GC-MS Characterization of the Volatile Fraction and Optimization of the Isolation Parameters. *Foods* **2021**, *10*, 2487. https://doi.org/10.3390/foods10102487

Academic Editors: Olga Escuredo and M. Carmen Seijo

Received: 28 September 2021
Accepted: 15 October 2021
Published: 17 October 2021

Publisher's Note: MDPI stays neutral with regard to jurisdictional claims in published maps and institutional affiliations.

Copyright: © 2021 by the authors. Licensee MDPI, Basel, Switzerland. This article is an open access article distributed under the terms and conditions of the Creative Commons Attribution (CC BY) license (https://creativecommons.org/licenses/by/4.0/).

Abstract: For long heather honey has been a special variety due to its unique organoleptic characteristics. This study aimed to characterize and optimize the isolation of the dominant volatile fraction of Greek autumn heather honey using solid-phase microextraction (SPME) followed by gas chromatography-mass spectrometry (GC-MS). The described approach pointed out 13 main volatile components more closely related to honey botanical origin, in terms of occurrence and relative abundance. These volatiles include phenolic compounds and norisoprenoids, with benzaldehyde, safranal and *p*-anisaldehyde present in higher amounts, while ethyl 4-methoxybenzoate is reported for the first time in honey. Then, an experimental design was developed based on five numeric factors and one categorical factor and evaluated the optimum conditions (temperature: 60 °C, equilibration time: 30 min extraction time: 15 min magnetic stirrer velocity: 100 rpm sample volume: 6 mL water: honey ratio: 1:3 (*v*/*w*)). Additionally, a validation test set reinforces the above methodology investigation. Honey is very complex and variable with respect to its volatile components given the high diversity of the floral source. As a result, customizing the isolation parameters for each honey is a good approach for streamlining the isolation volatile compounds. This study could provide a good basis for future recognition of monofloral autumn heather honey.

Keywords: autumn heather honey; *Erica manipuliflora* Salisb.; volatiles; gas chromatography-mass spectrometry; solid-phase microextraction; optimization; response surface methodology

1. Introduction

Honey bees (*Apis mellifera* L.) are primary pollinators with an important role in ecosystem conservation [1], offering many services and products, such as honey. Honey through the centuries has always been a vital food for humans, with many health properties [2,3]. The Mediterranean region, specifically Greece pronounces a set of several common and rare monofloral honeys in international markets [4]. Additionally, nowadays few rare honeys, like heather, have become increasingly well-known for their special characteristics and have received several awards in national and international food quality or taste competitions [5]. The term "heather" is used for plant species belonging to *Erica* and *Calluna* genera. However, this term is used to describe the honey produced from *Calluna vulgaris* (L.) Hull and not from other Ericaceae botanical sources [6]. In relation to honey from common species, including *Erica arborea* L., *Erica carnea* L., and *Erica cinerea* L., the given names are "Tree heath", "Spring heather", and "Bell heather", respectively [6].

Greek flora includes four Ericaceae nectar-secretion bee plants. Two of them are spring flowering species including *Erica arborea* L., and *Rhododendron* sp. while the other two (*Erica manipuliflora* Salisb. and *Arbutus unedo* L.) bloom in autumn. *Erica manipuliflora* is indigenous in Greece and is known as "autumn heather", while the traditional term used is "sousoura". However, honey from *E. manipuliflora* should not be confused with other heather honeys produced during autumn, including from *C. vulgaris*, and *Erica multiflora* L. Monofloral autumn heather honey can be quite easily produced [7], as its collection period does not coincide with the blooming of other bee plants, with the exception of *A. unedo* honey, which blooms in late autumn and its blooming period follows that of *E. manipuliflora*.

Greek autumn heather honey is well-known for its extraordinary aroma profile, characterized by perfume reminiscent "caramel" notes, which is worth studying since data for this honey variety are scarce. In the last twenty years, just two studies [8,9] have dealt with the volatile fraction of *E. manipuliflora* honey. However, there are numerous studies concerning heather honey [7,10–17]. As shown in a review study [18], the above studies refer to different botanical species, geographical origin, number of samples, isolation, and analysis procedures.

The volatile isolation method is usually followed by gas chromatography-mass spectrometry (GC-MS), and plays a significant role in the qualitative and quantitative determination of volatiles. Solid-phase microextraction (SPME) as a volatile fraction extraction methodology constitutes a simple procedure with no pre-treatment of samples and environmentally friendly solvents [19]. The main factors, including temperature, equilibration time, extraction time, sample volume, water-honey ratio, and magnetic stirring velocities contribute simultaneously to the isolation of volatiles, sometimes synergistically [20]. For this reason, it is necessary to study all-factors-at-a-time, in terms of their effectiveness of volatiles isolation. This may be possible by using multivariate statistic techniques, like response surface methodology (RSM) [20,21].

The aim of the present study was the identification and semi-quantification of the volatile fraction of indigenous monofloral Greek autumn heather honey from *E. manipuliflora*. The main SPME factors were simultaneously examined for their potential to isolate the dominant volatile fraction and each molecule separately using RSM.

2. Materials and Methods

2.1. Honey Samples

The analyses of volatiles were carried out to 25 honey samples provided directly by Greek beekeepers. Samples were produced during the 2019–2021 harvest period. The botanical origin was assessed by the beekeepers and then confirmed by melissopalynological [22], and physicochemical analysis [23,24], as previously described [25]. Floral origin was confirmed firstly according to European [26] and secondly according to the more strict Greek [27] legislation (sum of fructose and glycose not less than 60% w/w; sucrose content not more than 5% w/w; moisture content not more than 20% w/w; electrical conductivity not more than 800 ($\mu S\ cm^{-1}$); diastase activity (Schade scale) not less than 8; HMF not more than 40 mg kg^{-1}; heather pollen not less than 45%). Honey samples were kept in the dark at 4 °C in hermetically closed glass bottles until further analysis.

2.2. Experimental Design

A central composite design (CCD) was used combined with RSM methodology by Box and Wilson [28]. A flexible design structure was constructed to accommodate a custom model, with numeric and categorical independent factors and irregular constrained regions. Five numeric factors (A, B, C, D, and E) and one categorical factor (F) were analyzed by a quadratic design domain. A total of 38 runs were determined by a selection criterion chosen during the experimental design (Table 1).

Table 1. Independent experimental factors and design layout runs.

Run	A: Temperature	B: Equilibration Time	C: Extraction Time	D: Magnetic Stirrer Velocity	E: Sample Volume	F: Water: Honey Ratio
Units	°C	min	min	rpm	mL	v/w
1	30.0	5.0	15.0	700.0	4.0	1:3
2	30.0	5.0	60.0	400.0	2.0	1:1
3	30.0	5.0	15.0	100.0	6.0	1:1
4	30.0	5.0	30.0	700.0	6.0	3:1
5	30.0	5.0	15.0	100.0	2.0	1:3
6	30.0	15.0	15.0	100.0	4.0	3:1
7	30.0	15.0	60.0	700.0	6.0	1:3
8	30.0	15.0	15.0	700.0	2.0	1:1
9	30.0	30.0	30.0	400.0	6.0	1:3
10	30.0	30.0	60.0	100.0	2.0	1:3
11	30.0	30.0	60.0	700.0	6.0	1:1
12	30.0	30.0	60.0	100.0	6.0	3:1
13	30.0	30.0	15.0	700.0	6.0	3:1
14	30.0	30.0	30.0	100.0	4.0	1:1
15	30.0	30.0	60.0	700.0	2.0	3:1
16	45.0	5.0	60.0	100.0	6.0	1:3
17	45.0	5.0	60.0	700.0	2.0	1:3
18	45.0	5.0	15.0	400.0	4.0	1:1
19	45.0	15.0	60.0	100.0	2.0	1:1
20	45.0	15.0	60.0	700.0	6.0	3:1
21	45.0	30.0	15.0	100.0	6.0	1:3
22	45.0	30.0	30.0	100.0	2.0	3:1
23	45.0	30.0	15.0	700.0	2.0	1:3
24	60.0	5.0	60.0	100.0	2.0	3:1
25	60.0	5.0	15.0	100.0	6.0	3:1
26	60.0	5.0	60.0	700.0	6.0	1:1
27	60.0	5.0	30.0	100.0	2.0	1:1
28	60.0	5.0	15.0	700.0	6.0	1:3
29	60.0	5.0	60.0	400.0	6.0	3:1
30	60.0	5.0	15.0	700.0	2.0	3:1
31	60.0	15.0	30.0	400.0	2.0	1:3
32	60.0	30.0	60.0	100.0	6.0	1:1
33	60.0	30.0	30.0	700.0	4.0	3:1
34	60.0	30.0	15.0	100.0	2.0	1:1
35	60.0	30.0	15.0	700.0	6.0	1:1
36	60.0	30.0	60.0	700.0	6.0	1:3
37	60.0	30.0	60.0	700.0	2.0	1:1
38	60.0	30.0	15.0	100.0	4.0	1:3

The responses of the volatile compounds expressed as chromatographic area (%) were used as dependent variables. For this purpose, a randomly selected sample was used for response prediction. The model's fitness was confirmed by analysis of variance (ANOVA) and the determination coefficient (R^2) using p-values. Dependent variables were also confirmed by the Box-Cox, correlations, and normality of residuals statistical tests. All possible optimized solutions, for (a) volatile profile and (b) each volatile molecule separately were evaluated by maximizing desirability indices. The robustness of the model was validated with response data of 24 samples according to the optimum SPME solution.

Statistical analysis was carried out using Desing-Expert 11.0.5.0 (Stat-Ease, Inc., Minneapolis, MN, USA).

2.3. Isolation and Analysis of Volatile Compounds

Isolation of the volatile fraction was done based on experimental design layout run (Table 1) using a manual holder with triple-phase divinylbenzene/carboxen/polydimethylsiloxane (DVB/CAR/PDMS) fiber 50/30 µm (Supelco, Bellefonte, PA, USA) with length of 1 cm. Before each analysis, fibers were conditioned at 270 °C. Moreover, a blank sample was performed for cleaning from previous volatile residues. Then, a predetermined volume ratio of water: Honey (v/w) was transferred in 15 mL screw top (22.7 × 86 mm) vials with PTFE/silicone septa and a portion of 20 µL (300 µg mL^{-1} in methanol) of benzophenone (Alfa Aesar, Kandal, Germany) was added as an internal standard.

RSM experiments were performed using a Trace Ultra gas chromatograph (GC) (Thermo Scientific Inc., Waltham, MA, USA), coupled with a mass spectrometer (MS) (DSQII, Thermo Scientific Inc., Waltham, MA, USA). GC-MS was performed with a Restek Rtx-5MS (30 m × 0.25 mm i.d., 0.25 µm film thickness) chromatography column with helium as carrier gas at a 1 mL min^{-1} rate. The chromatography conditions and temperature program have been previously described [29]. In brief, the GC inlet temperature 260 °C in the splitless mode for 3 min, with a 0.8 mm injector liner (SGE International Pty Ltd., Ringwood, Australia). Oven temperature was adapted to 40 °C for 6 min, then increased to 120 °C at a rate of 5 °C min^{-1}, followed by an increment of 3 °C min^{-1} up to 160 °C and up to 250 °C with a step of 15 °C min^{-1}. Finally, the temperature of 250 °C was kept constant for 1 min. The transfer line and injector temperatures were maintained at 290 and 220 °C, respectively. Electron impact was 70 eV, and mass spectra were recorded at the 35–650 mass range.

The peak identification was achieved with the Wiley 275 mass spectra library, and the arithmetic index provided by Adams [30]. Retention Index (RI) values of volatile compounds were calculated using n-alkane (C8–C20) standards (Supelco, Bellefonte, PA, USA). The isolated compounds were semi-quantified against the internal standard (benzophenone) and expressed as mg kg^{-1} of honey. All samples were analyzed in triplicate.

3. Results and Discussion

3.1. Evaluation of Isolated Volatile Compounds

In total, 49 volatile compounds were identified, including esters, hydrocarbons, alcohols, aldehydes, ketones, acids, terpenoids, and others (Table 2).

Esters have been encountered almost always in all blossom and honeydew honeys with some of them being dominant volatiles [18]. In our results, most esters were detected in small amounts except for methyl nonanoate. However, methyl nonanoate has been reported at much higher concentrations in honeydew honey, like fir and pine [29]. This presence could occur in the collection period of pine honey by the bees in October. Moreover, methyl octanoate and methyl dodecanoate can be related to the above conjecture [29]. Ethyl 4-methoxybenzoate was a derivative coming from p-anisic acid which has been reported in *Erica arborea* L. honey [10] by a Likens-Nickerson steam distillation (L-N) isolation technique. However, ethyl 4-methoxybenzoate was worth studying as it has not been detected in other botanical sources yet.

Table 2. Volatile compounds isolated from headspace of autumn heather honey.

No.	Volatile Compounds	CAS Number	RT [a]	RI [b]	Min (mg kg^{-1})	Max (mg kg^{-1})	Average (mg kg^{-1})
		Esters					
1	methyl benzoate	93-58-3	17.3	1093	0.00	0.33	0.02
2	methyl octanoate	111-11-5	18.3	1124	0.00	0.17	0.06
3	ethyl benzoate	93-89-0	19.6	1165	0.00	1.68	0.11
4	methyl 2-phenylacetate	101-41-7	19.8	1179	0.00	0.32	0.04
5	methyl 2-hydroxybenzoate (methyl salicylate)	119-36-8	20.4	1192	0.00	0.54	0.06
6	methyl nonanoate	1731-84-6	21.3	1222	0.06	0.44	0.16
7	methyl decanoate	110-42-9	24.3	1322	0.00	0.10	0.05
8	ethyl 4-methoxybenzoate (Ethyl anisate)	94-30-4	28.7	1458	0.00	0.24	0.02
9	methyl dodecanoate	111-82-0	30.8	1521	0.00	0.06	0.01
10	bis(2-methylpropyl) benzene-1,2-dicarboxylate	84-69-5	39.0	1859	0.00	0.07	0.02
		Hydrocarbons					
11	octane	111-65-9	6.3	800	0.00	0.18	0.06
12	nonane	111-84-2	10.3	898	0.00	0.16	0.03
13	undecane	1120-21-4	17.6	1101	0.10	0.52	0.21
14	dodecane	112-40-3	20.7	1201	0.00	0.18	0.02
		Alcohols					
15	oct-1-en-3-ol	3391-86-4	13.4	981	0.00	0.26	0.02
16	2-ethylhexan-1-ol	104-76-7	15.1	1029	0.00	0.16	0.03
17	5-(3,3-dimethyloxiran-2-yl)-3-methylpent-1-en-3-ol (cis-linalool oxide)	5989-33-3	16.6	1072	0.00	0.35	0.07
18	2-phenylethan-1-ol	60-12-8	17.9	1114	0.00	0.34	0.06
19	4-methyl-1-(prop-1-en-2-yl)cyclohex-3-en-1-ol (1,8-methadien-4-ol)	3419-02-1	20.0	1183	0.00	0.54	0.02
20	3,4,5-trimethylphenol	527-54-8	24.0	1314	0.00	0.93	0.08
21	4,6,10,10-tetramethyl-5-oxatricyclo[4.4.0.01,4]dec-2-en-7-ol	97371-50-1	29.3	1476	0.00	0.11	0.01
22	6,6-dimethyl-5-methylenebicyclo[2.2.1]heptan-2-ol (6-camphenol)	3570-04-5	30.5	1510	0.00	0.38	0.02
		Aldehydes					
23	furan-2-carbaldehyde (furfural)	98-01-1	7.4	826	0.01	2.61	1.14
24	benzaldehyde	100-52-7	12.6	959	0.02	1.44	0.18
25	octanal	124-13-0	14.2	1001	0.00	0.15	0.05
26	2-phenylacetaldehyde	122-78-1	15.6	1041	0.00	0.85	0.16
27	nonanal	124-19-6	17.7	1104	0.07	0.46	0.19

Table 2. Cont.

No.	Volatile Compounds	CAS Number	RT [a]	RI [b]	Min (mg kg^{-1})	Max (mg kg^{-1})	Average (mg kg^{-1})
28	2,6,6-trimethylcyclohexa-1,3-diene-1-carbaldehyde (safranal)	116-26-7	20.6	1198	0.00	0.54	0.12
29	decanal	112-31-2	20.8	1205	0.00	0.35	0.15
30	4-methoxybenzaldehyde (p-anisaldehyde)	123-11-5	22.4	1261	0.00	1.36	0.23
	Ketones						
31	1-(furan-2-yl)ethan-1-one	1192-62-7	10.6	907	0.07	0.34	0.16
32	cyclohex-2-en-1-one	930-68-7	14.7	1015	0.00	0.15	0.01
33	3,5,5-trimethylcyclohex-2-en-1-one (a-isophorone)	78-59-1	18.2	1120	0.01	4.16	0.43
34	2,6,6-trimethylcyclohex-2-ene-1,4-dione (4-oxoisophorone)	1125-21-9	18.9	1143	0.01	0.89	0.13
35	2-hydroxy-3,5,5-trimethylcyclohex-2-en-1-one (2-hydroxyisophorone)	4883-60-7	19.0	1145	0.00	0.29	0.09
36	1-(1,4-dimethylcyclohex-3-en-1-yl)ethanone	43219-68-7	19.1	1149	0.00	0.22	0.02
37	(E)-1-(2,6,6-trimethylcyclohexa-1,3-dien-1-yl)but-2-en-1-one (β-damascenone)	23726-93-4	26.0	1377	0.00	0.16	0.05
38	(E)-4-(2,4,4-trimethylcyclohexa-1,5-dien-1-yl)but-3-en-2-one	187519 [c]	27.6	1420	0.00	0.10	0.01
39	(E)-1,6,6-trimethyl-7-(3-oxobut-1-en-1-yl)-3,8-dioxatricyclo[5.1.0.02,4]octan-5-one	192009 [c]	28.0	1437	0.00	0.23	0.05
40	(E)-4-(2,6,6-trimethylcyclohexa-1,3-dien-1-yl)but-3-en-2-one (Dehydro-beta-ionone)	1203-08-3	29.2	1474	0.00	0.10	0.01
41	1-(4-(tert-butyl)-2,6-dimethylphenyl)ethan-1-one	2040-10-0	33.3	1584	0.00	0.14	0.04
42	(E)-3,5,5-trimethyl-4-(3-oxobut-1-en-1-yl)cyclohex-2-en-1-one	20194-68-7	35.0	1654	0.00	0.09	0.02
	Acids						
43	nonanoic acid	112-05-0	23.2	1288	0.00	0.27	0.11
	Terpenoids						
44	1-methyl-4-propan-2-ylbenzene (p-cymene)	99-87-6	14.9	1022	0.00	0.15	0.01
45	1-methyl-4-(prop-1-en-2-yl)benzene (p-cymenene)	1195-32-0	17.2	1090	0.00	0.18	0.02
46	1-methoxy-4-propylbenzene (4-propylanisole)	104-45-0	23.5	1299	0.00	0.83	0.04

Table 2. Cont.

No.	Volatile Compounds	CAS Number	RT [a]	RI [b]	Min (mg kg^{-1})	Max (mg kg^{-1})	Average (mg kg^{-1})
	Others						
47	(2S,8aR)-2,5,5,8a-tetramethyl-3,5,6,8a-tetrahydro-2H-chromene	41678-29-9	23.9	1306	0.00	0.09	0.01
48	1,1,5-trimethyl-1,2-dihydronaphthalene	357258 [c]	25.2	1352	0.00	0.19	0.08
49	8-isopropyl-1-methyl-1,2,3,4-tetrahydronaphthalene	81603-43-2	31.5	1535	0.00	0.11	0.05

[a] RT: Retention time (min); [b] RI: Experimental retention index; [c] NIST#.

Hydrocarbons were detected in most samples with undecane having the highest average compared to the rest. This class of volatiles is very common among honeys [18].

The chemical group of alcohols including 5-(3,3-dimethyloxiran-2-yl)-3-methylpent-1-en-3-ol (syn: *cis*-Linalool oxide) [11,17] and 2-phenylethan-1-ol [10,13,19] have been previously identified as dominant volatiles compounds of citrus, acacia, chestnut, and thyme honeys [18]. The compound, 3,4,5-trimethylphenol, has been previously described as one of the major volatile compounds of heather honey from Poland [14]. Furthermore, 4,6,10,10-tetramethyl-5-oxatricyclo[4.4.0.01,4]dec-2-en-7-ol and 6,6-dimethyl-5-methylenebicyclo[2.2.1]heptan-2-ol (6-camphenol) have been reported in *Erica* spp. honeys from Iberian Peninsula [17]. Nevertheless, the latter was not identified in all of our samples.

Aldehydes were detected in all samples, in smaller or larger amounts. Octanal, nonanal, and decanal were present in small quantities and are considered as important components of honeydew honey volatile profile [18]. Benzaldehyde [11,19] and 2-phenylacetaldehyde [13,16,17] were detected in all samples at a remarkable concentration. In addition, 2,6,6-trimethylcyclohexa-1,3-diene-1-carbaldehyde (safranal) [7]; 4-methoxybenzaldehyde (p-anisaldehyde) [10], and furan-2-carbaldehyde (furfural) [13] have been attributed to heather honey.

Ketones include many degraded carotenoids related to heather honey. Some of these compounds, such as 3,5,5-trimethylcyclohex-2-en-1-one (a-isophorone); 2,6,6-trimethylcyclohex-2-ene-1,4-dione (4-oxoisophorone); 2-hydroxy-3,5,5-trimethylcyclohex-2-en-1-one (2-hydroxyisophorone); (E)-1-(2,6,6-trimethylcyclohexa-1,3-dien-1-yl)but-2-en-1-one (β-damascenone); (E)-1,6,6-trimethyl-7-(3-oxobut-1-en-1-yl)-3,8-dioxatricyclo[5.1.0.02,4]octan-5-one; and (E)-4-(2,6,6-trimethylcyclohexa-1,3-dien-1-yl)but-3-en-2-one [10,13,14,31] are known heather honey compounds, all of which have been detected in our samples. Notably, 1-(furan-2-yl)ethan-1-one was found in all samples.

Terpenoids, acids, and other compounds do not include significant volatile compounds of heather honey, except for 1,1,5-trimethyl-1,2-dihydronaphthalene and 8-isopropyl-1-methyl-1,2,3,4-tetrahydronaphthalene, that have been identified in another study as well [17].

Other studies refer to hotrienol, *cis*-linalool oxide, and 2-phenylacetaldehyde as the main volatile compounds of *Erica* spp. honey [13,17,19]. However, this is not confirmed by our samples. Hotrienol was not detected in any of our samples. Oxide of cis-linalool was linked with hive atmospheres or combustion of wood/vegetation during beekeeping activity [32], and 2-phenylacetaldehyde also had been reported in relevant concentrations, while some studies attribute this molecule to long-term storage by enzymatic catalysis of phenylalanine or heat treatment [33]. Furan derivatives identified in some of our samples, emanate from thermal processing and/or prolonged storage [34–36] and cannot be related to honey botanical origin.

3.2. Optimization of Each Dominant Volatile Compound

Several SPME conditions (A: Temperature; B: Equilibration time; C: Extraction time; D: Magnetic stirrer velocity; E: Sample volume; F: water: honey ratio) were investigated to determine the most suitable conditions for each volatile compound. A total of 13 volatile compounds were chosen for optimization (responses R1-R13) (Table 3). These compounds were selected as they constitute dominant and characteristic responses of autumn heather honey.

Table 3. Dominant volatile compounds (responses R1-R13).

Response	Volatile Compound	Min (%Area)	Max (%Area)	Mean (%Area)	Std. Dev.
R1	benzaldehyde	1.59	6.70	4.52	1.22
R2	3,5,5-trimethylcyclohex-2-en-1-one	0.00	1.11	0.59	0.36
R3	2,6,6-trimethylcyclohex-2-ene-1,4-dione	0.00	2.90	1.32	0.75
R4	2-hydroxy-3,5,5-trimethylcyclohex-2-en-1-one	0.53	2.88	1.42	0.55
R5	2,6,6-trimethylcyclohexa-1,3-diene-1-carbaldehyde	0.98	4.57	2.37	0.81
R6	4-methoxybenzaldehyde	0.00	13.52	6.25	4.07
R7	3,4,5-trimethylphenol	0.00	0.72	0.22	0.29
R8	1,1,5-trimethyl-1,2-dihydronaphthalene	0.00	6.09	2.00	1.53
R9	(E)-1-(2,6,6-trimethylcyclohexa-1,3-dien-1-yl)but-2-en-1-one	0.00	1.80	1.13	0.40
R10	(E)-1,6,6-trimethyl-7-(3-oxobut-1-en-1-yl)-3,8-dioxatricyclo[5.1.0.02,4]octan-5-one	0.00	1.38	0.63	0.53
R11	ethyl 4-methoxybenzoate	0.00	3.96	1.39	1.30
R12	(E)-4-(2,6,6-trimethylcyclohexa-1,3-dien-1-yl)but-3-en-2-one	0.00	2.18	0.37	0.48
R13	4,6,10,10-tetramethyl-5-oxatricyclo[4.4.0.01,4]dec-2-en-7-ol	0.00	1.35	0.75	0.40

Prior to undertaking the processing steps, data for each volatile compound were confirmed by normal distribution, Box-Cox test, determination of coefficient (R^2) and ANOVA (Table 4). The condition number of coefficient matrix (<10) did not indicate multicollinearity. Additionally, all responses followed the normal distribution. Box-Cox test provides a guideline for selecting the correct power law transformation. If the 95% confidence interval around this lambda includes 1.00, it does not require a specific transformation. Table 4 shows the lambda values at the 95% confidence range, as well as the current lambda. R-square (R^2) constitutes a measure of the amount of variation around the mean explained by the model. The ANOVA in this case confirms the adequacy of the model (p-value < 0.05) and indicated whether the model terms were significant. Significant model terms may have a real effect on the response.

ANOVA results showed many considerable independent SPME conditions, while some of them could contribute in combination. At the same time, equations were developed in terms of coded factors that can be used to make predictions about the response for given levels of each factor. By default, the high levels of the factors are coded as +1 and the low levels are coded as −1. The coded equation is useful for identifying the relative impact of the factors by comparing the factor coefficients. However, these equations should be considered with caution because it is not safe to use them as panacea for modeling future responses. In this case, all these results are presented in Table 5.

Table 4. ANOVA, Box-Cox and determination of coefficient (R^2) of each response subjected to the model.

Response	ANOVA (p-Value < 0.05)						Box-Cox			R^2
	A *	B *	C *	D *	E *	F *	CI Low [a]	Current Lambda	CI High [a]	
R1	0.19	0.50	0.40	0.35	0.60	0.00	0.26	1.00	3.02	0.988
R2	0.23	0.84	0.11	0.05	0.04	0.01	−0.04	1.00	1.09	0.979
R3	0.09	0.12	0.54	0.19	0.44	0.00	0.28	1.00	1.24	0.988
R4	0.00	0.30	0.33	0.44	0.11	0.04	−1.63	1.00	2.28	0.932
R5	0.09	0.53	0.73	0.36	0.17	0.07	−2.22	1.00	2.35	0.895
R6	0.00	0.53	0.00	0.64	0.74	0.00	0.33	1.00	2.35	0.997
R7	0.39	0.05	0.74	0.52	0.35	0.00	−0.79	1.00	1.23	0.978
R8	0.00	0.41	0.37	0.07	0.59	0.00	0.00	1.00	1.88	0.985
R9	0.66	0.95	0.35	0.18	0.12	0.00	0.70	1.00	2.85	0.956
R10	0.00	0.39	0.97	0.24	0.32	0.09	−0.17	1.00	1.78	0.991
R11	0.00	0.32	0.01	0.70	0.23	0.00	−0.11	1.00	0.83	0.988
R12	0.11	0.31	0.66	0.18	0.98	0.72	−0.61	1.00	1.40	0.913
R13	0.29	0.04	0.08	0.13	0.37	0.01	0.30	1.00	2.02	0.986

[a] 95% confidence interval level. * A: Temperature; B: Equilibration time; C: Extraction time; D: Magnetic stirrer velocity; E: Sample volume; F: water: honey ratio.

Table 5. Contingent combinations of the SPME conditions and final equation in terms of coded factors.

R1	MT [a]	F	AD	DF	B^2	D^2		
	CE [b]	+0.69	+0.31	+0.39	+0.75	−0.81		
R2	MT	E	F	AC	AF			
	CE	+0.08	+0.21	−0.13	−0.28			
R3	MT	F	AC	AF	CF	DE	A^2	D^2
	CE	+0.56	−0.20	+0.14	+0.20	+0.13	−0.42	−0.43
R4	MT	A	F					
	CE	−0.47	+0.21					
R5	MT							
	CE		No significant model terms					
R6	MT	A	C	F				
	CE	+0.99	+0.53	−0.14				
R7	MT	F	AB	AC	AD	BD	CF	A^2	E^2
	CE	−0.18	+0.10	+0.09	+0.08	−0.09	−0.06	+0.17	+0.31
R8	MT	A	F	AF				
	CE	−0.88	−0.12	+0.60				
R9	MT	F	A^2					
	CE	+0.21	−0.39					

Table 5. Cont.

R10	MT	A	A²					
	CE	+0.50	−0.42					
R11	MT	A	C	F	AF			
	CE	+0.96	+0.27	+0.60	+0.40			
R12	MT	No significant model terms						
	CE							
R13	MT	B	F	AB	AF	BE	DE	E²
	CE	−0.09	+0.16	−0.12	+0.13	+0.12	−0.11	−0.21

[a] MT: model term; [b] CE: coded equation.

After these steps, optimization models were developed based on each volatile molecule. The results are presented in Table 6 and were evaluated by desirability indices. A high level of ideal cases is coded as 1 and low level as zero. A predicted mean for each volatile response is also included.

Table 6. Optimum conditions, desirabilities and predicted mean for each dominant volatile compound.

Response	A *	B *	C *	D *	E *	F *	Desirability	Predicted Mean (%Area)
R1	60	5	60	700	6	1:1	1.000	4.07 ± 0.36
R2	45	30	15	700	2	1:3	1.000	0.96 ± 0.18
R3	60	30	15	100	4	1:3	1.000	1.63 ± 0.22
R4	60	30	15	100	4	1:3	1.000	0.85 ± 0.19
R5	45	30	15	100	6	1:3	1.000	1.97 ± 0.71
R6	60	15	15	100	2	1:3	1.000	12.61 ± 0.64
R7	45	30	15	100	6	1:3	1.000	0.11 ± 0.02
R8	60	5	15	700	6	1:3	1.000	0.64 ± 0.11
R9	60	15	30	400	2	1:3	1.000	0,77 ± 0.23
R10	60	30	30	100	4	1:3	1.000	0.84 ± 0.22
R11	60	30	15	100	2	1:1	1.000	2.48 ± 0.39
R12	60	30	60	700	6	1:3	1.000	1.09 ± 0.39
R13	60	30	15	700	6	1:1	1.000	0.98 ± 0.27

* A: Temperature; B: Equilibration time; C: Extraction time; D: Magnetic stirrer velocity; E: Sample volume; F: water: honey ratio.

Experimental findings showed that optimum conditions of some volatiles required the maximum value of model terms. However, this conclusion is overturned by extraction time and magnetic stirrer velocity. As previously described, extraction time was a significant parameter, along with magnetic stirrer velocity, which in some cases, allowed better isolation of some compounds [20], whilst usually shortened the equilibration time.

3.3. Optimization and Validation of Dominant Volatile Compounds

The optimum conditions proposed for dominant volatile compounds of autumn heather honey were A: 60 °C B: 30 min C: 15 min D: 100 rpm E: 6 mL F: 1:3 (v/w). Predicted mean (% Area) was estimated for benzaldehyde (4.53%), 3,5,5-trimethylcyclohex-2-en-1-one (0.88%), 2,6,6-trimethylcyclohex-2-ene-1,4-dione (1.64%), 2-hydroxy-3,5,5-trimethylcyclohex-2-en-1-one (0.80%), 2,6,6-trimethylcyclohexa-1,3-diene-1-carbaldehyde (1.29%), 4-methoxybenzaldehyde (12.21%), 3,4,5-trimethylphenol (0.12%), 1,1,5-trimethyl-1,2-dihydronaphthalene (0.66%), (E)-1-(2,6,6-trimethylcyclohexa-1,3-dien-1-yl)but-2-en-1-one (0.15%), (E)-1,6,6-trimethyl-7-(3-oxobut-1-en-1-yl)-3,8-dioxatricyclo[5.1.0.02,4]octan-5-one (0.71%), ethyl 4-methoxybenzoate (2.88%), (E)-4-(2,6,6-trimethylcyclohexa-1,3-dien-1-

yl)but-3-en-2-one (0.77%), and 4,6,10,10-tetramethyl-5-oxatricyclo[4.4.0.01,4]dec-2-en-7-ol (0.97%). Moreover, the desirability of optimized model was calculated at 1.000.

The validation of the above results was carried out with a test set of 24 samples. All responses (R1-R13) were isolated in all samples with the confirmed optimum conditions of the proposed method. Data mean (% Area) was estimated for benzaldehyde (2.62%), 3,5,5-trimethylcclohex-2-en-1-one (6.14%), 2,6,6-trimethylcyclohex-2-ene-1,4-dione (1.90%), 2-hydroxy-3,5,5-trimethylcyclohex-2-en-1-one (1.29%), 2,6,6-trimethylcyclohexa-1,3-diene-1-carbaldehyde (1.75%), 4-methoxybenzaldehyde (3.62%), 3,4,5-trimethylphenol (0.98%), 1,1,5-trimethyl-1,2-dihydronaphthalene (1.21%), (E)-1-(2,6,6-trimethylcyclohexa-1,3-dien-1-yl)but-2-en-1-one (0.84%), (E)-1,6,6-trimethyl-7-(3-oxobut-1-en-1-yl)-3,8-dioxatricyclo[5.1.0.02,4]octan-5-one (0.60%), ethyl 4-methoxybenzoate (0.43%), (E)-4-(2,6,6-trimethylcyclohexa-1,3-dien-1-yl)but-3-en-2-one (0.37%), and 4,6,10,10-tetramethyl-5-oxatricyclo[4.4.0.01,4]dec-2-en-7-ol (0.37%).

The extraction temperature indicated a notable effect on total volatility (Figure 1). The ideal temperature for the isolation of compounds with lower molecular weight and high volatility was 30 °C. Contrariwise, 60 °C was better for molecules with lower volatility (Table 6). In our case, the temperature of 60 °C was selected. Considering the above results, the higher the extraction temperature the larger the partition coefficients of compounds [14]. Nonetheless, this relation was not linear because higher temperatures may lead to the formation of by-products or thermal decomposition [20].

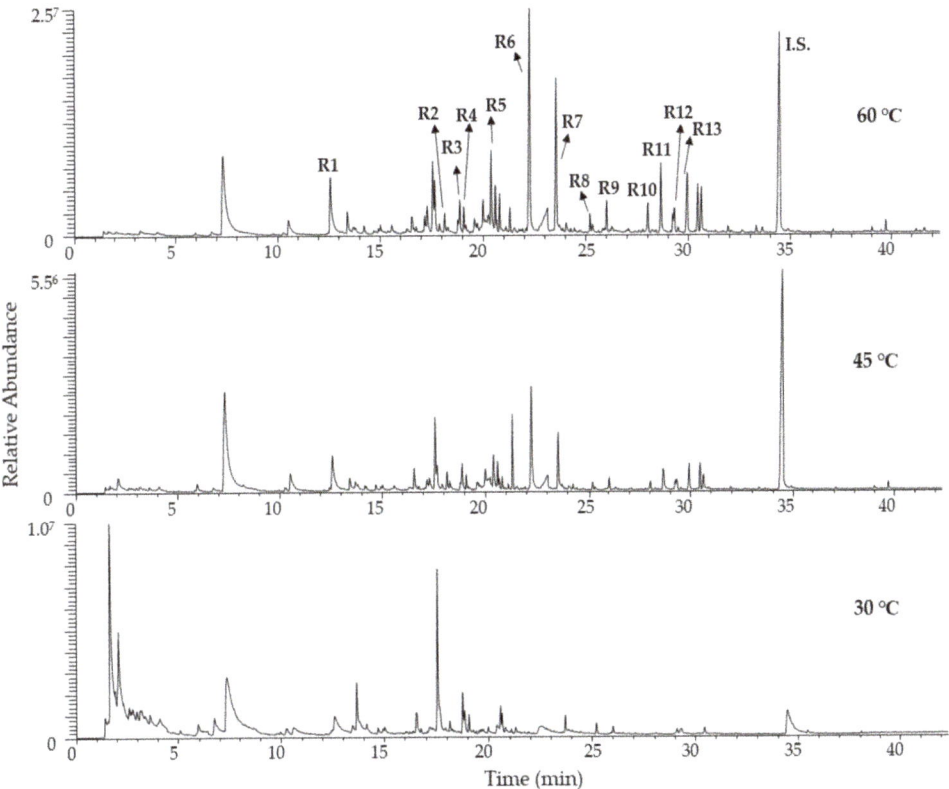

Figure 1. Chromatograms of the same sample at different temperatures (30, 45, and 60 °C).

The equilibration time, also known as "thermostating time", showed an uncommon high value at 30 min. Figure 2 presents the desirability surface area of all responses.

Due to the nature and molecular structure variability of autumn heather honey volatiles, compounds with short equilibration time can be displaced from the headspace of the vial gradually, by compounds with higher equilibration time. However, it is not a factor that significantly affects the efficiency of the system [14], since it reacts with other parameters at a relative moderate impact, as observed from the ANOVA test (Table 5).

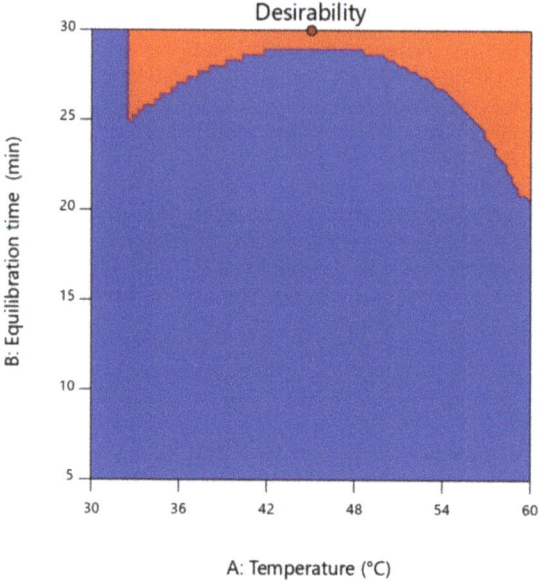

Figure 2. Desirability contour of equilibration time as a function of temperature of all responses.

Extraction time is critical for the sample to establish equilibrium with the SPME fiber coating. A typically 15 min extraction time showed a good performance. This parameter depends on the interactions of the molecules and can be lower when using the headspace technique and interact with high concentration samples. At the same time, it should be considered that many volatiles encumber the overall sensitivity and drive the specific compounds out of the fiber, which is easier to happen at prolonged extraction. On the other side, some studies indicated that the efficiency increases together with the extension of extraction time [14]. Obviously, the assessment is very difficult due to the intrinsic variability of honeys.

The magnetic stirrer speed, as previously referred to by Xagoraris [20], was confirmed to interact with most of the responses. However, this contribution is not always important as shown by the coded equation. Although one can assume that compounds with lower volatility require greater velocities, and in our case, 100 rpm gave satisfactory results. Thus, maintaining consistent agitation improves the accuracy and precision of the system.

The optimum sample volume was 6 mL, while this parameter had minimum impact on the isolation. Nevertheless, sensitivity is better when the headspace volume is small and fiber extracts faster compounds with higher volatility [37].

Finally, the ratio of water: Honey was proven a significant parameter. Honey is highly viscous and the addition of water enables the sample agitation, while also water evaporation drifts more easily the volatiles from honey. However, the excessive addition of water tends to dilute the honey reducing the concentration of specific molecules. The most favorable water: Honey ratio was 1:3 (v/w).

Reviewing the literature, in other botanical origin honey samples, different optimization conditions are reported. Ceballos [38] suggested the analytical conditions of the optimized SPME method by RSM as 60 μm PDMS/DVB fiber, 6 g honey, 3 mL water, and

20% w/w sodium chloride, 20 min for thermostatic time, 30 min for extraction at 60 °C. Plutowska [14] referenced the following conditions: CAR/PDMS/DVB fiber, 5:1 w/w honey to water ratio, 3 g sample, equilibration time 10 min, extraction time 30–60 min, and temperature at 60 °C. Bianchi [39] examined four different sets of conditions for volatile fraction isolation from thistle honey. Bianchin [40] proposed a new optimization strategy based on the use of three different extraction temperatures (60, 40, and 30 °C) followed by equilibration time (60, 36, and 6 min), respectively, in a single assay. Robotti, [41] reported as optimal extraction conditions for multi-floral honeys (extraction temperature: 70 °C; extraction time: 60 min; salt percentage: 27.50% w/w). Da Costa [42] reported the optimum condition for extraction of volatile compounds were as follows: equilibration time of 15 min, extraction time of 45 min, and extraction temperature of 45 °C.

On the basis of the above considerations, it is difficult to predict the factors that can affect the isolation of each volatile compound of honey. In a previous study, the optimized combination of isolation conditions of thyme honey was different with temperature (60 °C), equilibration time (15 min), extraction time (30 min), magnetic velocity speed (700 rpm), sample volume (6 mL) and water honey ratio (1:3 v/w) [20]. Each honey has its unique composition thus requiring different optimization conditions regarding volatile compounds.

4. Conclusions

In terms of this research, the volatile fraction of 25 honey samples from indigenous monofloral autumn heather honey was investigated. The most important compounds indicating the botanical origin of this honey are benzaldehyde, 3,5,5-trimethylcclohex-2-en-1-one, 2,6,6-trimethylcyclohex-2-ene-1,4-dione, 2-hydroxy-3,5,5-trimethylcyclohex-2-en-1-one, 2,6,6-trimethylcyclohexa-1,3-diene-1-carbaldehyde, 4-methoxybenzaldehyde, 3,4,5-trimethylphenol, 1,1,5-trimethyl-1,2-dihydronaphthalene, (E)-1-(2,6,6-trimethylcyclohexa-1,3-dien-1-yl)but-2-en-1-one, (E)-1,6,6-trimethyl-7-(3-oxobut-1-en-1-yl)-3,8-dioxatricyclo[5.1.0.02,4]octan-5-one, ethyl 4-methoxybenzoate, (E)-4-(2,6,6-trimethylcyclohexa-1,3-dien-1-yl)but-3-en-2-one, and 4,6,10,10-tetramethyl-5-oxatricyclo[4.4.0.01,4]dec-2-en-7-ol. These compounds were identified in almost all samples in significant concentrations, with few exceptions. Obviously, the assessment of a quantitative reference value is very difficult to be set due to endogenous or exogenous factors. Some of the above volatiles were previously reported in heather honey. However, the autumn heather honey from *E. manipuliflora* has not been previously investigated. The main volatile compounds were analyzed using a well-suited RSM methodology and predictive models were created to evaluate each volatile separately. Moreover, preconized optimum conditions (A: 60 °C B: 30 min C: 15 min D: 100 rpm E: 6 mL F: 1:3 (v/w)) were proposed for all dominant volatiles. In addition, a validation set amplified the results by responsiveness. This study reinforces the more reliable characterization of the volatile profile of autumn heather honey, aiming at the assessment of its botanical origin. In addition, it investigates the most common isolation factors, in terms of their ability to isolate their aroma fraction with relative abundance.

Author Contributions: Conceptualization, M.X.; methodology, M.X., E.A. and P.-K.R.; software, M.X., F.C., P.-K.R. and P.A.T.; validation, M.X., F.C. and C.S.P.; formal analysis, M.X. and F.C.; investigation, M.X., P.-K.R. and C.S.P.; resources, E.A., P.A.T. and P.-K.R.; data curation, M.X. and C.S.P.; writing—original draft preparation, M.X.; writing—review and editing, P.-K.R., P.A.T., E.A. and C.S.P.; visualization, M.X. and F.C.; supervision, C.S.P.; project administration, C.S.P. and P.A.T.; funding acquisition, E.A. and P.A.T.; All authors have read and agreed to the published version of the manuscript.

Funding: This research was partly financed by the Emblematic Action "The Honeybee Roads" of the Greek Public Investments Program (P.I.P.) of General Secretariat for Research and Technology (GSRT) (project code: 2018ΣE01300000).

Institutional Review Board Statement: Not applicable.

Informed Consent Statement: Not applicable.

Data Availability Statement: Not applicable.

Acknowledgments: The authors would like to thank the Laboratory of Sericulture and Apiculture of Aristotle University of Thessaloniki and the Laboratory of Analytical Chemistry of the Mediterranean Agronomic Institute of Chania (M.A.I.Ch.) for verifying the botanical origin of the samples.

Conflicts of Interest: The authors declare no conflict of interest.

References

1. Vercelli, M.; Novelli, S.; Ferrazzi, P.; Lentini, G.; Ferracini, C. A qualitative analysis of beekeepers' perceptions and farm management adaptations to the impact of climate change on honey bees. *Insects* **2021**, *12*, 228. [CrossRef]
2. Scepankova, H.; Saraiva, J.A.; Estevinho, L.M. Honey health benefits and uses in medicine. In *Bee Products—Chemical and Biological Properties*; Springer: Cham, Switzerland, 2017; pp. 83–96. ISBN 9783319596891.
3. Cianciosi, D.; Forbes-Hernández, T.Y.; Afrin, S.; Gasparrini, M.; Reboredo-Rodriguez, P.; Manna, P.P.; Zhang, J.; Lamas, L.B.; Flórez, S.M.; Toyos, P.A.; et al. Phenolic compounds in honey and their associated health benefits: A review. *Molecules* **2018**, *23*, 2322. [CrossRef]
4. Popescu, A.; Dinu, T.A.; Stoian, E.; Șerban, V. A Statistical Approach. *Sci. Pap. Ser. Manag. Econ. Eng. Agric. Rural Dev.* **2021**, *21*, 383–392.
5. The Best Greek Honey You'll ever Taste—Greece Is. Available online: https://www.greece-is.com/madeingreece/the-best-greek-honey-youll-ever-taste/ (accessed on 22 September 2021).
6. Persano Oddo, L.; Piana, L.; Bogdanov, S.; Bentabol, A.; Gotsiou, P.; Kerkvliet, J.; Martin, P.; Morlot, M.; Ortiz Valbuena, A.; Ruoff, K.; et al. Botanical species giving uniflora honey in Europe. *Apidologie* **2004**, *35*, S82–S93. [CrossRef]
7. Boi, M.; Llorens, J.A.; Cortés, L.; Lladó, G.; Llorens, L. Palynological and chemical volatile components of typically autumnal honeys of the western Mediterranean. *Grana* **2013**, *52*, 93–105. [CrossRef]
8. Tzitsa, E.; Tzakou, O.; Loukis, A. Volatile constituents of erica manipuliflora salisb. from Greece. *J. Essent. Oil Res.* **2000**, *12*, 67–68. [CrossRef]
9. Aliferis, K.A.; Tarantilis, P.A.; Harizanis, P.C.; Alissandrakis, E. Botanical discrimination and classification of honey samples applying gas chromatography/mass spectrometry fingerprinting of headspace volatile compounds. *Food Chem.* **2010**, *121*, 856–862. [CrossRef]
10. Guyot, C.; Scheirman, V.; Collin, S. Floral origin markers of heather honeys: Calluna vulgaris and Erica arborea. *Food Chem.* **1999**, *64*, 3–11. [CrossRef]
11. de la Fuente, E.; Martínez-Castro, I.; Sanz, J. Characterization of Spanish unifloral honeys by solid phase microextraction and gas chromatography-mass spectrometry. *J. Sep. Sci.* **2005**, *28*, 1093–1100. [CrossRef] [PubMed]
12. Soria, A.C.; Martínez-Castro, I.; Sanz, J. Some aspects of dynamic headspace analysis of volatile components in honey. *Food Res. Int.* **2008**, *41*, 838–848. [CrossRef]
13. Castro-Vázquez, L.; Díaz-Maroto, M.C.; González-Viñas, M.A.; Pérez-Coello, M.S. Differentiation of monofloral citrus, rosemary, eucalyptus, lavender, thyme and heather honeys based on volatile composition and sensory descriptive analysis. *Food Chem.* **2009**, *112*, 1022–1030. [CrossRef]
14. Plutowska, B.; Chmiel, T.; Dymerski, T.; Wardencki, W. A headspace solid-phase microextraction method development and its application in the determination of volatiles in honeys by gas chromatography. *Food Chem.* **2011**, *126*, 1288–1298. [CrossRef]
15. Yang, Y.; Battesti, M.J.; Paolini, J.; Muselli, A.; Tomi, P.; Costa, J. Melissopalynological origin determination and volatile composition analysis of Corsican "erica arborea spring maquis" honeys. *Food Chem.* **2012**, *134*, 37–47. [CrossRef]
16. Seisonen, S.; Kivima, E.; Vene, K. Characterisation of the aroma profiles of different honeys and corresponding flowers using solid-phase microextraction and gas chromatography-mass spectrometry/olfactometry. *Food Chem.* **2015**, *169*, 34–40. [CrossRef] [PubMed]
17. Rodríguez-Flores, M.S.; Falcão, S.I.; Escuredo, O.; Seijo, M.C.; Vilas-Boas, M. Description of the volatile fraction of Erica honey from the northwest of the Iberian Peninsula. *Food Chem.* **2021**, *336*, 127758. [CrossRef]
18. Machado, A.M.; Miguel, M.G.; Vilas-Boas, M.; Figueiredo, A.C. Honey volatiles as a fingerprint for botanical origin—a review on their occurrence on monofloral honeys. *Molecules* **2020**, *25*, 374. [CrossRef]
19. Soria, A.C.; Sanz, J.; Martínez-Castro, I. SPME followed by GC-MS: A powerful technique for qualitative analysis of honey volatiles. *Eur. Food Res. Technol.* **2009**, *228*, 579–590. [CrossRef]
20. Xagoraris, M.; Skouria, A.; Revelou, P.K.; Alissandrakis, E.; Tarantilis, P.A.; Pappas, C.S. Response surface methodology to optimize the isolation of dominant volatile compounds from monofloral greek thyme honey using spme-gc-ms. *Molecules* **2021**, *26*, 3612. [CrossRef]
21. Bezerra, M.A.; Santelli, R.E.; Oliveira, E.P.; Villar, L.S.; Escaleira, L.A. Response surface methodology (RSM) as a tool for optimization in analytical chemistry. *Talanta* **2008**, *76*, 965–977. [CrossRef]
22. Louveaux, J.; Maurizio, A.; Vorwohl, G. Methods of Melissopalynology. *Bee World* **1978**, *59*, 139–157. [CrossRef]
23. AOAC. Method 988.05. In *Official Methods of Analysis*, 15th ed.; Helrich, K., Ed.; The Association of Official Analytical Chemists, Inc.: Arlington, TX, USA; Scientific Research Publishing: Arlington, VA, USA, 1990. Available online: https://www.scirp.org/(S(lz5mqp453edsnp55rrgjct55.))/reference/referencespapers.aspx?referenceid=66154 (accessed on 6 September 2021).

24. International Honey Commission. *Harmonides Methods of the International Honey Commission*; Bee Product Science; Bogdanov, S., Ed.; Swiss Bee Research Centre FAM: Bern, Switzerland, 2009. Available online: https://www.ihc-platform.net/ihcmethods2009.pdf. (accessed on 6 September 2021).
25. Xagoraris, M.; Revelou, P.K.; Alissandrakis, E.; Tarantilis, P.A.; Pappas, C.S. The use of right angle fluorescence spectroscopy to distinguish the botanical origin of greek common honey varieties. *Appl. Sci.* **2021**, *11*, 4047. [CrossRef]
26. Council of the European Union. Council of the European Union. Council Directive 2001/110/EC of relating to honey. *Off. J. Eur. Comm.* **2002**, *L10*, 47–52.
27. Government Gazette B-239/23-2-2005 Annex II Article 67 of Greek Food Code 2005. Available online: http://www.minagric.gr/images/stories/docs/agrotis/MeliMelissokomia/KYATaytopoiisi.pdf. (accessed on 6 September 2021).
28. Box, G.E.P.; Wilson, K.B. On the experimental attainment of optimum conditions. *J. R. Stat. Soc. Ser. B* **1951**, *13*, 1–38. [CrossRef]
29. Xagoraris, M.; Revelou, P.K.; Dedegkika, S.; Kanakis, C.D.; Papadopoulos, G.K.; Pappas, C.S.; Tarantilis, P.A. SPME-GC-MS and FTIR-ATR spectroscopic study as a tool for unifloral common greek honeys' botanical origin identification. *Appl. Sci.* **2021**, *11*, 3159. [CrossRef]
30. Adams, R.P.; Sparkman, O. Review of identification of essential oil components by gas chromatography/mass spectrometry. *J Am. Soc. Mass Spectrom.* **2007**, *18*, 803–806.
31. Tan, S.T.; Wilkins, A.L.; Holland, P.T.; McGhie, T.K. Extractives from New Zealand unifloral honeys. Degraded carotenoids and other substances from heather honey. *J. Agric. Food Chem.* **1989**, *37*, 1217–1221. [CrossRef]
32. Smith, G.C.; Bromenshenk, J.J.; Jones, D.C.; Alnasser, G.H. Volatile and semi-volatile organic compounds in beehive atmospheres. *Honey Bees Estim. Environ. Impact Chem.* **2002**, 12–41.
33. Jerković, I.; Kuś, P.M. Terpenes in honey: Occurrence, origin and their role as chemical biomarkers. *RSC Adv.* **2014**, *4*, 31710–31728. [CrossRef]
34. D'Arcy, B.R.; Rintoul, G.B.; Rowland, C.Y.; Blackman, A.J. Composition of Australian honey extractives. Norisoprenoids, monoterpenes, and other natural volatiles from blue gum (*Eucalyptus leucoxylon*) and yellow box (*Eucalyptus melliodora*) honeys. *J. Agric. Food Chem.* **1997**, *45*, 1834–1843. [CrossRef]
35. Wootton, M.; Edwards, R.A.; Faraji-Haremi, R.; Williams, P.J. Effect of accelerated storage conditions on the chemical composition and properties of australian honeys 3. Changes in volatile components. *J. Apic. Res.* **1978**, *17*, 167–172. [CrossRef]
36. Visser, F.R.; Allen, J.M.; Shaw, G.J. The effect of heat on the volatile flavour fraction from a unifloral honey. *J. Apic. Res.* **1988**, *27*, 175–181. [CrossRef]
37. Supelco; Sigma-Aldrich. Solid Phase Microextraction Troubleshooting Guide. 1998. Available online: https://www.sigmaaldrich.com/US/en/deepweb/assets/sigmaaldrich/marketing/global/documents/306/877/t101928.pdf (accessed on 6 September 2021).
38. Ceballos, L.; Pino, J.A.; Quijano-Celis, C.E.; Dago, A. Optimization of a hs-spme/gc-ms method for determination of volatile compounds in some cuban unifloral honeys. *J. Food Qual.* **2010**, *33*, 507–528. [CrossRef]
39. Bianchi, F.; Mangia, A.; Mattarozzi, M.; Musci, M. Characterization of the volatile profile of thistle honey using headspace solid-phase microextraction and gas chromatography-mass spectrometry. *Food Chem.* **2011**, *129*, 1030–1036. [CrossRef]
40. Bianchin, J.N.; Nardini, G.; Merib, J.; Dias, A.N.; Martendal, E.; Carasek, E. Screening of volatile compounds in honey using a new sampling strategy combining multiple extraction temperatures in a single assay by HS-SPME-GC-MS. *Food Chem.* **2014**, *145*, 1061–1065. [CrossRef] [PubMed]
41. Robotti, E.; Campo, F.; Riviello, M.; Bobba, M.; Manfredi, M.; Mazzucco, E.; Gosetti, F.; Calabrese, G.; Sangiorgi, E.; Marengo, E. Optimization of the extraction of the volatile fraction from honey samples by SPME-GC-MS, experimental design, and multivariate target functions. *J. Chem.* **2017**, *2017*, 6437857. [CrossRef]
42. da Costa, A.C.V.; Sousa, J.M.B.; Bezerra, T.K.A.; da Silva, F.L.H.; Pastore, G.M.; da Silva, M.A.A.P.; Madruga, M.S. Volatile profile of monofloral honeys produced in Brazilian semiarid region by stingless bees and key volatile compounds. *LWT-Food Sci. Technol.* **2018**, *94*, 198–207. [CrossRef]

Article

The Occurrence of Skeletons of Silicoflagellata and Other Siliceous Bioparticles in Floral Honeys

Donát Magyar [1,*], Paulian Dumitrica [2], Anna Mura-Mészáros [3], Zsófia Medzihradszky [4], Ádám Leelőssy [5] and Simona Saint Martin [6]

1. National Public Health Center, 1097 Budapest, Hungary
2. Institute of Earth Sciences, Université de Lausanne, 1015 Lausanne, Switzerland; Paulian.Dumitrica@unil.ch
3. Faculty of Biological Sciences, Friedrich Schiller University Jena, 07743 Jena, Germany; muram.anna@gmail.com
4. Museum and Library of Hungarian Agriculture, 1146 Budapest, Hungary; medzihradszky.zsofia@mmgm.hu
5. Department of Meteorology, Eötvös Loránd University, 1053 Budapest, Hungary; leelossyadam@gmail.com
6. Centre de Recherche en Paléontologie, Muséum National d'Histoire Naturelle, Sorbonne Université, 75006 Paris, France; simona.saint-martin@mnhn.fr
* Correspondence: magyar.donat@gmail.com

Citation: Magyar, D.; Dumitrica, P.; Mura-Mészáros, A.; Medzihradszky, Z.; Leelőssy, Á.; Saint Martin, S. The Occurrence of Skeletons of Silicoflagellata and Other Siliceous Bioparticles in Floral Honeys. *Foods* **2021**, *10*, 421. https://doi.org/10.3390/foods10020421

Academic Editor: Paweł Kafarski, Olga Escuredo and M. Carmen Seijo

Received: 23 December 2020
Accepted: 9 February 2021
Published: 14 February 2021

Publisher's Note: MDPI stays neutral with regard to jurisdictional claims in published maps and institutional affiliations.

Copyright: © 2021 by the authors. Licensee MDPI, Basel, Switzerland. This article is an open access article distributed under the terms and conditions of the Creative Commons Attribution (CC BY) license (https://creativecommons.org/licenses/by/4.0/).

Abstract: Siliceous marine microfossils were unexpectedly discovered during the analysis of flower honey samples from Poland and Tunisia. The microfossils were represented by protist with siliceous skeletons: silicoflagellates, diatoms, and endoskeletal dinoflagellates. This is the first record of such microfossils in honeys. Based on the high percent of anemophilous pollen grains and spores in the sample, it was hypothesized that silicoflagellates were deposited from the air onto the nectariferous flowers, then bees harvested them with the nectar. Based on the comparison of pollen content of honeys and flowering calendar of Tunisia, the harvest time of honey was identified as a period between 1 April and 31 May 2011. Trajectory analysis of air masses in this period confirmed that siliceous microfossils could be aerosolized by wind from the rocks of the so-called Tripoli Formation of Messinian age (6–7 Ma). Similar to the Tunisian case, the Polish trajectory simulation also supports the hypothesis of atmospheric transport of silicoflagellates from outcrops of Oligocene age in the Polish Outer Carpathians. In the case of diatom content of honey, however, the source can be both natural (wind) and artificial (diatomaceous earth filters). For a correct determination, natural sources of siliceous bioparticles, such as wind transport from nearby outcrops should be also considered. Silicoflagellates could be used as complementary indicators of the geographical origin of honeys collected in areas characterized by diatomite outcrops, supporting the results obtained with other methods; thus, such indicators merit further studies within the area of honey authenticity.

Keywords: honey; Silicoflagellata; diatoms; pollen; spores

1. Introduction

The identification of the origin of food is one of the most important issues in food quality control [1]. Depending on its geographical origin (the region where the beehives are located and the surrounding environment), honey can acquire different characteristics and properties. Therefore, geographical origin is an important parameter with respect to honey differentiation and valorization.

The determination of the geographical origin of honey relies on microscopical examination of its pollen profile if it is specific enough in the area of interest. Because of the limitations of this method (being expensive, time-consuming, and strongly dependent on the qualifications and judgement of the analyst), there is a tendency to replace pollen analysis by finding other markers for honey discrimination. Minerals and trace elements [1], and fungal spore content [2] are some of the parameters that have been examined for the recognition of the origin of honeys. However, when microscopical analysis of honeys is performed, one can see a great variety of particles, other than pollen grains or spores. Some of them belong to insect parts, most commonly bee hair, tracheae and, especially in *Pinus* honeydew honeys, wax produced by Pseudococcidae. Surprisingly, some particles come from microfauna as well (Acari, Rotifera, eggs of Tardigrade). Plant trichomes, starch, and phytoliths are also common components of honey samples. As a sign of human activity, microplastics can also be observed in honey under the microscope. These particles are often overlooked during the routine melissopalyonolgical analysis of samples. As no attention is paid to them, they remain unidentified, although they would be useful in the analysis.

During our research to find new indicators of the origin of the honeys, unusual multi-radiate structures were found in honey samples that originated from Poland and Tunisia. Therefore, we aimed to identify these particles and other accompanying components in honey to find any indication of their origin.

2. Materials and Methods

The honeys were purchased from food shops in Poland (according to the information on the product label, the honey was harvested from *Fagopyrum esculentum* in Stróze, near Nowy Sącz, 2013) and Tunisia (mixed floral honey, Nabeul, 2011). Ten grams were taken from 250 g of previously homogenized honey, dissolved in 20 mL of distilled water at 40 °C, centrifuged for 5 min at 2500 rpm, and allowed to settle. The sediment was recovered in 10 mL of distilled water and again centrifuged. The sediment was then collected with a Pasteur pipette and dried onto microscope slides at 40 °C. It was then mounted in glycerine-gelatine and covered [3,4]. The entire surface of the preparation was scanned under 600× magnification of an Olympus CX 31 microscope. Preliminary identification indicated that multiradiate particles may belong to extinct microscopic organisms, occurring as fossils.

To see the frequency of occurrence of these particles, we studied samples from a collection containing 106 honeys, prepared according to the method mentioned above. These samples were listed in the Supplement 1 with their collection code, type, botanical origin, and location.

Because the Polish (P) and the Tunisian (T) honeys were particularly rich in the investigated multiradiate particles, further analyses were performed. To identify the geographical origin of the multiradiate particles, a combination of methods was applied. First, fungal spore and pollen composition were determined.

Among fungal spores, honeydew indicators were not found [2], but indicators of floral origin (*Metschnikowia reukaufii*, P,T) and common airborne fungi were present (*Alternaria* sp. P,T, *Aspergillus/Penicillium* P,T, *Bipolaris spicifera* T, *Botrytis* sp. P, *Chaetomium* sp. P,T, *Cladosporium* spp. P,T, *Coprinus* sp. P,T, *Curvularia* sp. T, *Diplodia frumenti* T, *Drechslera biseptata* T, *Drechslera/Helminthosporium* T, *Ellisembia* sp. T, *Epicoccum nigrum* P,T, *Ganoderma* sp. P, *Leptosphaeria* spp. P, *Melampsoridium* sp. T, *Paraphaeosphaeria michotii* P, *Periconia* sp. T, Peronosporaceae P, *Pithomyces chartarum* T, *Polythrincium trifolii* P, Pucciniaceae T, *Rhizopus* sp. T, *Stemphylium* sp. P,T, Telephoraceae P, *Torula* sp. P,T, *Trichothecium roseum* P, *Tripospermum* spp. P, Ustilaginomycetes P,T, other Ascomycota). Pollen content was expressed as percentage of pollen grains (N = 300) [5]: P: *Brassica* 34%, *Centaurea cyanus* 5%, *Fagopyrum* 5%, *Trifolium* 2%, Ericaceae and *Tilia* < 2%; T (in descending order of frequency; data are shown on Figure 1): Poaceae spp., Brassicaceae (*Brassica* cf. *napus*), *Eucalyptus* sp., *Myrtus* sp., *Acacia* sp., Ericaceae sp., *Carex* sp., Caryophyllaceae, Chenopodiaceae, Compositae-Tubuliflorae, *Trifolium* sp., Umbelliferae, *Vicia* sp., *Zea mays*, Boraginaceae, Compositae-Liguliflorae, *Convolvulus* sp., *Echium* sp., Labiatae, Polygalaceae sp., *Rumex* sp. [6,7]. Percentage of pollen grains also showed that dominant taxa are anemophilic.

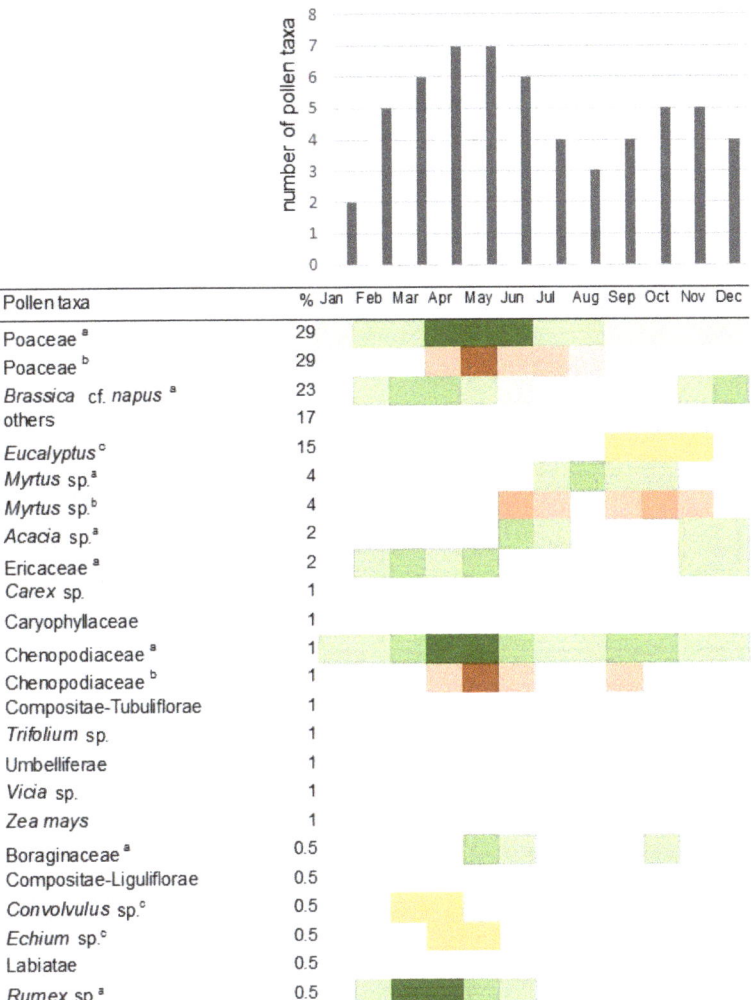

Figure 1. Combined aerobiological and phenological calendar of pollen taxa found in Tunisian honey. [a]: Flowering time from aerobiological data [8] covering northern Tunisia shown in green; deep green indicates the main pollination period. [b]: Flowering time from another aerobiological dataset [8] covering Tunisia shown in red. Dark shades mean flowering peaks, according to the original illustrations. [c]: Flowering time from phenological data [9] shown in yellow. 'Pollen taxa %' means the relative abundance of pollen taxa in the Tunisian honey. In [8], taxa are referred to as: Amaranthaceae (instead of Chenopodiaceae), *Erica* (instead of Ericaceae), *Borago* (instead of Boraginaceae), Brassicaceae (instead of *Brassica* cf. *napus*), and Myrtaceae (instead of *Myrtus*), in [10]: 'Graminees' (instead of Poaceae), 'Myrtacees' instead of *Myrtus*), in [9]: *Eucalyptus gomphocephalla* (instead of *Eucalyptus*). Polygalaceae, found in low numbers (<0.5%) are not shown, as flowering data is not available for this region.

Consequently, we hypothesized that multiradiate particles found in the honey samples might also have airborne origin. To test this hypothesis, we searched for the possible source of multiradiate particles. The source of airborne particles can be identified with the calculation of wind trajectory of air masses carrying particles from long distances. An important information for calculations is the time (year and month) of honey harvest. However, in case of the Polish honey, only the year (2013) was known. Harvesting period was determined to be July–August according to the blooming of *Fagopyrum esculentum* [11]. In the case of Tunisian honey, the year 2011 was shown on the product's label. The month of acquisition was known as well (31 July). Therefore, the month of harvest had to be identified. With this aim, a forensic palynological method was used [12,13]. To collect information on pollination of local melliferous plants, phenological calendars were reviewed [9]. Because the dominant pollen taxa in the honey belonged to anemophilous plants, data from aerobiological literature were also considered [8] (data from the year of 2011) [10]. Number of pollen taxa found in the honey was summarized by month and illustrated on a histogram (Figure 1). According to this analysis, the honey was most probably harvested during the months of April and May.

To investigate whether atmospheric conditions supported the transport of aerosol particles from the suspected source area to the harvesting region, an atmospheric dispersion model was applied for the flowering periods (P: July–August 2013, T: April–May 2011). The dust emission flux was estimated to be a cubic function of the friction velocity, according to the dust emission model presented by Bagnold [14], discussed more recently by Xuan [15], and applied as described in a previous study [16]. The threshold friction velocity was set to 0.5 ms^{-1}, a medium value within the range of experimental results of Marticorena and Bergametti [17]. If the friction velocity was higher than the threshold friction velocity and no precipitation occurred, 1000 particles were released in every hour from each of 20 levels between 1–1000 m above ground; and their atmospheric trajectories were simulated for 48 h. Meteorological data was obtained from the GDAS FNL (Global Data Assimilation System—Final Analysis) database [18] with 3 h temporal and 0.25° spatial resolution. Atmospheric dispersion was simulated with the Lagrangian particle dispersion model RAPTOR that calculates advection, turbulent dispersion, and deposition [19,20]. As the extent and amount of mobilizable dust is unknown, sensitivity maps were produced with unit m^{-3}, normalized to a total sensitivity of 1 over the entire domain. This way, the spatial and temporal pattern of the dispersion could be investigated while the amount of deflated dust remains unknown.

3. Results and Discussion

Our investigations have shown that honey from Tunisia (T), Morocco (M), Africa (A), Greece (G) Poland (P), and Romania (R) contained silica skeletons of planktonic marine Silicoflagellata belonging to *Dictyocha fibula* (T), *Distephanopsis crux* (T), *Stephanocha speculum* (A,G,P,R), *Stephanocha* cf. *speculum* (M), *Stephanocha speculum speculum* (T), *Stephanocha speculum speculum* f. *notabilis* (T) (Figure 2, Table 1). Diatoms, e.g., *Actinocyclus divisus*, *Coscinodiscus marginatus*, *Coscinodiscus* (?) sp., *Fragilaria* (?) sp., *Hantzschia amphioxys*, *Mastogloia* (?) sp., *Melosira* sp., *Nitzschia* (?) sp., *Thalassionema nitzschioides*, and very rare endoskeletal siliceous dinoflagellates belonging to *Actiniscus pentasterias* were also found (Figure 3, Table 1). To our knowledge, this is the first record in the literature of silicoflagellates and other protists with siliceous skeletons occurring in honey.

Silicoflagellates are planktonic marine chloroplast-bearing protists with a flagellum and a siliceous skeleton formed of distally closed hollow bars known to have existed starting from the mid-Cretaceous (Albian) to recent. Their skeletons usually comprise 1–2% of the siliceous component of marine sediments [21] and in some cases, as for example in some Sarmatian deposits from Romania, they are so abundant that practically these rocks could be called silicoflagellitites. Their skeleton has a rather simple geometrical form and consists usually of two parts: a basal ring and an apical structure, both interconnected by bars. All these elements have a special descriptive nomenclature [21,22]. *Dictyocha fibula* is a species characteristic of warm water, whereas *Stephanocha speculum* is much more frequent in colder waters. *Distephanopsis crux* is a Miocene and Pliocene species that became extinct at the base of the Pleistocene [23]. All these three species are common in the diatomites of the so-called Tripoli Formation of Messinian age that can be visible in outcrops and found also in cored sediments [24–27] in the Mediterranean area. They were deposited 6–7 million years ago before the period of the closing of the Mediterranean Sea, which determined the famous "Messinian Salinity Crisis" that lasted until the Pliocene [28]. These diatomites are well known especially in Spain, Italy, Crete, Cyprus, and in the northwestern part of Africa, in Morocco and Algeria. The only mining of Messinian diatomites, formerly active in the Oran region in Algeria [29], has long been abandoned.

Table 1. Silicoflagellates, diatoms, and cyanobacteria found in honey samples.

Name	Chloroplast	Major Group	Occurrence	Habitat	Sample Code	Country	Source
Dictyocha fibula Ehrenberg	no	Silicoflagellata	fossil	marine	FH29Af	Tunisia	floral
Distephanopsis crux (Ehrenberg)	no	Silicoflagellata	fossil	marine	FH29Af	Tunisia	floral
Stephanocha speculum (Ehrenberg)	no	Silicoflagellata	fossil	marine	FH25Af	Africa	floral
Stephanocha speculum	no	Silicoflagellata	fossil	marine	HP09Gr	Greece	honeydew, *Pinus*
Stephanocha speculum	no	Silicoflagellata	fossil	marine	FH33Po	Poland	floral, *Fagopyrum*
Stephanocha speculum	no	Silicoflagellata	fossil	marine	UK02Ro	Romania	unknown
Stephanocha cf. *speculum* (Ehrenberg)	no	Silicoflagellata	fossil	marine	FH28Af	Morocco	floral
Stephanocha speculum speculum (Ehrenberg)	no	Silicoflagellata	fossil	marine	FH29Af	Tunisia	floral
Stephanocha speculum speculum f. *notabilis* Locker & Martini	no	Silicoflagellata	fossil	marine	FH29Af	Tunisia	floral
Achnanthes sp.	no	Diatom	*	freshwater	UK05Cz	Czech Republic	unknown
Achnanthes sp.	living	Diatom	*	freshwater	UK05Cz	Czech Republic	unknown
Achnanthidium sp.	living	Diatom	*	freshwater	HH19It	Italy	honeydew
Actiniscus pentasterias Ehrenberg	no	dinoflagellates	fossil and actual	marine	FH29Af	Tunisia	floral
Actinocyclus divisus (Grunow) Hustedt	no	Diatom	fossil and actual	marine	FH29Af	Tunisia	floral
Aulacodiscus sp.	no	Diatom	*	marine	UK02Ro	Romania	unknown
Aulacoseira distans (Ehrenberg) Simonsen	no	Diatom	fossil and actual	freshwater	UK04Ge	Germany	unknown
Aulacoseira distans (Ehrenberg) Simonsen	no	Diatom	fossil and actual	freshwater	HC04It	Italy	floral, *Castanea*
Aulacoseira distans (Ehrenberg) Simonsen	no	Diatom	fossil and actual	freshwater	UK02Ro	Romania	unknown
Aulacoseira cf. *distans* (Ehrenberg) Simonsen	no	Diatom	fossil and actual	freshwater	FH33Po	Poland	floral, *Fagopyrum*
Aulacoseira sp.	no	Diatom	*	freshwater	UK02Ro	Romania	unknown
Aulacoseira sp.	no	Diatom	*	freshwater	HA05Gr	Greece	honeydew, *Abies*

Table 1. Cont.

Name	Chloroplast	Major Group	Occurrence	Habitat	Sample Code	Country	Source
Chroococcus sp.	living	Cyanobacteria	*	mainly freshwater	HA05Gr	Greece	honeydew, *Abies*
Chroococcus sp.	living	Cyanobacteria	*	mainly freshwater	UK07Sv	Switzerland	unknown
Chroococcus sp.	living	Cyanobacteria	*	mainly freshwater	HH29Sl	Slovakia	honeydew
Chroococcus sp.	living	Cyanobacteria	*	mainly freshwater	UK04Ge	Germany	unknown
Chroococcus sp.	living	Cyanobacteria	*	mainly freshwater	FH14Mx	Mexico	floral
Chroococcus sp.	living	Cyanobacteria	*	mainly freshwater	HH04It	Italy	honeydew
Chroococcus sp.	living	Cyanobacteria	*	mainly freshwater	HH04It	Italy	honeydew
Chroococcus sp.	living	Cyanobacteria	*	mainly freshwater	UK05Cz	Czech Republic	unknown
Coscinodiscus ? sp.	no	Diatom	fossil and actual	marine	FH33Po	Poland	floral, *Fagopyrum*
Coscinodiscus ? sp.	no	Diatom	fossil and actual	marine	FH29Af	Tunisia	floral
Coscinodiscus marginatus Ehrenberg	no	Diatom	fossil and actual	marine	FH29Af	Tunisia	floral
Coscinodiscus sp.	no	Diatom	fossil and actual	marine	UK02Ro	Romania	unknown
Cyclostephanos dubius Hustedt (Round)	no	Diatom	fossil and actual	freshwater	HA04Gr	Greece	honeydew, *Abies*
Cyclostephanos dubius Hustedt (Round)	no	Diatom	fossil and actual	freshwater	HA05Gr	Greece	honeydew, *Abies*
Cyclotella sp.	no	Diatom	fossil and actual	freshwater	HH31Hu	Hungary	honeydew
Fragilaria ? sp.	?	Diatom	?	freshwater	FH29Af	Tunisia	floral
Fragilaria intermedia? Grunow (Grunow)	living	Diatom	fossil? recent to actual	freshwater	UK05Cz	Czech Republic	unknown
Fragilaria? sp.	no	Diatom	*	freshwater	HH30Hu	Hungary	honeydew
Hantschia amphioxys (Ehrenberg) Grunow	no	Diatom	fossil and actual	freshwater	FH33Po	Poland	floral, *Fagopyrum*
Hantzschia amphioxys Ehrenberg (Grunow)	no	Diatom	fossil and actual	freshwater	HH31Hu	Hungary	honeydew
Mastogloia ? sp.	living	Diatom	*	freshwater/brackish	FH29Af	Tunisia	floral
Nitzschia ? sp.	no	Diatom	?	marine?	FH29Af	Tunisia	floral
Nitzschia paleacea Grunow in van Heurck	living	Diatom	*	freshwater	HH30Hu	Hungary	honeydew
Nitzschia sp.	living	Diatom	*	freshwater	FH20Cu	Cuba	floral
Nitzschia sp.	living	Diatom	*	freshwater	UK07Sv	Switzerland	unknown
Nitzschia sp.	living	Diatom	*	freshwater	FH31Hu	Hungary	floral, *Foeniculum*

Table 1. Cont.

Name	Chloroplast	Major Group	Occurrence	Habitat	Sample Code	Country	Source
Nitzschia sp.	living	Diatom	*	freshwater	FH31Hu	Hungary	floral, *Foeniculum*
Oscillatoria sp.	living	Cyanobacteria	*	mainly freshwater/marine	HA17Gr	Greece	honeydew, *Abies*
Scenedesmus sp.	living	green algae	*	freshwater	HA15Gr	Greece	honeydew, *Abies*
Thalassionema nitzschioides (Grunow) Mereschkowsky	no	Diatom	fossil and actual	marine	FH29Af	Tunisia	floral
centric sp.	no	Diatom	*	?	HH04It	Italy	honeydew
pennate sp.	living	Diatom	*	freshwater	HH08It	Italy	honeydew
pennate sp.	living	Diatom	*	freshwater	HA03Gr	Greece	honeydew, *Abies*

* for the living taxa not designated to species level, the geological record is not indicated.

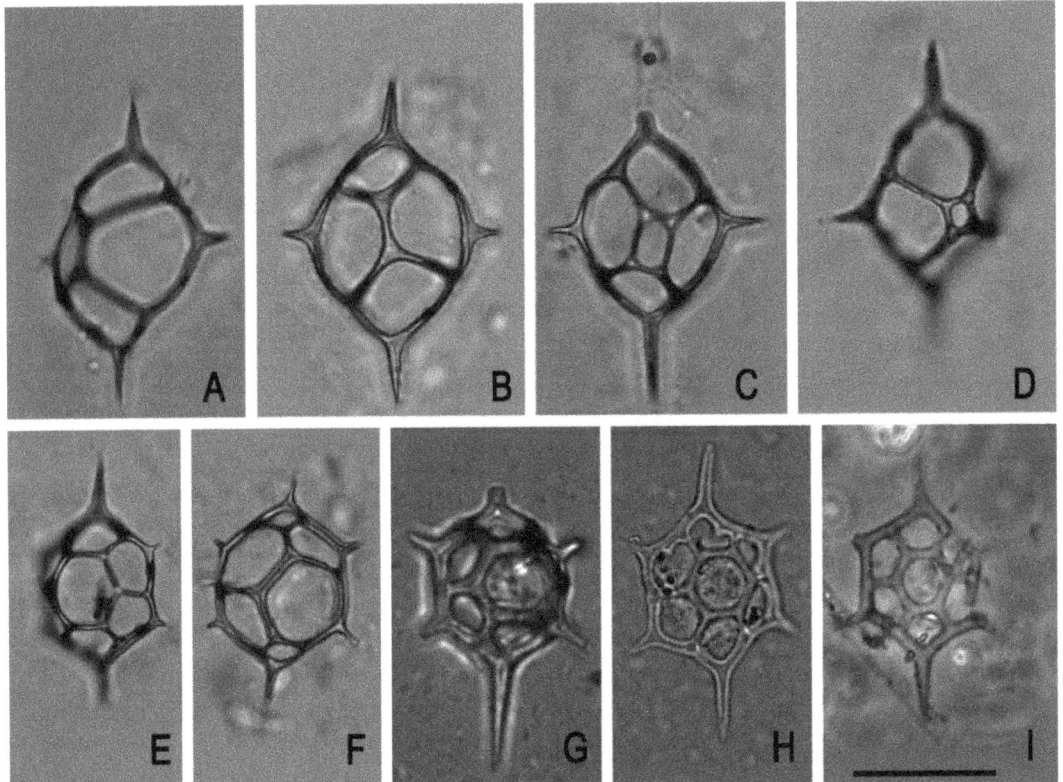

Figure 2. Silicoflagellates and diatoms found in floral honey. (**A,B**). *Dictyocha fibula*. (**C,D**). *Distephanopsis crux*. (**E**). *Stephanocha speculum speculum*. (**F**). *Stephanocha speculum speculum* f. *notabilis*. (**G–I**). *Stephanocha speculum*. (**A–F**) from Tunisia, (**G**) Poland, (**H**) Romania, (**I**) Greece. Scale bar: 20 μm.

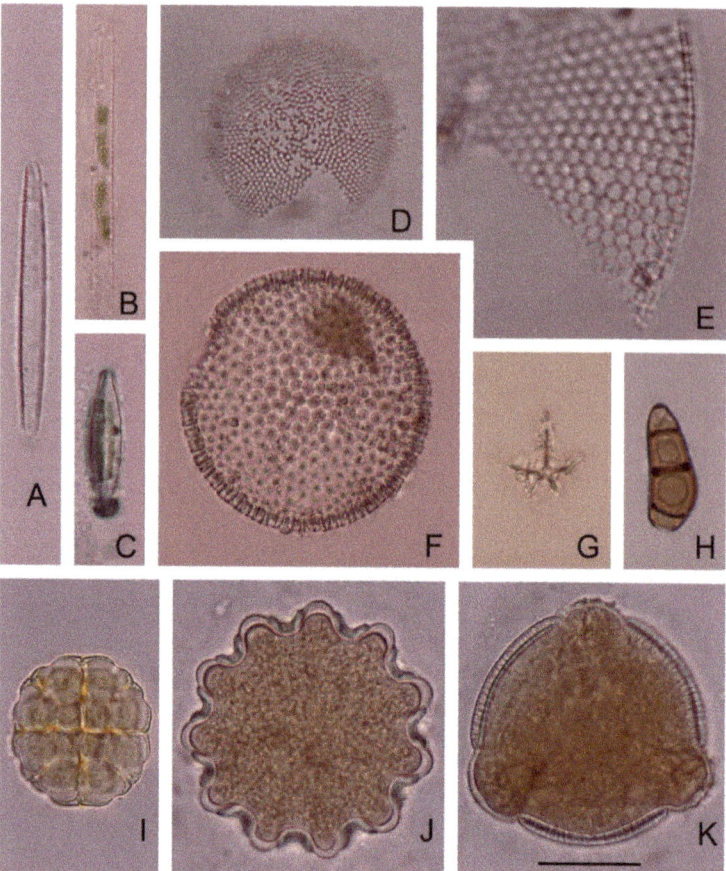

Figure 3. Diatoms, dinoflagellates, fungal spores, and pollen grains found in floral honey from Tunisia. (**A,B**). *Thalassionema nitzschioides*. (**C**). *Mastogloia* sp., (**D**). *Actinocyclus divisus*. (**E**). fragment of *Coscinodiscus*. (**F**). *Coscinodiscus marginatus*, (**G**). *Actiniscus pentasterias*. (**H**). *Curvularia* sp., (**I**). *Acacia* sp., (**J**). Polygalaceae sp., (**K**). *Convolvulus* sp., Scale bar: 20 μm.

The diatomites are porous rocks of marine or lacustrine origin. The marine diatomites originated in zones of high planktonic fertility. The lacustrine diatomites were especially formed in lakes of volcanic craters or in zones with volcanic tufs. Although they are very light due to the porosity of diatom frustules, they may contain up to 3000 frustules/mm³ [30]. Therefore, we hypothesized that these marine siliceous microfossils found in honey are of airborne origin. They may have been deposited as airborne dust on the flowers from where the bees collected them with the nectar. The presence of other particles (pollen grains, fungal spores, dinoflagellates, and diatoms) corroborate this hypothesis.

Diatoms can become airborne from outcrops (e.g., by mining activity), by deflation from dried lakebeds, or after swamp fires and storms [31] or via sea-foam and bursting bubbles [32]. Geissler and Gerloff [33] showed that the species composition of airborne diatoms above the city of Berlin is identical to the composition of diatoms in Berlin lakes and rivers. Folger [34] as well as many others, found *Melosira granulata* and *Stephanodiscus astrea* to be the most common diatoms in airborne dust samples from the Equatorial Atlantic. *Cyclotella* and *Stephanodiscus* spp. were found in high quantities as house dust in a building constructed on a dried lakebed in Hungary (Magyar, unpublished observation). Specimens of *Corethron*, another marine diatom, were recently detected in Late Cretaceous amber [35]; the authors considered that the amber forest grew in a nearshore environment where wind introduced the marine diatoms into the terrestrial realm. The occurrence of airborne algae in the atmosphere has been recorded as early as the middle of the 19th century. In 1833, aboard the famous vessel Beagle, Charles Darwin observed airborne diatoms in the dust from North Africa deposited on the board when it was near Cape Verde Islands [36]. Ehrenberg [37] reported 18 species of freshwater diatoms from the dust samples sent by Darwin. Since then, North African dust particles associated with diatoms were frequently observed [38]. Diatoms from the Bodélé Depression (once part of Mega-Lake Chad, North Africa) are the main source material for the dust [39]. Direct sampling of the atmosphere in various environments (e.g., terrestrial, marine, and freshwater) provided evidence that airborne algae are naturally occurring in the aerial biota [40]. In the Tunisian honey we investigated, recent diatoms were present. *Mastogloia* sp. and *Thalassionema nitzschioides* contained chloroplasts, thus they could be recent, originating from sea spray or high tide and wind.

Actiniscus pentasterias is a Miocene to recent species of endoskeletal dinoflagellates. Its star-like specimens with five, rarely four arms are frequent in marine sediments with other siliceous microfossils [41]. Its stellate arched structure, although rather small, is very easily recognized in microscope slides. Each spicule represents one of the two spicules present in a living cell and disposed symmetrically face to face with their concave sides opposed in the ovoid cell. Usually, they are separated in fossil material, but interconnected specimens by the end of the rays can be encountered when the sample is not treated too much with hydrochloric or other acids that can dissolve the points of interconnection. Rather neglected in the fossil samples, it was studied in detail by Dumitrica [41] who described several species and tried to make an order in this group.

There are no diatomites and therefore no mining of diatomites in Tunisia. However, the closest Messinian diatomites are those of Sicily about 250 km to 300 km from Nabeul (the location of honey harvest) to the northeast [26,27].

Atmospheric dispersion simulations performed in the period April–May 2011 revealed a situation when atmospheric conditions supported the transport of dust from Sicily to the region of Nabeul. Dust emission was assumed only when the threshold friction velocities exceeded 0.5 ms^{-1} and no precipitation occurred. Among these, 21–22 May was characterized by northerly-northeasterly winds in Sicily, brought by a Mediterranean cyclone marking the end of an 18-day long drought. Dry surface conditions with approaching thunderstorms were ideal for gust fronts and evaporative cooling, a well-known pattern for deflation [42]. According to WMO synop reports from the hilltop meteorological station of Enna (20 km to the northeast and 350 m above Caltanissetta) on the night of 21–22 May 2011, repeated thunderstorms with or without precipitation occurred, although yielding a total precipitation of only 3 mm/12 h. This confirms the potential for evaporative cooling and the formation of gust fronts. Thunder with no rain and the 1-h mean wind velocity reaching 30 km/h was reported at 3 UTC. Continuous rain inhibiting further deflation initiated at approximately 6 UTC; however, the total precipitation remained relatively low during the day (12 mm/24 h). Atmospheric dispersion maps of particles released from Caltanissetta between 18–6 UTC on 21–22 May 2011 (Figure 4) confirm the potential of the deflated dust to reach Tunisia. In the flow of the cyclone, the dust would have travelled in a moist environment to North Africa and deposited efficiently with rain onto the surface.

It might also be noted that in the previous week (10–18 May 2011), a documented Saharan dust event had occurred in Portugal [43], related to the ongoing shallow cyclonic activity over the Mediterranean. While dust transport typically occurs northward on the leading edge of cyclones, similar dust transport potential is present on the rear edge of a cyclone towards North Africa.

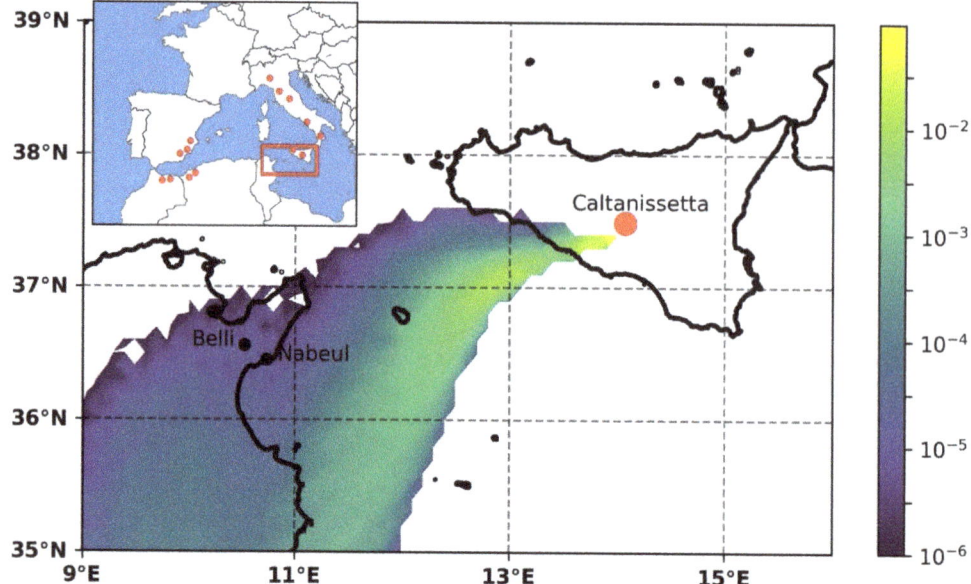

Figure 4. An episode of air mass possibly transporting microfossils from the emission area (Caltanissetta, Sicily) to the honey harvesting area shown on an atmospheric dispersion sensitivity map [m^{-3}] for the dust deflated between 18–6 UTC, 21–22 May 2011. Red dots in the small map show the distribution of Messinian diatomites.

Therefore, trajectory simulations support the hypothesis of atmospheric transport of silicoflagellates from Sicily to Nabeul.

Stephanocha speculum was found in a Polish product labelled as *Fagopyrum* honey. Since this silicoflagellata occurs from the Miocene to Recent, one can suppose that it comes from the diatomite intercalated in the Middle Miocene from the Silezian Basin [44]. Outcrops of Oligocene age diatomites in the Carpathians are exposed on the surface (Figure 5); thus, particles as small as silicoflagellate skeletons can be lifted from outcropping on the surface of soft sediments and transported by winds, being a plausible source of silicoflagellates to explain our observations. It is possible that the source of these silicoflagellata is a diatomite in eastern part of the Polish Outer Carpathians, 60 km from the honey harvesting area [45]. A simulation study was performed with the atmospheric dispersion model for the harvesting period July–August 2013, i.e., the blooming of *F. esculentum* [11]. The source area was represented by the location 49.8 N and 22.6 E and sensitivity maps were produced for each day to investigate whether the atmospheric conditions were suitable to deflate particles and transport them to the harvesting location near Stróze, Poland. It was found that in the beginning and the end of the harvesting period, e.g., on 4 July and 26 August, atmospheric conditions supported the potential transport of microfossils to the honey harvesting area (Figure 6). Meteorological observations reported from Nowy Sącz, located in a distance of 20 km from Stróze, confirmed that between 10–19 UTC on 4 July and 9–16 UTC on 26 August 2013, easterly-northeasterly winds dominated the area, potentially transporting microfossils from the upwind direction. No precipitation but trace had been reported for the previous four days. Similar to the Tunisian case, the Polish trajectory simulation also supports the hypothesis of atmospheric transport of silicoflagellates from outcrops of Oligocene age (possibly quarries).

Figure 5. A quarry located in the Carpathians, Sibiciu de Sus, Romania—a possible source of airborne microfossils. Photo courtesy of Emilia Tulan.

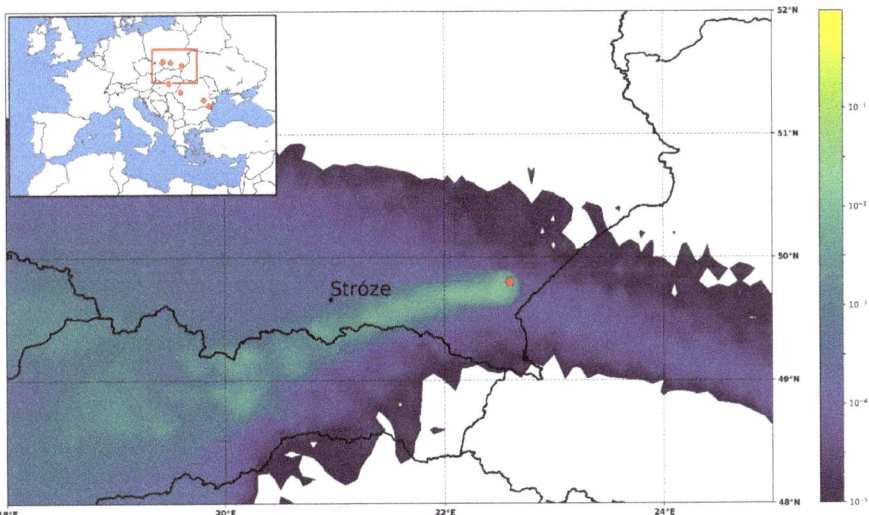

Figure 6. An episode of air mass possibly transporting microfossils from the emission area (red dot) to the honey harvesting area near Stróze, Poland; shown on an atmospheric dispersion sensitivity map [m^{-3}] for the dust deflated on 26 August 2013. Red dots on the small map show the distribution of Oligocene-Miocene diatomites.

The dominance of the Poaceae pollen in the Tunisian honey sample corroborates our hypothesis that the particles were deposited from the atmosphere. Poaceae produce typical anemophilous pollen grains, and their presence in honey is rare and incidental in North African honeys [46]. Fungal spores were also the common members of the airspora.

Surprisingly, we did not find any scientific report on the presence of silicoflagellata in air samples, possibly because aerobiological networks focus on the monitoring of pathogenic bioaerosols, e.g., allergenic pollen grains and spores, rather than other particles [47]. Further studies are needed to study the presence of silicoflagellates and diatoms in air samples and honeys harvested within the area of diatomite outcrops.

In our honey collection, a total of 21 countries are represented by samples, 14 of them from Europe (most of them were from Italy, Greece and Hungary). Silicoflagellates were observed in three of our European honey samples (*Pinus* honeydew honey from Greece; *Fagopyrum esculentum* honey from Poland, Stróze; and honeydew honey from Romania, Odorheiu Secuiesc). No silicoflagellates were found in eight other *Pinus* honeydew honey samples collected from Greece, and four *Fagopyrum esculentum* samples collected from Poland, Stróze or nearby (Królów, Lipowy and Wiśniowa). Other two samples (*Robinia pseudo-acacia* honey and a honeydew honey) collected from Stróze or nearby (Pogorzany) were analysed, but again, with negative results.

In case of North African honeys, samples from Egypt, Morocco, and Tunisia were available for analysis. Silicoflagellates were found in the honeys originating from the latter two countries. For Morocco, there is a possible source from Messinian diatomite outcrops known in Boudinar and Melilla basins and for Tunisia the closest possible source may be the Messinian diatomites outcrops in Sicily [48,49]. Another positive sample was found from Africa, but more precise information on the origin of the honey is not available. Because data on the composition of the North-African honeys are uncommon [9], information on new indicators of geographical origin of honeys in this region is useful. According to our findings, the occurrence of silicoflagellates may be expected in honey harvested near diatomites in Algeria, Crete, Hungary, Poland, Romania, Sicily, and Spain. We propose further studies on the presence of microfossils in honeys of other areas of the world as well.

The identification of the origin of food is one of the most important issues in food quality control [1,50]. Considering the increasing global trade and owing to the higher economic value of specific honeys (e. g. those having protected geographical indication), such products are targets of adulteration through incorrect labelling and fraudulent admixing with honey of lower value and quality. Thus, in order to promote fair competition among producers, and protect consumers, there is a growing need to assess the authenticity of honey, particularly with regard to geographical origins [1]. Microfossil identification could be compared with other microscopy-based analytical methods, such as melissopalynology. These methods allow a good differentiation of honeys, however, are not suitable for application in the case of filtered honeys. Melissopalynology has a limitation in its application in honey adulterated by pollen addition [1]. Silicoflagellates can be used as non-quantitative indicators, as the mere presence of their distinct siliceous skeletons can indicate the geographical source of honeys. Because no particle counting is needed, the analysis is less time-consuming than quantitative microscopical methods. (Similar, non-quantitative indicators of honey origin were previously proposed on the basis of biodiversity of fungal spores [2].) Our findings encourage the confirmation of honey origin also by recording the occurrence of microfossil elements during routine melissopalynological analysis. Silicoflagellata skeletons are characteristic multiradiate particles and it is easy to observe them in honey samples, thus they seem to be good candidates of indicators of geographical origin of honeys in food analysis. The most frequent species was *Stephanocha speculum* (1 particle/g honey), followed by *Dictyocha fibula* and *Distephanopsis crux* (both having 0.3/g; data from the Tunisian sample). Only *Stephanocha speculum* was found in the other honey samples from Africa and Europe. The limitation of our method is the low frequency of microfossils in honeys. Observation of the presence of silicoflagellates can be used in a complementary way. Because of their low frequency in honey, only positive findings can be interpreted as indicators of geographical origin. Apparently, silicoflagellates in honeys are not common, but if detected, can provide a strong evidence of geographical origin of the honey.

Marine species of diatoms (*Coscinodiscus marginatus*, *Actinocyclus divisus* and *Thalassionema nitzschioides*) have a large geological range and are also known in actual assemblages. They are known in Messinian diatomites in the Mediterreanean area, but since they have a very large geological time span distribution, they cannot be considered as markers for Messinian [26,48]. Thus, silicoflagellates only support a possible Messinian age. Consequently, diatoms cannot be used as indicators of the honey origin. It should be mentioned that fossil diatoms can have another source in the honeys: diatomite earth filters. The use of diatomite earth filters is largely known in food industry. The presence of fossil diatoms was already observed in honey, and is explained by the use of diatomite filters in order to: (a) filtration to obtain pure honey [1] (b) forge [51,52]. Forgers combine two kinds of honey: a local honey with another kind of honey that is much cheaper. Diatomaceous earth filter aids completely remove the pollen, and thus prevent any identification of source by analysis of pollen. In our samples, three categories of diatoms were found: fossil marine diatoms, freshwater diatoms, and living freshwater pennate diatoms.

1. Fossil marine diatoms (ex. *Coscinodiscus marginatus*, *Actinocyclus divissus*) together with fossil silicoflagellates. Their occurrence can be explained:

   (a) By wind transport from fossil diatomite (e.g., Tripoli formation);

   (b) Contamination from filters made from diatomite earth. Diatomite used for the fabrication of filters may be made from marine diatomites that contain marine diatom genera like *Coscinodiscus* and *Actinocyclus*.

2. Freshwater centric diatoms (e.g., *Aulacoseira distans*, *Cyclotella*), known from Miocene, Pliocene up to day in freshwater and lacustrine assemblages. Their occurrence might be explained:

   (a) By wind transport from fossil diatomite from strata of fossil freshwater diatoms.

   (b) Contamination from filters made from diatomite earth. For example, in France, there is known exploitation of fossil lacustrine diatomite Miocene in age from Massif Central that might be used for the fabrication of filters [53]. This lacustrine diatomite contains *Aulacoseira distans* and *Cyclotella*. Recovery of *Aulacoseira* sp. was reported from only pressed honey sample in Nigeria, Africa, and interpreted as an indication of secondary contamination during processing [46].

3. Living freshwater pennate diatoms that present the chloroplasts and cellular content, together with living cyanobacteria (*Chroococcus* and *Oscillatoria*) and green algae (*Scenedesmus*). Obviously, living diatoms do not originate from diatomite earth filters. Their occurrence might be explained:

   (a) By wind transport. *Nitzschia* and *Chroococcus* were common living diatom and cyanobacteria in our samples (especially in Cuba, France, Greece, Hungary and Italy). These genera were reported to be airborne [54,55].

   (b) The presence of green microalgae (e.g., *Scenedesmus*) is characteristic of honeydew honeys as well [56]. When honey bees collect honeydew, they may also collect other attached structures such as algae that grow on plants. In our samples, such algae were not associated with silicoflagellata, but were found in honeys of honeydew origin.

   (c) Freshwater, as a source cannot be excluded either. Adding water to honey is a known authentication technique to increase the volume. Food analytical tests, such as microscopic yeast count and analysis of fermentation products are available to indirectly detect this type of honey adulteration [1].

Each of the above-mentioned hypotheses can be considered as true alone, but they are not exclusive (i.e., diatoms and silicoflagelates transported by wind and living diatoms were added using freshwater).

In case of the Tunisian honey, matching evidences (wind trajectory analysis and high percentage of anemophilous pollen and spores) indicate that the source of siliceous bioparticles is possibly not the result of filtration, but air. In the unknown African honey, a high percentage of anemophilous pollen (Chenopodiaceae) was found too, leading to similar hypothesis. The Polish product was labelled as *Fagopyrum* honey. Our pollen analysis confirmed the presence of this pollen. Here, again both trajectory analysis and pollen composition suggested natural sources of particles in the honey. In the Romanian honey, fungal indicators (*Metschnikowia*, *Retiarius* and *Tripospermum*) suggested that the origin is mixed floral and honeydew honey, but of natural (i.e non-forged) source [2]. Similar evidences are available in the honey sample from Altenst, Germany, where conidia of a honeydew-indicator hypomycete (*Retiarius*) were found with the freshwater diatom *Aulacoseira distans*. Consequently, the presence of siliceous bioparticles in honey does not necessarily indicate the manipulation of the product. For correct determination, natural sources of siliceous bioparticles, such as wind transport from nearby outcrops, should also be considered. According to a world map showing main directions of atmospheric transport of diatoms, presented by Harper and McKay [57], the occurrence of windborne microfossils in honeys might be a world-wide phenomenon.

4. Conclusions

Silica skeletons of planktonic marine silicoflagellates were found in honey samples from Greece, Morocco, Poland, Romania, and Tunisia. In Tunisia, the source of silicoflagellate content of honey is suspected to be the wind erosion of microfossils from the Tripoli Formation of Messinian age in nearby Sicily. In Poland, the source is similarly shown in the Polish Outer Carpathians, but with low diversity of silicoflagellata species. Atmospheric trajectory analysis confirmed the possibility of atmospheric transport of deflated grains in the harvesting period. Therefore, silicoflagellates could be used as indicators of the geographical origin of honeys collected in areas characterized by diatomite outcrops. It was demonstrated that the diatom content of honey can have both natural (wind) and artificial (filters) sources.

Supplementary Materials: The following are available online at https://www.mdpi.com/article/10.3390/foods10020421/s1, Supplement 1: The studied honey samples. Data are organized by collection code, botanical origin, type and location.

Author Contributions: Conceptualization, observation of silicoflagellates, research concept, supervision, writing—original draft preparation, photos, spore identification: D.M.; identification of silicoflagellates and writing: P.D.; laboratory works, writing—review and editing: A.M.-M.; identification of pollen grains: Z.M.; trajectory analysis, funding acquisition: Á.L.; identification of diatoms, conceptualization and writing: S.S.M. All authors have read and agreed to the published version of the manuscript.

Funding: The atmospheric transport analysis presented in this research was supported by the National Research, Development and Innovation Office of Hungary (No. K128818).

Institutional Review Board Statement: Not applicable.

Informed Consent Statement: Not applicable.

Data Availability Statement: Data is contained within the article or supplementary material.

Acknowledgments: The authors are thankful to Szilvia Babinszkyné Nagy (Hungarian Institute for Forensic Sciences, Budapest, Hungary) for gifting the Tunisian honey and for the translation of French and Arabic texts; Francesca Grillenzoni (Consiglio per la Ricerca in Agricoltura e l'Analisi dell'Economia Agraria, Centro di Ricerca Agricoltura e Ambiente, Italy) for gifting several honey samples; Magdalena Wójcik (University of Rzeszów, Poland) for gifting a honey sample from Poland; Jakub Alexandrowicz (Geology and Palaeogeography Unit, Faculty of Geosciences University of Szczecin, Poland) for the consultation on diatomites in Poland; and Ágnes Schütz and Zsuzsanna Udvardi (National Public Health Center, Budapest, Hungary) for their kind technical help in the laboratory.

Conflicts of Interest: The authors declare no conflict of interest.

References

1. Soares, S.; Amaral, J.S.; Oliveira, M.B.P.; Mafra, I. A comprehensive review on the main honey authentication issues: Production and origin. *Compr. Rev. Food Sci. Food Saf.* **2017**, *16*, 1072–1100. [CrossRef]
2. Mura-Mészáros, A.; Magyar, D. Fungal honeydew elements as potential indicators of the botanical and geographical origin of honeys. *Food Anal. Methods* **2017**, *10*, 3079–3087. [CrossRef]
3. Louveaux, J.; Maurizio, A.; Vorwohl, G. Methods of melissopalynology. *Bee World* **1978**, *51*, 139–157. [CrossRef]
4. Magyar, D.; Mura-Mészáros, A.; Grillenzoni, F. Fungal diversity in floral and honeydew honeys. *Acta Bot. Hung.* **2016**, *58*, 145–166. [CrossRef]
5. Von Der Ohe, W.; Oddo, L.P.; Piana, M.L.; Morlot, M.; Martin, P. Harmonized methods of melissopalynology. *Apidologie* **2004**, *35*, S18–S25. [CrossRef]
6. Beug, H.J. *Leitfaden der Pollenbestimmung für Mitteleuropa und Angrenzende Gebiete*; Dr.Friedrich Pfeil: München, Germany, 2004.
7. Reille, M. *Pollen et Spores d'Europe et d'Afrique du Nord*; Laboratoire de Botanique Historique et Palynologie: Marseille, France, 1992.
8. Hamda, S.H.; Dhiab, A.B.; Galán, C.; Msallem, M. Pollen spectrum in northern Tunis, Tunisia. *Aerobiologia* **2017**, *33*, 243–251. [CrossRef]
9. Jilani, B.; Paul Schweitzer, P.; Larbi Khouja, M.; Zouaghi, M.; Ghrabi, Z. Physicochemical properties and pollen spectra of honeys produced in Tunisia (Southwest Kef). *Apiacta* **2008**, *43*, 38–48.
10. El Gharbi, B.; Charpin, H.; Aubert, J.; Renard, M.; Mallea, M.; Soler, M. Le calendrier pollinique de Tunis. *Rev. Française D'allergologie D'immunologie Clin.* **1976**, *16*, 25–31. [CrossRef]
11. Alekseyeva, E.S.; Bureyko, A.L. Bee visitation, nectar productivity and pollen efficiency of common buckwheat. *Fagopyrum* **2000**, *17*, 77–80.
12. Zavada, M.S.; McGraw, S.M.; Miller, M.A. The role of clothing fabrics as passive pollen collectors in the north-eastern United States. *Grana* **2007**, *46*, 285–291. [CrossRef]
13. Mercuri, A.M. Applied palynology as a trans-disciplinary science: The contribution of aerobiology data to forensic and palaeoenvironmental issues. *Aerobiologia* **2015**, *31*, 323–339. [CrossRef]
14. Bagnold, R.A. *The Physics of Blown Sand and Desert Dunes*; Methuen: London, UK, 1941.
15. Xuan, J. Turbulence factors for threshold velocity and emission rate of atmospheric mineral dust. *Atmos. Environ.* **2004**, *38*, 1777–1783. [CrossRef]
16. Leelőssy, Á.; Lagzi, I.; Mészáros, R. Spatial and temporal pattern of pollutants dispersed in the atmosphere from the Budapest Chemical Works industrial site. *Időjárás* **2017**, *121*, 101–115.
17. Marticorena, B.; Bergametti, G. Modeling the atmospheric dust cycle: 1. Design of a soil-derived dust emission scheme. *J. Geophys. Res. Atmos.* **1995**, *100*, 16415–16430. [CrossRef]
18. NOAA NCEP GDAS/FNL 0.25 Degree Global Tropospheric Analyses and Forecast Grids. In *Research Data Archive at the National Center for Atmospheric Research*; Computational and Information Systems Laboratory: Boulder, CO, USA, 2015. [CrossRef]
19. Mészáros, R.; Leelőssy, Á.; Kovács, T.; Lagzi, I. Predictability of the dispersion of Fukushima-derived radionuclides and their homogenization in the atmosphere. *Sci. Rep.* **2016**, *6*, 1995. [CrossRef] [PubMed]
20. Leelőssy, Á.; Mészáros, R.; Kovács, A.; Lagzi, I.; Kovács, T. Numerical simulations of atmospheric dispersion of iodine-131 by different models. *PLoS ONE* **2017**, *12*, e0172312. [CrossRef]
21. McCartney, K. Silicoflagellates. In *Fossil Prokaryotes and Protists*; Lipps, J.H., Ed.; Blackwell Scientific: Boston, MA, USA, 1993; pp. 143–154.
22. Dumitrica, P. Double skeletons of silicoflagellates: Their reciprocal position and taxonomical and paleobiological values. *Revue Micropaléontologie* **2014**, *57*, 57–74. [CrossRef]
23. Dumitrica, P. Paleocene, late Oligocene and post-Oligocene silicoflagellates in southwestern Pacific sediments cored on DSDP Leg 21. In *Initial Reports of the Deep Sea Drilling Project, Vol. 21*; Burns, R.E., Andrews, J.E., van der Lingen, G.J., Churkin, M., Jr., Galehouse, J.S., Packham, G.H., Davies, T.A., Kennett, J.P., Dumitrica, P., Edwards, A.R., et al., Eds.; U.S. Government Printing Office: Washington, DC, USA, 1973; Volume 21, pp. 837–883.
24. Dumitrica, P. Miocene and Quaternary silicoflagellates in sediments from Mediterranean Sea. In *Initial Reports of the Deep Sea Drilling Project, Vol. 13*: 902–933; Ryan, W.B.F., Hsü, K.J., Cita, M.B., Dumitrica, P., Lort, J.M., Mayne, W., Nesteroff, W.D., Pautot, G., Stradner, H., Wezel, F.C., et al., Eds.; U.S. Government Printing Office: Washington, DC, USA, 1972.
25. Rouchy, J.M. La genèse des évaporites messiniennes de Méditerranée. *Mémoires Muséum Natl. D'histoire Nat. Sér. C-Sci. Terre* **1982**, *50*, 267.
26. Pestrea, S.; Blanc-Valleron, M.M.; Rouchy, J.M. Les assemblages de diatomées des niveaux infra-gypseux du Messinien de Méditerranée (Espagne, Sicile, Chypre). *Geodiversitas* **2002**, *24*, 543–583.
27. Pellegrino, L.; Dela Pierre, F.; Natalicchio, M.; Carnavale, G. The Messinian diatomite deposition in the Mediterranean region and its relationships to the global silica cycle. *Earth-Sci. Rev.* **2018**, *178*, 154–176. [CrossRef]
28. Krijgsman, W.; Hilgen, F.J.; Raffi, I.; Sierro, F.J.; Wilson, D.S. Chronology, causes and progression of the Messinian salinity crisis. *Nature* **1999**, *400*, 652–655. [CrossRef]
29. Deflandre, G. *La vie créatrice de roches. coll. Que sais-je? nr. 20*; Presses Univ. Fr.: Paris, France, 1967.
30. De Wever, P.; Cornée, A. *Roches à tout faire*; EDP Sci: Paris, France, 2020.
31. Fenner, J.; Houben, G.; Kaufhold, S.; Lechner-Wiens, H.; Adams, F.; Baez, J. Flying diatoms a key to the path and origin of a dust storm. In *Abstracts of the 22nd International Diatom Symposium*; VLIZ Special Publication: Ghent, Belgium, 2012; Volume 58, p. 43.

32. Schlichting, H.E. A preliminary study of the algae and protozoa in sea foam. *Bot. Mar.* **1971**, *14*, 24–28.
33. Geissler, U.; Gerloff, J. Das Vorkommen von Diatomeen in menschlichen. Organen und in der Luft. *Nova Hedwig.* **1965**, *10*, 565–577.
34. Folger, D.W. Wind transport of land-derived mineral, biogenic, and industrial matter over the North Atlantic. *Deep-Sea Res. Oceanogr. Abstr.* **1970**, *17*, 337–342. [CrossRef]
35. Saint Martin, S.; Saint Martin, J.P.; Schmidt, A.R.; Girard, V.; Néraudeau, D.; Perrichot, V. The intriguing marine diatom genus Corethron in Late Cretaceous amber from Vendée (France). *Cretac. Res.* **2015**, *52*, 64–72. [CrossRef]
36. Gregory, P.H. *The Microbiology of the Atmosphere*; Leonard Hill: London, UK, 1961; pp. 12–13.
37. Ehrenberg, G.G. Bericht über die zur Bekanntmachung geeigneten Verhandlungen. *Königlich-Preuss. Akad. Der Wiss. Berl.* **1844**, *9*, 194–197.
38. Romero, O.E.; Lange, C.B.; Swap, R.; Wefer, G. Eolian-transported freshwater diatoms and phytoliths across the equatorial Atlantic record: Temporal changes in Saharan dust transport patterns. *J. Geophys. Res.* **1999**, *104*, 3211–3222. [CrossRef]
39. Washington, R.; Todd, M.C.; Engelstaedter, S.; Mbainayel, S.; Mitchell, F. Dust and the low-level circulation over the Bodele Depression. *J. Geophys. Res.* **2006**, *111*, D03201.
40. Sharma, N.K.; Rai, A.K.; Singh, S.; Brown Jr, R.M. Airborne algae: Their present status and relevance 1. *J. Phycol.* **2007**, *43*, 615–627. [CrossRef]
41. Dumitrica, P. Cenozoic endoskeletal dinoflagellates in Southwestern Pacific sediments cored during Leg 21 of the DSDP. In *Initial Reports of the Deep Sea Drilling Project, Vol. 21*; Burns, R.E., Andrews, J.E., van der Lingen, G.J., Churkin, M., Jr., Galehouse, J.S., Packham, G.H., Davies, T.A., Kennett, J.P., Dumitrica, P., Edwards, A.R., et al., Eds.; U.S. Government Printing Office: Washington, DC, USA, 1973; pp. 819–835.
42. Gläser, G.; Knippertz, P.; Heinold, B. Orographic effects and evaporative cooling along a subtropical cold front: The case of the spectacular Saharan dust outbreak of March 2004. *Mon. Weather Rev.* **2012**, *140*, 2520–2533. [CrossRef]
43. Monteiro, A.; Fernandes, A.P.; Gama, C.; Borrego, C.; Tchepel, O. Assessing the mineral dust from North Africa over Portugal region using BSC-DREAM8b model. *Atmos. Pollut. Res.* **2015**, *6*, 70–81.
44. Alexandrowicz, J.; University of Szczecin, Geology and Palaeogeography Unit, Faculty of Geosciences, Szczecin, Poland. Personal communication, 2019.
45. Figarska-Warchoł, B.; Stańczak, G.; Rembiś, M.; Toboła, T. Diatomaceous rocks of the Jawornik deposit (the Polish Outer Carpathians): Petrophysical and petrographical evaluation. *Geol. Geophys. Environ.* **2015**, *41*, 311–331. [CrossRef]
46. Adeonipekun, P.A. Palynology of honeycomb and a honey sample from an apiary in Lagos, Southwest Nigeria. *Asian J. Plant Sci. Res.* **2012**, *2*, 274–283.
47. Buters, J.T.; Antunes, C.; Galveias, A.; Bergmann, K.C.; Thibaudon, M.; Galán, C.; Oteros, J. Pollen and spore monitoring in the world. *Clin. Transl. Allergy* **2018**, *8*, 9. [CrossRef] [PubMed]
48. Saint Martin, S.; Conesa, G.; Saint Martin, J.P. Signification paléoécologique des assemblages de diatomées du Messinien dans le bassin de Melilla-Nador (Rif Nord-Oriental, Maroc). *Rev. Micropaléontologie* **2003**, *46*, 161–190. [CrossRef]
49. El Ouahabi, F.Z.; Saint Martin, S.; Benmoussa, A.; Saint Martin, J.P. Les assemblages de diatomées du bassin messinien de Boudinar (Maroc nord-oriental). *Rev. Micropaléontologie* **2007**, *50*, 149–167. [CrossRef]
50. Stanimirova, I.; Üstün, B.; Cajka, T.; Riddelova, K.; Hajslova, J.; Buydens, L.M.C.; Walczak, B. Tracing the geographical origin of honeys based on volatile compounds profiles assessment using pattern recognition techniques. *Food Chem.* **2010**, *118*, 171–176. [CrossRef]
51. Benmbarek, M.; Bonhomme, C.; Boussalem, Z.; Landbeck, T. Mieux communiquer sur le miel, vers une nouvelle approche apiculteur-consommateur. In *Unpublished Memory MASTER 2 Management Administration des Entreprises*; Université de Strasbourg: Strasbourg, France, 2018; p. 99.
52. Borneck, R.; Gauthron, R.; Guiraute, F.; Horguelin, P.; Loveaux, J.; Pedelucq, A. Les techniques de conditionnement et de commercialisation du miel au Canada et aux USA. *Ann. L'abeille* **1964**, *7*, 103–159.
53. Saint Martin, J.-P.; Métais, G.; Saint Martin, S.; Sen, S. La diatomite du Coiron et son lagerstate. *Géochronique* **2017**, *141*, 57–68.
54. Genitsaris, S.; Kormas, K.A.; Moustaka-Gouni, M. Airborne algae and cyanobacteria: Occurrence and related health effects. *Front. Biosci.* **2011**, *3*, 772–787.
55. Tesson, S.V.; Skjøth, C.A.; Šantl-Temkiv, T.; Löndahl, J. Airborne microalgae: Insights, opportunities, and challenges. *Appl. Environ. Microbiol.* **2016**, *82*, 1978–1991. [CrossRef]
56. Seijo, C.M.; Escuredo, O.; Fernández-González, M. Fungal diversity in honeys from northwest Spain and their relationship to the ecological origin of the product. *Grana* **2011**, *50*, 55–62. [CrossRef]
57. Harper, M.A.; McKay, R.M. Diatoms as markers of atmospheric transport. In *The Diatoms*; Smol, J.P., Stoermer, E.F., Eds.; Cambridge University Press: Cambridge, UK, 2010; pp. 552–559.

Article

Prediction of Physicochemical Properties in Honeys with Portable Near-Infrared (microNIR) Spectroscopy Combined with Multivariate Data Processing

Olga Escuredo *, María Shantal Rodríguez-Flores, Laura Meno and María Carmen Seijo

Department of Vegetal Biology and Soil Sciences, Faculty of Sciences, University of Vigo, 32004 Ourense, Spain; mariasharodriguez@uvigo.es (M.S.R.-F.); laura.meno@uvigo.es (L.M.); mcoello@uvigo.es (M.C.S.)
* Correspondence: oescuredo@uvigo.es

Abstract: There is an increase in the consumption of natural foods with healthy benefits such as honey. The physicochemical composition contributes to the particularities of honey that differ depending on the botanical origin. Botanical and geographical declaration protects consumers from possible fraud and ensures the quality of the product. The objective of this study was to develop prediction models using a portable near-Infrared (MicroNIR) Spectroscopy to contribute to authenticate honeys from Northwest Spain. Based on reference physicochemical analyses of honey, prediction equations using principal components analysis and partial least square regression were developed. Statistical descriptors were good for moisture, hydroxymethylfurfural (HMF), color (Pfund, L and b* coordinates of CIELab) and flavonoids (RSQ > 0.75; RPD > 2.0), and acceptable for electrical conductivity (EC), pH and phenols (RSQ > 0.61; RPD > 1.5). Linear discriminant analysis correctly classified the 88.1% of honeys based on physicochemical parameters and botanical origin (heather, chestnut, eucalyptus, blackberry, honeydew, multifloral). Estimation of quality and physicochemical properties of honey with NIR-spectra data and chemometrics proves to be a powerful tool to fulfil quality goals of this bee product. Results supported that the portable spectroscopy devices provided an effective tool for the apicultural sector to rapid in-situ classification and authentication of honey.

Keywords: honey; quality control; authentication; botanical origin; NIR; modified partial least squares; linear discriminant analysis

Citation: Escuredo, O.; Rodríguez-Flores, M.S.; Meno, L.; Seijo, M.C. Prediction of Physicochemical Properties in Honeys with Portable Near-Infrared (microNIR) Spectroscopy Combined with Multivariate Data Processing. *Foods* **2021**, *10*, 317. https://doi.org/10.3390/foods10020317

Academic Editors: Christopher John Smith

Received: 5 January 2021
Accepted: 30 January 2021
Published: 3 February 2021

Publisher's Note: MDPI stays neutral with regard to jurisdictional claims in published maps and institutional affiliations.

Copyright: © 2021 by the authors. Licensee MDPI, Basel, Switzerland. This article is an open access article distributed under the terms and conditions of the Creative Commons Attribution (CC BY) license (https:// creativecommons.org/licenses/by/ 4.0/).

1. Introduction

Healthy habits are part of our daily life. In recent years, there is a greater concern for health taking care of the daily diet and physical activity. Honey is a healthful food produced by honeybees from floral nectar or secretions of plants or some kind of aphids. Increasing honey consumption can be attributed to the consumer interest in natural foods with health benefits [1,2]. The properties derive from its particular chemical composition, as well as its sensorial characteristics and depend mainly on the flowers and biogeographical regions involved in its production [1–6]. Although they are also affected by processing, manipulation, packaging, and storage time [3,4,7], hence ensuring product quality is essential.

The quality and authenticity of its geographical and botanical origins remain as important factors in reliable marketing [2,8]. This bee product can be classified according to their botanical source as unifloral honey (if arising predominantly from a single plant species), multifloral honey (obtained from multiples plant species), or honeydew when was from secretions in plants. In north-western Spain, unifloral honeys of eucalyptus, chestnut, heather, and blackberry are produced, and there is also a good production of honeydew honey [9]. Since 2007, these unifloral honeys in European countries were recognized in the Protected Geographical Indication (PGI) *Miel de Galicia* (Commission Regulation (EC) No 868/2007 of 23 July 2007). Honeydew honeys are currently in revision process to include them in this designation. The increasing demand for unifloral honeys, with protected designation of origin (PDO) and PGI, generally perceived as high-quality products, produced an increase in their commercial value and, at the same time, an increase in counterfeiting [10]. Undoubtedly, food products with high value-added require exhaustive controls examining a large number of samples that guarantee quality [11].

Microscopic pollen analysis is commonly used to determine the botanical origin of honey, and in some cases, the geographical origin [1,12]. Pollen profile is a useful tool in the identification of the main pollen of honey sample, as a result of bees collecting nectar. Accurate identification is aided by standards and pollen databases [13]. Nevertheless, this approach is time-consuming and strongly dependent on the qualifications and experience of the analyst [1,6,13]. Although for exact determination of honey origin, sensory, and physicochemical properties are also needed [12,14,15].

To assess honey quality standard methods are used, including spectrophotometric, refractometric, titration and melissopalynological methods [12,16]. Some quality parameters (such as water content, pH, electrical conductivity, acidity, hydroxymethylfurfural (HMF) content, diastase activity, or reducing sugars) inform about the botanical origin and confirm the adequate manipulation and storage of honey [2,3,16,17].

On the other hand, the color is one of the most common commercial attributes of honey. Consumers have preferences and the particular tonalities in honey depend on the botanical origin, deriving from certain chemical compounds such as the polyphenols, carotenoids, or minerals [14,18–21]. Consequently, the correct classification of botanical origin of honeys based on color allows beekeepers and exporters to determine the most advantageous market destination for this apicultural product [14].

In the food industry, the concern for public health and detection of possible frauds regarding the labelling of commercial products have led many regulatory bodies to demand rigorous inspections. Hence, raw material identification or verification is a common quality-control practice. However, common analytical chemistry determinations such as high-performance liquid chromatography typically take a long time to complete, is expensive and destructive for samples. Recently, non-destructive characterization tools for quality control based on the principles of spectroscopy were introduced [11]. More specifically, the near-infrared (NIR) technique has expanded its scope of application in the past decade, through the rapid integration of technologies from various fields [7]. Large amounts of spectral data, characterized by an intercorrelation among the recorded spectral variables, can be processed [11]. Unfortunately, this technology is heavily dependent on reference conventional methods to develop a calibration model and to their validation [13,22]. The assessment and interpretation of spectroscopy instrumentation are usually not straightforward, and the application of chemometric approach is crucial to guarantee the success of this technique. Major advantages of NIR spectroscopy are that it requires little to no preparation, its rapidness, safety to the analyst, non-destructiveness, and multicomponent remote analysis [13,23].

NIR technology has been described as a successful tool in analytical instrumentation and quality control in various fields such as food adulteration, authenticity control, the assessment of physicochemical attributes, rheological, or technological properties [1,13,23–26], as well as in petrochemical industries, plastic contaminants, pharmaceuticals, cosmetics, and medical applications [22,27]. This method has been widely used in the field of honey quality detection, such as in the determination of 5-hydroxymethylfurfural [24,28–30], diastase activity [7], moisture and reducing sugar content [23,24,28,31], compounds with antioxidant activity such polyphenols or minerals [1,25], floral origin discrimination [5,6,32], and honey adulteration [33,34].

A high standard of quality for honey is achieved by developing alternative methods that are simple, fast, cheap, and reliable. In this context, the objectives pursued with the present study were (a) assessment the potential of portable microNIR and chemometric techniques for the prediction of main physicochemical parameters in the honey; (b) assessment the ability to discriminate different honey types combining the linear discriminate analysis with main physicochemical attributes and botanical profile. The physicochemical data by conventional methods (moisture, pH, EC, HMF content, diastase index, color, phenols, and flavonoids) of 100 samples from Northwest Spain for the development NIR regression models and discrimination of honey samples were included.

2. Materials and Methods

2.1. Honey Samples

A total of 100 honey samples of the label Protected Geographical Indication Miel de Galicia (northwest Spain) were used for this study. The samples were collected directly from the beekeepers and stored refrigerated at 4 °C until further analysis. All the physicochemical determinations were conducted in duplicate, after the homogenization of the samples. NIR models were developed using 84 samples for the calibration group and 16 samples for external validation group. Each honey sample were registered by triplicate and the mean value was used for the chemometric treatment.

2.2. Palynological Analysis

The pollen analysis of the honey samples was performed based on the method proposed by Louveaux, et al. [35], with some adaptations [9]. The honey samples (10 g) were diluted in distilled water and centrifuged at 4500 rpm for 10 min (first centrifugation), the obtained sediment was re-dissolved and centrifuged for 5 min (second centrifugation). The slide for the microscopical analysis was prepared by duplicate with 100 µL of the sediment. The pollen grains were identified using a Nikon Optiphot II microscope (Nikon UK Ltd., London, UK) at 400× or 1000×. The results were expressed in percentage considering the total number of pollen grains counted.

2.3. Determination of Quality Parameters

The moisture was measured with a digital refractometer (ABBE URA-2WAJ-325; Auxilab S.L., Navarra, Spain) on a drop of honey sample. The refractive index values at 20 °C, were converted to moisture content using the Chataway table.

Then, 5 g honey dissolved in 25 mL bi-distilled water were used to determine the pH and the electrical conductivity (EC). The pH was measured directly using a pH meter (Crison micro pH 2001; Crison Instruments S.A., Barcelona, Spain). EC was measured in a honey solution considering the moisture content of sample with a portable conductivity meter (Knick Portamess 913 Conductivity, Beuckestr, Berlin). Results were expressed in µs/cm.

HMF content and the diastase activity were determined following the methodology proposed by Bogdanov et al. [36]. The HMF content (expressed in mg/100 g) was estimated using the White spectrophotometric method. This method considers the difference between the UV absorbance at 284 nm of a honey solution and the same solution after adding bisulphite. The HMF level was calculated after subtraction of the background absorbance at 336 nm using a UV-Visible Spectrophotometer (Jenway 6305 UV-Visible Spectrophotometer, Staffordshire, UK).

The diastase activity was calculated through the hydrolysis rate of the starch solution by the α-amylase present in a honey buffer solution at 40 °C (Schade method). The amount of starch converted was determined measuring the absorbance of the honey solution at 660 nm for different time points until an endpoint when the absorbance was less than 0.235. An UV-Visible spectrophotometer (Jenway 6305 UV-Visible Spectrophotometer, Staffordshire, UK) was used for this purpose. Diastase activity was expressed as diastase number (DN) or grams of starch hydrolyzed each hour per 100 g honey at 40 °C.

2.4. Determination of Color

The color of samples was determined with a portable Minolta Chroma Meter CR-210 colorimeter (Konica Minolta, Tokyo, Japan). The CIE L a*b* coordinates were registered, where L is the luminance component (ranging from 0 to 100), while a* and b* are chromatic coordinates related with the gradient red/green and yellow/blue, respectively.

A HANNA Honey Color C221 colorimeter was also used to determine the color in mm Pfund. The honey must be fluid to be analyzed correctly. For crystallized or weakly fluid honeys, they were heated up to 45 °C in a thermostatic bath until the complete fluidification. Approximately 4 mL of sample was placed in a cuvette and sonicated to eliminate bubbles. The measure was taken after calibration with glycerin (Glycerol HANNA instruments).

2.5. Determination of Total Phenol and Flavonoid Content

Folin–Ciocalteu spectrophotometric method adapted to honey was used for the determination of the phenol content [37]. This method is based on the oxidation of the phenolic compounds forming a bluish complex. The samples (0.1 g/mL) were mixed with the reactive of Folin–Ciocalteu and the absorbance at 765 nm was measured using a UV–Vis spectrophotometer (Jenway 6305, UK). Gallic acid solutions (0.01–0.50 mg/mL) as a reference standard were used. Finally, the total phenolic content was expressed as gallic acid equivalents in mg/100 g honey.

The total flavonoid content was measured by spectrophotometry using the Dowd method adapted by Arvouet-Grand et al. [38]. The method uses a solution of aluminum chloride that reacts with the flavonoids present in the honey solution (0.33 g/mL). Solutions developed a yellow color for which absorbance was determined spectrophotometrically at 425 nm. Quercetin solutions (0.002–0.01 mg/mL) as reference standard were used. The results were expressed as equivalents of quercetin in mg/100 g honey.

2.6. Near Infrared Spectroscopy Measurement

A portable MicroNIR Pro v2.5 equipment (MicroNIR 1700 ES, VIAVI, Santa Rosa, CA, USA) was used for acquisitions of NIR spectra data on honey samples (Figure 1). The samples properly homogenized were placed to the spectrometer housed in 4 mm borosilicate glass vials with measurements performed through the bottom of the vials. The vials were coupled directly with an adapter to the system NIR, and each sample a minimum of three times was scanned. The NIR spectrophotometer includes an instrument designed to measure diffuse reflectance in the NIR region of the electromagnetic spectrum [39]. A tungsten light bulb composes the system as the radiation source, and a linear-variable filter (LVF) connected to a linear indium gallium arsenide (InGaAs array detector) are integrated into the equipment itself. This equipment used a Spectralon® ceramic tile as a white reference (100% reflectance) of politetrafluoroetilen (~99%). Therefore, a 99% diffuse reflectance panel was used for the 100% reference value, and the 0% reference value was taken by leaving the tungsten Lampson with an empty support (known as dark Current Scan). Spectra were recorded using the instrument acquisition software MicroNIR™ Pro v.2.2 (VIAVI, Santa Rosa, CA, USA). The measuring of reflectance in the NIR zone was recorded in a range between 900 and 1700 nm at intervals of 6 nm in the spectra. To minimize sampling errors, all the samples were analyzed in triplicate and the mean was used in statistical treatments. The diffuse reflectance signal of the NIR spectrum is referred to as reflectance (R), using log (1/R) values for performing chemometric analyses [40]. The advantage of portable MicroNIR systems is its easy handling and size. The MicroNIR dimensions are 45 mm in diameter and 42 mm in height, weighing about 60 g, and it is equipped with a 128-pixel detector array. In addition, the MicroNIR can be directly connected to a USB port of any laptop [27].

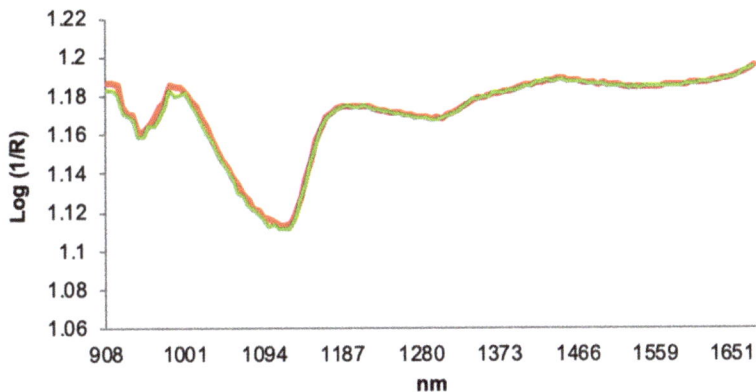

Figure 1. Near-Infrared (NIR) spectra of honey samples recorded by portable MicroNIR.

2.7. NIR-Chemometric Analyses

Chemometric techniques for the calibration of the main physicochemical parameters of honey were applied. Spectral data of samples corresponding to the samples of the calibration set were analyzed by principal component analysis (PCA) [41]. Anomalous spectra were detected by applying the Mahalanobis distance (H-statistic). Considering an H-value greater than 3 (the spectra not belonging to the population), the equations are not used to make any prediction. The modified partial least squares (MPLS) regression method was used to obtain the NIR equations. Partial least squares (PLS) regression is similar to principal component regression (PCR) but uses both reference data and spectral information to form the factors useful for fitting purposes. Using the $T \geq 2.5$ criterion, samples that presented high residual values when they were predicted were eliminated from the set. Therefore, statistical parameters of the calibration were obtained for each of the components after removing the samples for spectral (H criterion) or chemical (T criterion) reasons. To optimize the multivariate regression equations, the spectral scattering effects were taken into account with several mathematical treatments: multiplicative scatter correction (MSC), standard normal variate (SNV), D-trend (DT), and SNV-DT [42]. A nomenclature using 4 digits was used (1,4,4,1), in which the first digit is the number of the derivate, the second is the gap over which the derivative is calculated, the third is the number of data points in a running average or smoothing, and the fourth is the second smoothing.

Cross-validation is recommended to select the optimal number of factors and to avoid over fitting [43]. The calibration set is divided into several groups for the cross-validation. Each group is then validated using a calibration developed on the other group of samples. Validation errors generated are combined into a root mean square error of cross-validation (RMSECV). This statistic is considered the best single estimate for the prediction capability of the equations [44]. Cross-validation was performed by splitting the population into eight groups for all cases.

The performance of the models was determining by the squared correlation coefficient for predicted versus measured quantified in cross-validation and the ratio of standard deviation (SD) to SECV of the data set. RPD (ratio of performance to deviation) is the relation between SD and RMSEC, and it is desired to be larger than 2 for a good calibration, and an RPD ratio less than 1.5 indicates poor predictions and the model cannot be used for further prediction [44]. The statistics used to select the best equation for each physicochemical parameter were the highest RSQ (multiple correlation coefficients) and the lowest SECV (standard error of cross-validation) [25]. The software used for chemometric analysis was WinISI II version 1.50 (Infrasoft International, LLC, Silver Spring, Maryland, MD, USA).

2.8. Linear Discriminant Analysis

Linear discriminant analysis (LDA) is a supervised classification technique, which uses a class member known for the analysis. In this case, the known variable was honey type determined by palynological analysis. Considering the pollen profile, six honey groups were characterized: heather, chestnut, eucalyptus, blackberry, honeydew, and multifloral. LDA was applied to the collected reference data set (physicochemical and botanical data) to determine a linear combination of these groups of subjects. LDA is considered as a dimensional reduction method to determine a lower dimension hyperplane on which the points will be projected from the higher dimension space [10]. A linear function of the variables is sought which maximizes the ratio of between-class variance and minimizes the ratio of within-class variance. STATGRAPHICS Centurion XVI software (Statpoint Technologies, Inc., The Plains, VA, USA) was used for treatment of data.

3. Results

3.1. Physicochemical Properties of Honeys: Reference Values

The descriptive analysis of physicochemical data obtained by reference methods are summarized in (Table 1). The data are expressed in function of two groups established for the NIR treatment (calibration and validation).

Table 1. Reference physicochemical data of honey samples selected for the calibration and validation sets.

	Calibration Set (N = 84)				Validation Set (N = 16)			
	Mean	SD	Min	Max	Mean	SD	Min	Max
Moisture (%)	17.9	1.1	15.5	20.6	18.0	1.0	16.2	19.6
EC (µs/cm)	869.2	307.6	302.5	1649.5	825.5	296.8	277.0	1334.5
pH	4.2	0.3	3.7	4.8	4.2	0.3	3.7	4.7
HMF (mg/100 g)	0.1	0.2	0.0	1.7	0.1	0.2	0.0	0.5
Diastase index	29.0	9.3	10.1	44.0	27.9	8.0	13.6	36.7
Pfund (mm)	98.2	31.2	41.0	150.0	91.0	35.5	36.0	150.0
L	69.5	10.4	51.9	84.5	74.5	16.7	53.2	88.0
a*	7.4	5.2	−3.1	14.6	4.8	8.3	−3.9	14.2
b*	23.1	9.2	2.8	36.2	22.4	10.7	2.3	33.5
Phenols (mg/100 g)	126.3	51.9	36.3	254.5	120.3	63.8	24.6	233.3
Flavonoids (mg/100 g)	6.7	3.2	3.0	16.7	6.1	3.3	2.8	15.7

EC: electrical conductivity; HMF: hydroxymethylfurfural; N: number of samples; SD: standard deviation.

3.2. NIR Calibration Equations

The calibration process was performed with chemometric techniques using the spectra and the physicochemical data of the samples. The samples were split randomly in two groups: calibration group with 84 samples, and external validation group with 16 samples (Table 1). Firstly, a principal component analysis was carried out with the samples corresponding to the calibration group. The spectral variability explained ranged between 99.36% and 99.99%. The principal components required for each parameter were 8 for moisture, EC, HMF and L and b* coordinates, phenols and flavonoids; 5 for pH; 7 for diastase index; 9 for a* coordinate; and 6 for color (Pfund scale). During the cross-validation process, the identified outliers with the T and H criteria were removed of the calibration set. before of the development of equations. According to both criteria were deleted the following samples: 13 for moisture, 10 for EC, eight for pH, 29 for HMF, 11 for diastase index, 11 for Pfund, 15 for L, a* and b* coordinates, 11 for phenols, and 13 for flavonoids.

Calibrations were performed by MPLS using spectral data and the physicochemical data of honey. Table 2 shows the best mathematical treatment, the concentration range, standard deviations and the calibration descriptors for each parameter. The obtained results indicated that it was possible to predict most of the physicochemical parameters in honey samples with portable microNIR system. The degree to which the calibration fits the data set was calculated by considering the highest RSQ, and the lowest SEC and

SECV. Moisture, EC, HMF, Pfund, L and b* coordinates of CIELab scale, phenols, and flavonoids presented high RSQ coefficient (between 0.74 and 0.90). Values of RSQ lower than 0.70 had pH, diastase index and a* coordinate of CIELab. Although, the standard errors of calibration (SEC) and of cross-validation (SECV) were acceptable in all cases, presenting a minimum difference, which is an indicator that the NIR models obtained are suitable in the ranges indicated with the portable microNIR.

Table 2. Statistical descriptors of modified partial least squares (MPL)S calibration models for each physicochemical parameter.

Variable	Math Treatment	N	Mean	SD	Est. Min	Est. Max	SEC	RSQ	SECV	RPD
Moisture (%)	None 1,4,4,1	71	17.6	1.0	14.8	20.5	0.5	0.75	0.8	2.0
EC (µs/cm)	None 1,4,4,1	74	891.2	301.5	0.0	1795.8	155.1	0.74	217.9	1.9
pH	None 0,0,1,1	76	4.2	0.3	3.4	5.0	0.2	0.61	0.2	1.6
HMF (mg/100 g)	None 1,4,4,1	55	0.1	0.1	0.0	0.3	0.0	0.83	0.1	2.4
Diastase index	Detrend only 1,4,4,1	73	30.3	8.6	4.6	56.1	5.8	0.54	8.0	1.5
Pfund (mm)	None 2,10,10,1	73	96.9	31.6	2.0	191.8	11.3	0.87	20.3	2.8
L	Detrend only 1,4,4,1	69	69.3	8.7	43.2	95.3	2.8	0.90	5.4	3.2
a*	None 2,4,4,1	69	7.5	4.6	0.0	21.4	3.0	0.57	3.9	1.5
b*	None 1,4,4,1	69	24.2	8.8	0.0	50.6	3.0	0.88	5.6	2.9
Phenols (mg/100 g)	Detrend only 2,8,6,1	73	120.2	47.3	0.0	262.2	24.5	0.73	38.7	1.9
Flavonoids (mg/100 g)	None 2,8,6,1	71	6.4	2.7	0.0	14.4	1.1	0.84	2.1	2.5

EC: electrical conductivity; HMF: hydroxymethylfurfural; N: number of samples; SD: standard deviation; RSQ: multiple correlation coefficients; SEC: standard error of calibration; SECV: standard error of cross-validation; RPD: ratio performance deviation.

3.2.1. Internal Validation of Models

Cross-validation method (internal validation) was used to study the predictive capacity of the obtained models. The samples of the calibration set were divided into a series of subsets. Six cross-validation sets in all cases were checked, one group for the results (prediction) and the other to construct the calibration model. The process was implemented as many times as there were groups, such that all of them passed through the calibration set and the prediction set. The results of internal validation (predicted values versus reference) for physicochemical variables in NIR is shown in (Figure 2). Considering the statistics SEP and SEP corrected (C) was deduced that the calibration models for moisture, EC, HMF, Pfund, a* and b* coordinates, phenols, and flavonoids were adequate. Therefore, the estimation of these physicochemical parameters was possible with good results.

The RPD statistic was used to determine the predictive capacity of reference methods for NIR calibration. The parameters of moisture, HMF, Pfund, a* and b*, and flavonoids had the higher values of RPD (>2.0). This is indicative of a good calibration of the data. While, EC, pH, and phenols showed an acceptable calibration, with values of RPD higher than 1.5. Diastase index and L coordinate had a RPD value of 1.5, resulting the poorest model.

3.2.2. External Validation of the Models

In external validation the solidity of the method is checked with 16 new honey samples which were not used in the calibration models (Table 1). The average of the sample spectra was taken, the equations obtained were applied, and the NIR values were compared with the reference in accordance with the residuals and the root mean square error (RMSE). In general, the results obtained for each physicochemical parameter analysed were satisfactory. The means of the residuals were between 0.12 for HMF and 188.37 for EC. RMSE values were between 0.18 and 270.36 for HMF and EC, respectively. The predicted values by the calibration models were compared with the reference data using the Student test for paired values ($p < 0.05$). The significance level showed that there were no differences between the results obtained (values were higher than 0.05), ranging the value between 0.07 for L coordinate and 0.92 for moisture (Table 3). Therefore, it can be concluded that the method provides significantly comparable data to the reference physicochemical data.

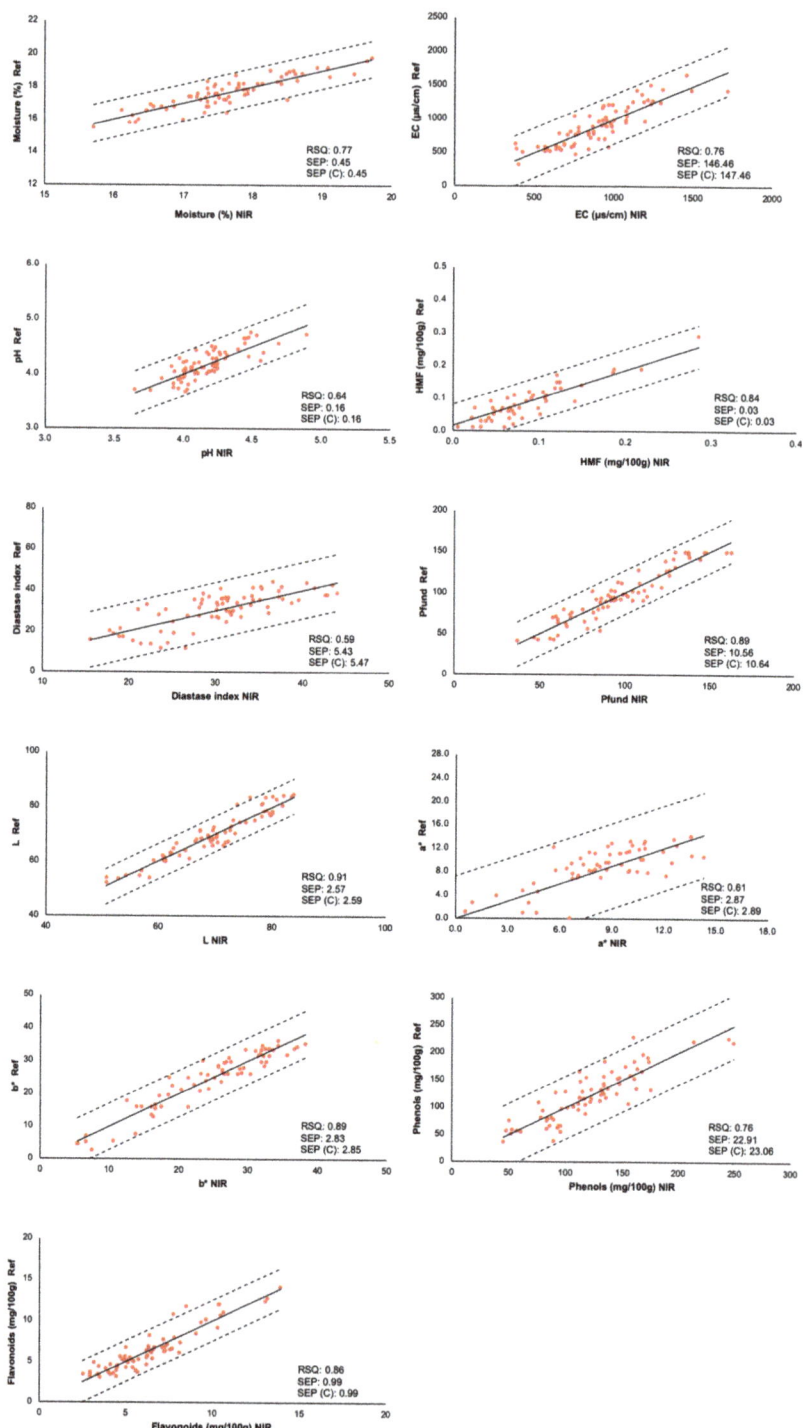

Figure 2. Measured references values versus the NIR values in the prediction set of honey samples.

Table 3. External validation of physicochemical properties in honeys by NIR (16 samples).

Constituent	Mean Residual	RMSE	p
Moisture (%)	0.87	1.02	0.92
EC (µs/cm)	188.37	270.36	0.16
pH	0.25	0.28	0.33
HMF (mg/100 g)	0.12	0.18	0.26
Diastase index	9.25	10.82	0.12
Pfund (mm)	24.81	29.12	0.18
L	10.21	14.76	0.07
a*	6.76	8.65	0.20
b*	6.09	7.68	0.58
Phenols (mg/100 g)	43.11	52.51	0.55
Flavonoids (mg/100 g)	1.92	2.58	0.83

EC: electrical conductivity; HMF: hydroxymethylfurfural; RMSE: root mean standard error.

3.3. Botanical Origin of Samples: Pollen Fingerprint

Eighty-four pollen types corresponding to 50 botanical families were identified in the honey samples. The best represented families were Fagaceae, Leguminosae, Ericaceae, Rosaceae, and Myrtaceae. According to their botanical origin, 10 samples were classified as blackberry honey (*Rubus*), 22 samples as chestnut honey (*Castanea sativa*), 9 as eucalyptus honey (*Eucalyptus*), 5 as heather honey (*Erica*), 18 as honeydew honey, and 36 as multifloral honey. The principal pollen types in each honey type are showed in (Table 4).

Table 4. Main pollen types identified in the samples by type honey.

Honey Type	N	Main Pollen Type	Secondary Pollen Types	Other Significant Pollen Types
Blackberry	10	Rubus (58.5%)	Castanea (26.4%)	Erica, Cytisus type, Eucalyptus Echium, Trifolium type
Chestnut	22	Castanea (76.1%)		Rubus, Cytisus type, Erica, Trifolium type, Echium
Eucalyptus	9	Eucalyptus (72.8%)		Cytisus type, Castanea, Erica, Rubus, Conium maculatum type, Salix
Heather	5	Erica (35.5%)	Castanea (37.9%)	Cytisus type, Eucalyptus, Rubus, Echium, Plantago
Honeydew	18	Castanea (52.3%)	Rubus (18.6%)	Cytisus type, Plantago, Salix, Erica
Multifloral	36	Castanea (41.3%)	Eucalyptus (21.0%)	Rubus, Cytisus type, Erica, Trifolium type, Salix, Echium Cynoglossum

Main pollen type: predominant pollen type in samples (frequently dominant pollen but not always). Secondary pollen types: 15–45% of pollen spectra; Other significant pollen types: <15% of pollen spectra.

The blackberry honeys had a mean value of 58.5% for *Rubus* pollen, while *Castanea* was secondary pollen (26.4%). Other important pollen types in sample were *Erica*, *Cytisus* type, *Eucalyptus*, *Echium*, and *Trifolium* type. The averaged percentage of *Castanea* in chestnut honeys was 76.1%. *Rubus* was also present in all samples with a mean value of 14.2%. Other important pollen types were *Cytisus* type, *Erica*, *Trifolium* type, and *Echium*. For eucalyptus honeys, *Eucalyptus* pollen had a mean value of 72.8%. Other significant pollen types were *Cytisus* type, *Castanea*, *Rubus*, *Erica*, *Conium maculatum* type, and *Salix*. In heather honeys, *Castanea* pollen was the pollen type with the highest mean value (37.9%) and a representation of 100% of the samples and the mean value for *Erica* was 35.5% corresponding to a slightly underrepresented pollen in samples. *Cytisus* type, *Eucalyptus*, and *Rubus* were also present in all samples.

The predominant pollen type in honeydew honey was *Castanea*, commonly found as dominant pollen (mean value of 52.3%). *Rubus* is usually secondary pollen and *Cytisus* type, *Erica*, *Plantago*, and *Salix* were also well represented pollen.

Multifloral honeys had diverse pollen spectra. Commonly *Castanea* was the main pollen type and *Eucalyptus* secondary pollen while, *Rubus*, *Cytisus* type, *Erica*, *Trifolium* type, *Salix*, *Echium*, and *Cynoglossum* appeared as other significant pollen.

3.4. Discrimination of the Samples by Honey Type

LDA was used to classify honey samples according the honey type. Main pollen types (*Castanea sativa*, *Eucalyptus*, *Erica*, and *Rubus*) and physicochemical characteristics (moisture, EC, pH, HMF, diastase index, color, phenols, and flavonoids) were included.

Five discriminant functions represented the 100% of variability of the data in the discriminant analysis (Table 5). The cumulative contribution rate of the first two linear discriminant functions accounted for 66.07%, which represented the largest fraction of overall variability in the dataset. In the first function a Wilks' Lambda = 0.01, Chi-Square = 432.3, DF = 75, $p < 0.01$ was formed; and in the second function a Wilks' Lambda = 0.04, Chi-Square = 296.4, DF = 56, $p < 0.01$. The significant value ($p < 0.05$) of Wilks' Lambda showed that the discriminant function was basic for the differentiation of the investigated groups. In addition, the higher values of eigenvalue and canonical correlations showed the high power of discrimination of the first two functions.

Table 5. Discriminant functions and statistics extracted of linear discriminant analysis.

Discriminant Function	Eigenvalue	Relative Percentage	Canonical Correlation	Wilks Lambda	Chi-Square	DF	p
1	3.64	38.77	0.89	0.01	432.27	75	<0.001
2	2.56	27.30	0.85	0.04	296.44	56	<0.001
3	1.73	18.37	0.80	0.13	183.97	39	<0.001
4	0.98	10.44	0.70	0.34	95.25	24	<0.001
5	0.48	5.13	0.57	0.67	34.79	11	<0.001

p: significance level.

Figure 3 shows the representation of the first two functions of discriminant analysis applied to all the honey samples. The graphical projection shown that honey types were satisfactorily differentiated. The overall correct classification rate was 88.1% for all samples (Table 6). The groups of unifloral samples (heather, eucalyptus, and blackberry) showed a correct classification (100% of samples) with the higher discrimination rate. Chestnut honeys and honeydew honeys were correctly classified with 83.4% and 83.3%, respectively. Finally, multifloral honeys had a correct classification of 83.3%. Regarding chestnut and honeydew honeys, 3 and 2 samples, respectively, were interchanged of group. This is possible due to the closeness of these samples, produced in same biogeographical area and from the same plant species but with different predominance. For this reason, some samples are difficult to classify being possible they are chestnut honeys with honeydew contributions.

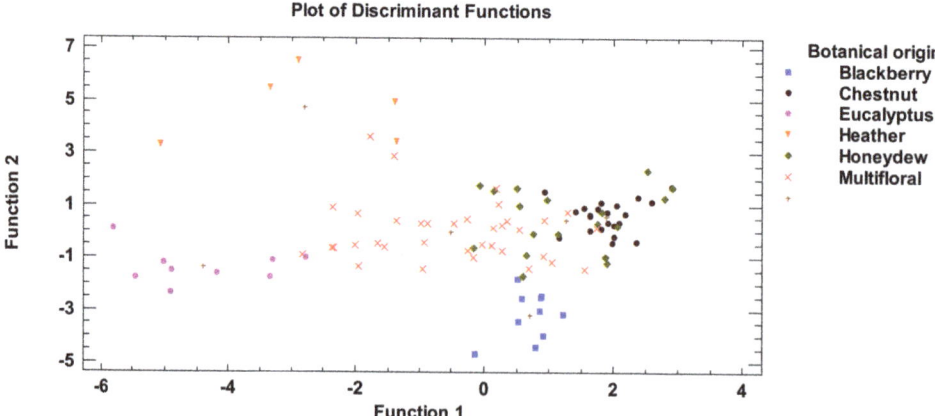

Figure 3. Plot of the first two functions of linear discriminant analysis based on main pollen types and physicochemical characteristics of honeys.

Table 6. Results of classification of honey samples considering the linear discriminant analysis.

Groups			Predicted Group Member					
Honey Type	N		Heather	Chestnut	Eucalyptus	Blackberry	Honeydew	Multifloral
Heather	5	N	5	0	0	0	0	0
		%	100	0	0	0	0	0
Chestnut	22	N	0	19	0	0	3	0
		%	0	83.4	0	0	13.6	0
Eucalyptus	9	N	0	0	9	0	0	0
		%	0	0	100	0	0	0
Blackberry	10	N	0	0	0	10	0	0
		%	0	0	0	100	0	0
Honeydew	18	N	0	2	0	0	15	1
		%	0	11.1	0	0	83.3	5.6
Multifloral	36	N	2	1	1	1	1	30
		%	5.6	2.8	2.8	2.8	2.8	83.3

Percent of cases correctly classified: 88.1%

N: number of samples; %: correct classification.

4. Discussion

The increasing consumption of natural products demands authenticity evaluations [33]. The first step in authenticity is the knowledge of a common pattern in samples with the same botanical and geographical origin. This is increasingly demanded by beekeepers, consumers, and by control entities. Furthermore, declarations of the botanical and geographical origin of honey are important to protect consumers from possible frauds and quality flaws.

In this context, the development of rapid and easy techniques to check quality and to typify many samples in a short time is a challenge. Some analytical methodologies aimed at foods authentication and traceability have been developed [11]. The best known is the NIR technology, which is enjoying increasing popularity in the fresh food and food processing industry. NIR spectroscopy represents an emerging analytical technique due to its low running costs, simple, non-destructive, environmentally friendly, rapid application, and allows several analytes to be detected simultaneously in large number of samples [7,23,26].

Quantitative or qualitative chemometric tools are necessary for the development of calibration or classification methods of groups of samples. The calibration of NIR models allows estimating with accurately desired parameters in unknown samples, but with a previous calibration with reference data obtained by conventional methods. PLS is applied in order to extract analytical information from the spectra and it has contributed much to spectral analysis works [7,26,40]. This method was used to develop quantitative calibration models on honeys matrix [1,7,23–25,29]. The correlation of the NIR spectral information with the main physicochemical parameters of reference in honey was analysed in the present study in order to assess the usefulness of portable microNIR in the analysis of quality and authenticity of this bee product. The use of this portable instrument constitutes an opportunity to honey industries to follow the progress of a product throughout the manufacturing process (traceability), to ensure food security movement, and to take trade decisions. The best calibration models were obtained for HMF, color (Pfund, L and b* coordinates by CieLAB) and flavonoid content, with a high capacity of prediction (RPD > 2). However, moisture, EC, pH, and phenols had lower values of RPD but acceptable (RPD > 1.5).

HMF and diastase content in honey are important parameters for assessing the quality and particularly its freshness. According to the Honey Quality and International Regulatory Standards, HMF must be absent or present in very low concentrations (with a maximum limit of 40 mg/kg), and the diastase activity must not be less than or equal to eight in fresh honey. All the honey samples analyzed in this study met these reference standards. Apriceno et al. [29] showed excellent results in Italian honeys based in NIR spectroscopy, with correlation coefficient of prediction of 0.98 and RPD higher than three. However, the prediction accuracy for HMF in honey by NIR was poor and unreliable in other studies [24,28]. Perhaps this is due good results are obtained with a large number of samples and it is sometimes difficult to work with quality samples knowing the provenance. Stöbener et al. [30] developed models for determining HMF content in honey using by Fourier transform attenuated total reflection infrared spectroscopy (ATR-FTIR), resulting coefficients of determination for calibration higher than 0.8. However, they observed for the proper development of a method used to determine trace compounds (such is the case of HMF), it is necessary to use a large number of calibration samples in a wide range of concentrations. Diastase is one the most important enzymes of honey, enriches the nutritional and therapeutic function of this bee product, and it is used as an important index to evaluate honey qualities [7]. In addition, the traditional chemical method for the determination of diastase content is complicated and time-consuming. Therefore, there is an urgent demand for a non-destructive and rapid method to measure this enzyme. However, the predictive capacity by NIR with the samples set selected in the present study for diastase activity was not satisfactory. Huang et al. [8] showed for diastase activity a prediction coefficient of 0.89 through a heating process based on visible and near-infrared spectroscopy. These researchers, in addition to the dependence of temperature and heating time on the diastase content, confirmed the relationship of the enzyme with the botanical origin of honey.

The range of estimation of phenols and flavonoids was similar to reported by a NIR benchtop equipment [25]. However, the results obtained in the calibration by PLS were better (with RSQ > 0.89 and RPD > 3). The polyphenols are secondary plant metabolites, which are determinants in the sensory and nutritional quality of honey, and they are recognized as having high scientific and therapeutic interest. The association of color with botanical origin and its important role in characterization of honey is well reported [9,14,19,21]. The highest polyphenolic content related with dark honeys (heather, chestnut or honeydew honeys), manifesting the importance of this parameter in the authenticity of honeys. According to Sipos et al. [13] the objective characterization of the color of different products and quantification of the differences between them is a fundamental area of research. Food color is the first sensation that consumers perceive, and it greatly influences their decision to purchase [45]. It is therefore important to distinguish small nuances in color that can differentiate similar honey as dark shades, as occurred with heather, chestnut and honeydew honeys. Therefore, the determination of different physicochemical characteristics combined with statistical techniques is essential to distinguish common patterns in samples.

As we have commented previously, different physicochemical properties have been used to authenticate the botanical origins of honey worldwide. Among the physicochemical parameters best differed the botanical origin of honeys were pH, EC, color, sugar profile, phenols, flavonoids, minerals, and volatile profile [2–4,14,16,17]. Among chemometry applications, LDA is used to differentiate groups of samples based on a common pattern. LDA confirmed that honeys of heather, eucalyptus and blackberry analysed in present study were 100% correctly classified. In the case of chestnut and honeydew honeys, the great similarity in some physicochemical characteristics complicates their differentiation, although the classification in this study resulted satisfactory (>83%). The high discrimination power of conventional physicochemical and color parameters by LDA has been previously reported for unifloral honeys. The variables with the greatest discriminatory power using LDA for unifloral honeys (*Citrus* and *Eucalyptus*) from southern Spain were water activity and EC [46]. *Citrus* honeys from Greece, Egypt, Morocco, and Spain correctly classified with rates of 97.3% [8]. Recently, combining physicochemical and botanical variables with LDA, correctly discriminated (97.6%) chestnut and honeydew honeys produced in the Northwest Spain [9].

Honey is the most researched bee product. However, some of the common determinations (such as those referring to quality parameters), beekeepers require qualified laboratories to guarantee the quality of the products in market. This implies more time for the commercialization of honey and an economic cost. Undoubtedly, the development of NIR technology with spectral-chemical information, as well as the application of portable equipment such as microNIR is an alternative for the beekeeping sector. The results obtained shows that portable microNIR equipment is a useful alternative that is comparable to the conventional technology for the determination of main physicochemical parameters in honey. The demand for increasingly versatile equipment by the food industry means that innovative and technological development is present. Development of portable vis/NIR systems including linear variable filter (LVF) of low-cost, with innovation in optical system design, miniaturization of equipment, applying it to non-professionals, which communicates through wireless technology [7,22] make them more attractive. The possibility of in situ identification of the quality parameters and physicochemical properties of honey by beekeepers in industries without destruction of sample is a desired advantage.

5. Conclusions

Versatility is a well-defined quality of portable NIR equipment and meets the needs of the primary sector in search of fast and competitive solutions. Results presented in this study showed that NIR spectroscopy by portable microNIR might be useful for the identification of floral origin of honeys from Northwest Spain. Although further experiments are proposed to build a robust database, which could support the use of this equipment as a quick alternative for honey authentication. Specifically, for HMF and diastase index, it would be interesting to establish temperature thresholds that allow adjusting the prediction with more reliable results in terms of its application.

Author Contributions: O.E. and L.M. conceived and designed methodology. O.E., and M.C.S. analyzed data. O.E. and M.C.S. writing original draft preparation. O.E., M.S.R.-F., and M.C.S. writing review and editing. M.C.S. project administration. All authors have read and agreed to the published version of the manuscript.

Funding: This study was supported by Xunta de Galicia (Rural Development Programme 2014/2020-FEADER 2018/054B, "Identification of specific markers to guarantee quality and authenticity of Galician honey".

Institutional Review Board Statement: Not applicable.

Informed Consent Statement: Not applicable.

Data Availability Statement: Data is contained within the article.

Conflicts of Interest: The authors declare no conflict of interest.

References

1. Escuredo, O.; González-Martín, M.I.; Rodríguez-Flores, M.S.; Seijo, M.C. Near infrared spectroscopy applied to the rapid prediction of the floral origin and mineral content of honeys. *Food Chem.* **2015**, *170*, 47–54. [CrossRef] [PubMed]
2. Adgaba, N.; Al-Ghamdi, A.A.; Getachew, A.; Tadesse, Y.; Belay, A.; Ansari, M.J.; Sharma, D. Characterization of honeys by their botanical and geographical origins based on physico-chemical properties and chemo-metrics analysis. *J. Food Meas. Charact.* **2017**, *11*, 1106–1117. [CrossRef]
3. Escuredo, O.; Dobre, I.; Fernández-González, M.; Seijo, M.C. Contribution of botanical origin and sugar composition of honeys on the crystallization phenomenon. *Food Chem.* **2014**, *149*, 84–90. [CrossRef] [PubMed]
4. Karabagias, I.K.; Badeka, A.; Kontakos, S.; Karabournioti, S.; Kontominas, M.G. Characterisation and classification of Greek pine honeys according to their geographical origin based on volatiles, physicochemical parameters and chemometrics. *Food Chem.* **2014**, *146*, 548–557. [CrossRef] [PubMed]
5. Ballabio, D.; Robotti, E.; Grisoni, F.; Quasso, F.; Bobba, M.; Vercelli, S.; Gosetti, F.; Calabrese, G.; Sangiorgi, E.; Orlandi, M.; et al. Chemical profiling and multivariate data fusion methods for the identification of the botanical origin of honey. *Food Chem.* **2018**, *266*, 79–89. [CrossRef] [PubMed]
6. Nasab, S.G.; Yazd, M.J.; Marini, F.; Nescatelli, R.; Biancolillo, A. Classification of honey applying high performance liquid chromatography, near-infrared spectroscopy and chemometrics. *Chemom. Intell. Lab. Syst.* **2020**, *202*, 104037. [CrossRef]
7. Huang, Z.; Liu, L.; Li, G.; Li, H.; Ye, D.; Li, X. Nondestructive Determination of Diastase Activity of Honey Based on Visible and Near-Infrared Spectroscopy. *Molecules* **2019**, *24*, 1244. [CrossRef]

8. Karabagias, I.K.; Louppis, A.P.; Karabournioti, S.; Kontakos, S.; Papastephanou, C.; Kontominas, M.G. Characterization and geographical discrimination of commercial *Citrus* spp. honeys produced in different Mediterranean countries based on minerals, volatile compounds and physicochemical parameters, using chemometrics. *Food Chem.* **2017**, *217*, 445–455. [CrossRef]
9. Rodríguez-Flores, M.S.; Escuredo, O.; Míguez, M.; Seijo, M.C. Differentiation of oak honeydew and chestnut honeys from the same geographical origin using chemometric methods. *Food Chem.* **2019**, *297*, 124979. [CrossRef]
10. Carabetta, S.; Di Sanzo, R.; Campone, L.; Fuda, S.; Rastrelli, L.; Russo, M. High-Performance Anion Exchange Chromatography with Pulsed Amperometric Detection (HPAEC–PAD) and Chemometrics for Geographical and Floral Authentication of Honeys from Southern Italy (Calabria region). *Foods* **2020**, *9*, 1625. [CrossRef]
11. Biancolillo, A.; Marini, F.; Ruckebusch, C.; Vitale, R. Chemometric Strategies for Spectroscopy-Based Food Authentication. *Appl. Sci.* **2020**, *10*, 6544. [CrossRef]
12. Persano-Oddo, L.; Bogdanov, S. Determination of honey botanical origin: Problems and issues. *Apidologie* **2004**, *35*, 1–3. [CrossRef]
13. Sipos, L.; Végh, R.; Bodor, Z.; Zaukuu, J.L.Z.; Hitka, G.; Bázár, G.; Kovacs, Z. Classification of Bee Pollen and Prediction of Sensory and Colorimetric Attributes—A Sensometric Fusion Approach by e-Nose, e-Tongue and NIR. *Sensors* **2020**, *20*, 6768. [CrossRef] [PubMed]
14. Escuredo, O.; Rodríguez-Flores, M.S.; Rojo-Martínez, S.; Seijo, M.C. Contribution to the chromatic characterization of unifloral honeys from Galicia (NW Spain). *Foods* **2019**, *8*, 233. [CrossRef]
15. Seijo, M.C.; Escuredo, O.; Rodríguez-Flores, M.S. Physicochemical properties and pollen profile of Oak honeydew and Evergreen Oak honeydew honeys from Spain: A comparative study. *Foods* **2019**, *8*, 126. [CrossRef]
16. International Honey Commission. World Network of Honey Science. Harmonised Methods of the International Honey Commission. Available online: https://www.ihc-platform.net/ihcmethods2009.pdf (accessed on 15 September 2019).
17. Puścion-Jakubik, A.; Borawska, M.H.; Socha, K. Modern Methods for Assessing the Quality of Bee Honey and Botanical Origin Identification. *Foods* **2020**, *9*, 1028. [CrossRef]
18. Escuredo, O.; Fernández-González, M.; Rodríguez-Flores, M.S.; Seijo-Rodríguez, A.; Seijo, M.C. Influence of the botanical origin of honey from North western Spain in some antioxidant components. *J. Apic. Sci.* **2013**, *57*, 5–14. [CrossRef]
19. Ciappini, M.C.; Gatti, M.B.; Di Vito, M.V. El color como indicador del contenido de flavonoides en miel. *Rev. Cienc. Tecnol.* **2013**, *19*, 53–63.
20. Combarros-Fuertes, P.; Valencia-Barrera, R.M.; Estevinho, L.M.; Dias, L.G.; Castro, J.M.; Tornadijo, M.E.; Fresno, J.M. Spanish honeys with quality brand: A multivariate approach to physicochemical parameters, microbiological quality, and floral origin. *J. Apic. Res.* **2019**, *58*, 92–103. [CrossRef]
21. Vela, L.; De Lorenzo, C.; Pérez, R.A. Antioxidant capacity of Spanish honeys and its correlation with polyphenol content and other physicochemical properties. *J. Sci. Food Agric.* **2007**, *87*, 1069–1075. [CrossRef]
22. Pasquini, C. Near infrared spectroscopy: A mature analytical technique with new perspectives—A review. *Anal. Chim. Acta* **2018**, *1026*, 8–36. [CrossRef] [PubMed]
23. Chen, G.Y.; Huang, Y.P.; Chen, K.J. Recent advances and applications of near infrared spectroscopy for honey quality assessment. *Adv. J. Food Sci. Technol.* **2014**, *6*, 461–467. [CrossRef]
24. Ruoff, K.; Luginbühl, W.; Bogdanov, S.; Bosset, J.-O.; Estermann, B.; Ziolko, T.; Kheradmandan, S.; Amad, R. Quantitative determination of physical and chemical measurands in honey by near-infrared spectrometry. *Eur. Food Res. Technol.* **2007**, *225*, 415–423. [CrossRef]
25. Escuredo, O.; Seijo, M.C.; Salvador, J.; González-Martín, M.I. Near infrared spectroscopy for prediction of antioxidant compounds in the honey. *Food Chem.* **2013**, *141*, 3409–3414. [CrossRef]
26. Porep, J.U.; Kammerer, D.R.; Carle, R. On-line application of near infrared (NIR) spectroscopy in food production. *Trends Food Sci. Technol.* **2015**, *46*, 211–230. [CrossRef]
27. Pakhomova, S.; Zhdanov, I.; van Bavel, B. Polymer Type Identification of Marine Plastic Litter Using a Miniature Near-Infrared Spectrometer (MicroNIR). *Appl. Sci.* **2020**, *10*, 8707. [CrossRef]
28. Qiu, P.Y.; Ding, H.B.; Tang, Y.K.; Xu, R.J. Determination of chemical composition of commercial honey by near-infrared spectroscopy. *J. Agric. Food Chem.* **1999**, *47*, 2760–2765. [CrossRef]
29. Apriceno, A.; Bucci, R.; Girelli, A.M.; Marini, F.; Quattrocchi, L. 5-Hydroxymethyl furfural determination in Italian honeys by a fast near infrared spectroscopy. *Microchem. J.* **2018**, *143*, 140–144. [CrossRef]
30. Stöbener, A.; Naefken, U.; Kleber, J.; Liese, A. Determination of trace amounts with ATR FTIR spectroscopy and chemometrics: 5-(hydroxymethyl)furfural in honey. *Talanta* **2019**, *204*, 1–5. [CrossRef]
31. Thamasopinkul, C.; Ritthiruangdej, P.; Kasemsumran, S.; Suwonsichon, T.; Haruthaithanasan, V.; Ozaki, Y. Temperature compensation for determination of moisture and reducing sugar of longan honey by near infrared spectroscopy. *J. Near Infrared Spectrosc.* **2017**, *25*, 36–44. [CrossRef]
32. Chen, L.; Wang, J.; Ye, Z.; Zhao, J.; Xue, X.; Heyden, Y.V.; Sun, Q. Classification of Chinese honeys according to their floral origin by near infrared spectroscopy. *Food Chem.* **2012**, *135*, 338–342. [CrossRef] [PubMed]
33. Leme, L.M.; Montenegro, H.R.; dos Santos, L.D.R.; Sereia, M.J.; Valderrama, P.; Março, P.H. Relation between near-infrared spectroscopy and physicochemical parameters for discrimination of honey samples from *Jatai weyrauchi* and *Jatai angustula* bees. *Food Anal. Methods* **2018**, *11*, 1944–1950. [CrossRef]

34. Ferreiro-Gonzalez, M.; Espada-Bellido, E.; Guillen-Cueto, L.; Palma, M.; Barroso, C.G.; Barbero, G.F. Rapid quantification of honey adulteration by visible-near infrared spectroscopy combined with chemometrics. *Talanta* **2018**, *188*, 288–292. [CrossRef] [PubMed]
35. Louveaux, J.; Maurizio, A.; Vorwohl, G. Methods of melissopalynology. *Bee World* **1978**, *59*, 139–157. [CrossRef]
36. Bogdanov, S.; Martin, P.; Lullmann, C. Harmonized methods of the international honey commission. *Apidologie* **1997**, extra issue. 1–59.
37. Singleton, V.L.; Orthofer, R.; Lamuela-Raventos, R.M. Analysis of total phenols and other oxidation substratesand antioxidants by means of Folin-Ciocalteu reagent. *Methods Enzymol.* **1999**, *299*, 152–178.
38. Arvouet-Grand, A.; Vennat, B.; Pourrat, A.; Legret, P. Standardization of propolis extract and identification of principal constituents. *J. Pharm. Belg.* **1994**, *49*, 462–468.
39. VIAVI Solutions Inc. *MicroNIR Pro v3.0. User Manual*; VIAVI Solutions Inc.: Santa Rosa, CA, USA, 2019; p. 308.
40. Herold, B.; Kawano, S.; Sumpf, B.; Tillmann, P.; Walsh, K.B. VIS/NIR spectroscopy. In *Optical Monitoring of Fresh and Processed Agricultural Crops*; Zude, M., Ed.; CRC Press: Boca Raton, FL, USA, 2009; pp. 141–249.
41. Lavine, B. A user-friendly guide to multivariate calibration and classification, tomas naes, tomas isakson, tom fearn and tony davies, NIR publications, Chichester. *J. Chemometr.* **2003**, *17*, 571–572. [CrossRef]
42. Dhanoa, M.S.; Sister, S.J.; Barnes, R.J. On the scales associated with near infrared reflectance difference spectra. *Appl. Spectrosc.* **1995**, *49*, 765–772. [CrossRef]
43. Shenk, J.; Westerhaus, M. Routine Operation, Calibration Development and Network System Management Manual for Near Infrared Instruments. Version 3.1; Infrasoft International: Silver Spring, MD, USA, 1995.
44. Williams, P.C.; Sobering, D. How do we do it: A brief summary of the methods we use in developing near infrared calibrations. In *Near Infrared Spectroscopy: The Future Waves*; Davies, A.M.C., Williams, P.C., Eds.; NIR Publications: Chichester, UK, 1996; pp. 185–188.
45. Pathare, P.B.; Opara, U.L.; Al-Said, F.A.J. Colour measurement and analysis in fresh and processed foods: A review. *Food Bioprocess. Technol.* **2013**, *6*, 36–60. [CrossRef]
46. Serrano, S.; Villarejo, M.; Espejo, R.; Jodral, M. Chemical and physical parameters of Andalusian honey: Classification of *Citrus* and *Eucalyptus* honeys by discriminant analysis. *Food Chem.* **2004**, *87*, 619–625. [CrossRef]

Article

Sensorial, Melissopalynological and Physico-Chemical Characteristics of Honey from Babors Kabylia's Region (Algeria)

Asma Ghorab [1,2,*], **María Shantal Rodríguez-Flores** [2], **Rifka Nakib** [2,3], **Olga Escuredo** [2], **Latifa Haderbache** [4], **Farid Bekdouche** [5] **and María Carmen Seijo** [2,*]

[1] Laboratoire d'Ecologie et Environnement, Faculté des Sciences de la Nature et de la Vie, Université A. Mira de Bejaia, Bejaia 06000, Algeria
[2] Department of Vegetal Biology and Soil Sciences, Facultade de Ciencias, Universidade de Vigo, 32004 Ourense, Spain; mariasharodriguez@uvigo.es (M.S.R.-F.); nakib.rifka@gmail.com (R.N.); oescuredo@uvigo.es (O.E.)
[3] Laboratory of Food Quality and Food Safety, University of Mouloud Mammeri, Tizi Ouzou 15000, Algeria
[4] Research Laboratory in Food Technology (LRTA), M'hamed Bougara University, Avenue de l'indépendance, Boumerdes 35000, Algeria; l.haderbache@univ-boumerdes.dz
[5] Department of Ecology and Environment, FSNV, University of Batna 2, Batna 05000, Algeria; bekdouche_21@yahoo.fr
* Correspondence: asma.ghorab@uvigo.es (A.G.); mcoello@uvigo.es (M.C.S.)

Citation: Ghorab, A.; Rodríguez-Flores, M.S.; Nakib, R.; Escuredo, O.; Haderbache, L.; Bekdouche, F.; Seijo, M.C. Sensorial, Melissopalynological and Physico-Chemical Characteristics of Honey from Babors Kabylia's Region (Algeria). *Foods* **2021**, *10*, 225. https://doi.org/10.3390/foods10020225

Academic Editor: Paweł Kafarski

Received: 29 December 2020
Accepted: 19 January 2021
Published: 22 January 2021

Publisher's Note: MDPI stays neutral with regard to jurisdictional claims in published maps and institutional affiliations.

Copyright: © 2021 by the authors. Licensee MDPI, Basel, Switzerland. This article is an open access article distributed under the terms and conditions of the Creative Commons Attribution (CC BY) license (https://creativecommons.org/licenses/by/4.0/).

Abstract: This study aimed to characterize the honeys of Babors Kabylia through sensory, melissopalynological and physico-chemical parameters. Thirty samples of honey produced in this region were collected over a period of two years and analyzed. All the samples presented physico-chemical parameters in conformity with legislation on honey quality, with few exceptions, linked mainly to beekeeping management. The pollen spectrum revealed a great diversity with 96 pollen types. The main pollen types were spontaneous species as Fabaceae (*Hedysarum*, *Trifolium*, Genisteae plants), Asteraceae plants, Ericaceae (*Erica arborea* L.) or *Myrtus* and *Pistacia*. The sensory properties of samples showed a high tendency to crystallization, the colors were from white to brown, but most of them had gold color. Smell and odor corresponded mainly to vegetal and fruity families and in taste perceptions besides sweetness highlighted sourness and saltiness notes. Seventeen samples were polyfloral, one was from honeydew and twelve were monofloral from heather, genista plants, sulla, blackberry or Asteraceae. Heather and the honeydew samples showed the darkest color, the highest electrical conductivity and phenol and flavonoid content. A statistical analysis based on the most representative pollen types, sensory properties and some physico-chemical components allowed the differentiation of honey samples in terms of botanical origin.

Keywords: honey; Babors Kabylia; sensorial properties; melissopalynology; quality parameters; multivariate analysis

1. Introduction

Honey is one of the apian products more linked to the territory in which was produced, due to plant communities of the area, climate, soil and apicultural practices drive its characteristics. In Algeria, beekeeping is considered an integral part of the agricultural and rural routine. It is practiced in several regions but has been more important in the north of the country thanks to the appropriate climatic conditions and the great floristic biodiversity that provides honey resources during most of the year [1]. There are more than 20,000 beekeepers with 700,000 hives throughout Algeria, mainly are modern hives (Langstroth type and lesser Dadant type) and rarely are traditional hives. About 90% are independent and amateur and only 10% are professionals.

Babors Kabylia's region is situated at the North east of Algeria being one of the most interesting regions for honey production in bio-geographical terms [2]. It has been considered as a biodiversity hotspot because of the richness of its flora and the presence of high number of endemic plants [3]. Within this large plant biodiversity, melliferous plants constitute an important part, so that it is possible to produce a wide variety of honey types [4–7]. According to [8,9], the plant species visited by the bees as well as the environment in which the honey was produced seem to have a strong influence on its quality and quantity; hence, it is possible to relate it to its geographical origin and on the other hand, honey could be a footprint of its environment.

The tellien sector of the Babors is made up of folded and scaled units. The soils are of a shisty and marly calcareous nature. Its Mediterranean-type climate is characterized by a rainy season mainly consisting of thunderstorms and torrential rains, concentrated during a very wet period from October to March with an average annual rainfall of nearly 900 mm/year, and a dry season between June and September. The vegetation is characterized by woodlands and shrubs, spontaneous plants, agricultural fields and other grasslands and hedgerows, forming a relatively heterogeneous landscape with numerous forage opportunities for honeybees.

The relief is from the sea level to high mountains (more than 1200 m). At the north slopes of these mountains, the forest vegetation is very dense with woodlands of resinous species such as *Cedrus atlantica* (Endl.) Carrière and *Abies numidica* de Lannoy ex Carrière. and caducifolia oaks such as *Quercus canariensis* Willd. and *Quercus afares* Pomel. The southern slopes are practically devoid of forest vegetation being shrubs such as *Calicotome spinosa* (L.) Link. and some herbaceous plants as *Ampelodesma mauritanicus* (Poir.) Durand & Schinz. At altitudes below 1200 m, appear some degraded green oak forests (*Quercus ilex* L.), but is the domain of cork oak forest (*Quercus suber* L.), with firstly the humid facies of *Cytisus villosus* Pourr. and then, at lower altitudes, the thermophilic facies of *Erica arborea* L. Most of the beekeeping is practiced in this area. The maquis constitutes an interesting plant community for apiculture; more particularly *Erica arborea* and *Pistacia lentiscus* L. association, which covers the slopes located at less than 600 m of altitude in inland regions. It includes, among others, *Cistus salviifolius* L., *Arbutus unedo* L., *Clinopodium vulgare* L., *Lavandula stoechas* L., *Daphne gnidium* L. and *Genista tricuspidata* Desf. Its degradation also promotes a great biodiversity of spontaneous species characteristics of stripped soils: *Cistus monspeliensis* L., *Bellis sylvestris* Cirillo, *Hypochaeris radicata* L., *Hedysarum coronarium* L., *Stachys ocymastrum* (L.) Briq. and Poaceae as *Ampelodesmos mauritanicus*, *Briza maxima* L., *Aira tenorei* Guss., *Festuca coerulescens* Desf. or *Cynosurus echinatus* L. [10].

On the side of Draa El Kaid, between 500 and 700 m of altitude, close to the villages, grows a shrub with *Retama sphaerocarpa* (L.) Boiss., *Calicotome spinosa*, *Thymus munbyanus* subsp. *ciliatus* (Desf.) Greuter & Burdet., *Capparis spinose* L., *Ziziphus lotus* (L.) Lam. and *Teucrium polium* L. On the North, the Gouraya National Park, have a vegetation characterized by a degraded shrub with *Pinus halepensis* Mill. dominated by *Quercus coccifera* L., *Erica arborea*, *Erica multiflora* L., *Stachys ocymastrum* and *Glebionis coronaria* (L.) Spach. [10]. Main agricultural crops are near to the coast and the rivers. The most common are Solanaceae as potatoes, tomatoes, peppers or Brassicaceae as cauliflower crops. Fruit arboriculture is represented by orchards of orange, lemon, apples and sometimes medlar. Deserving a special mention for viticulture are olives and fig trees.

Honey, the fruit of collaboration between the plant and animal worlds, has always been considered a sacred product because of its attributed nutritional and therapeutic benefits [11]. Furthermore, beekeeping is an environmentally friendly practice useful to promote local economy in areas with water scarcity and to facilitate pollination services in highly valued ecosystems. As occurs in many parts of the world, in Algeria, people prefer local beekeeping and consumers get their honey directly from beekeepers, trusting them for the quality and botanical origin of the honey. However, the characteristics of local productions are poorly studied and most Algerian honeys are mislabeled. In this context, increasing knowledge in local honeys contributes to their valorization and to avoid frauds for consumers. One of the main tasks is the authentication of the predominant botanical origin and quality. In this framework, sensory characteristics are the first attributes distinguished for consumers and together with melissopalynology deepen in the botanical and geographical origin of the honey [12]. Physico-chemical parameters complete the information to characterize local productions.

The objective of this study was to investigate the characteristics of the honey produced in one of declared Mediterranean biodiversity hotspots. For this purpose, thirty honey samples collected during the years 2018–2019, in the Babors Kabylia's region, a large geographical area of Northern Algeria, were analyzed.

2. Materials and Methods

2.1. Study Area and Honey Samples

The present study was conducted on 30 honey samples obtained from *Apis mellifera intermissa* apiaries situated throughout Babors Kabylia's region (North East of Algeria). The samples were collected during spring and summer seasons (2018–2019) and the honey extraction from the combs was by centrifugation (Table 1). Then the samples were stored in glass jars at −4 °C until its analysis.

Table 1. Geographical origin of honey samples and period of harvest.

Sample	Area	Harvest Period	Altitude (a.m.s.l.)
M01	Timsyet (Tizi N'berber)	Summer	312
M02	Draa El-Gaid	Summer	580
M03	Ighil Hassan	Summer	320
M04	Lota village	Summer	20
M05	Tahalaket (Tichy)	Summer	167
M06	Ijouyaze Sahel	Summer	205
M07	Adrar Oufarnou	Summer	280
M08	Ait Aissa, Aokas	Summer	185
M09	Djermana (Aokas)	Summer	250
M10	Agwni Oukouche (Melbou)	Summer	596
M11	Ouagaz Ichaabanen (Aokas)	Summer	245
M12	Annar Assam	Summer	335
M13	Annar Assam	Summer	335
M14	Timsyet (Tizi N'berber)	Summer	312
M15	Tahalaket	Spring	167
M16	Tizi N'berber	Summer	350
M17	Ouagaz Ichaabanen	Summer	245
M18	Mechta Ledjbel (Ziama)	Summer	590
M19	Draa El-Gaid	Summer	580
M20	Tabelout	Summer	305
M21	Tasabounet	Summer	220
M22	Zentout-Tamrijet	Summer	290
M23	Melbou	Summer	03
M24	Ait Idir-Adekar	Summer	500
M25	Tizi Ahmed	Summer	330
M26	Agounane-Derguina	Summer	300
M27	Djoa-Ajloh (Tichy)	Summer	260
M28	AiT Anane-Derguina	Summer	80
M29	Tababort	Summer	600
M30	Boukhlifa	Summer	280

During the harvest season, the main melliferous plants of the region were identified and reference slides of pollen were prepared [13], for comparison with the pollen types found in the honey samples.

The following determinations were carried out: sensorial analysis (color, smell, taste and aroma), melissopalynological analysis and physico-chemical analysis (quality parameters, phenol and flavonoid content and main mineral content).

2.2. Sensorial Analysis

Sensory analysis (visual, olfactory and gustatory characteristics) of collected honey samples was performed by a tasting panel constituted of a group of tasters (8 people) previously selected and trained according to international standards. The sensory test was performed in a sensory room under natural white light at room temperature. The samples were presented to the panel as 20 mL in small transparent glasses. Water was provided for rinsing the mouth between samples. Each honey sample was individually evaluated by descriptive grades using scales (1–10). The descriptors used for the evaluation can be seen in Table 2.

Table 2. Main descriptors used for sensorial analysis.

Sensory Perception	Descriptors
Estate	Liquid, Crystallized
Color	White, Straw, Gold, Orange, Brown
Smell	Vegetal, Floral, Animal, Chemical, Fruity, Degraded
Taste	Sweetness, Bitterness, Saltiness, Sourness
Aroma	Vegetal, Floral, Animal, Chemical, Fruity, Degraded
Tertiary attributes	Astringency, Spicy

2.3. Melissopalynological Analysis

Pollen analysis was performed for quantitative results (number of pollen grains per gram of honey) and qualitative results (pollen spectra of the honey samples).

2.3.1. Quantitative

The methodology is based on the methods of melissopalynology [13]. Ten grams of honey were weighed and fully dissolved in 40 mL of warm distilled water (not above 40 °C). The solution was centrifuged for 10 min at 4500 rpm and the supernatant was discarded. Afterwards, 40 mL of distilled water was added prior to centrifugation for 5 min. The supernatant was again discarded until a volume of 5 mL and then the sediment was vortexed. For the microscopical analysis two drops (10 µL) of this sediment were deposited in separate, over a slide. The total number of pollen grains in each drop were counted and the results were expressed as number of pollen grains per g of honey considering the mean value of both drops. Honeys are grouped considering the number of pollen grains per gram of honey (PG/G) into one of the following classes: Class I with less than 2000 pollen grains; Class II with 2000 to 10,000 pollen grains; Class III with 10,000 to 50,000 pollen grains; Class IV with between 50,000 and 100,000 pollen grains; and Class V with more than 100,000 pollen grains.

2.3.2. Qualitative

The obtained sediment for quantitative analysis was centrifuged again and the supernatant was discarded. After vortexing, two drops (100 µL) of sediment were placed separately on a slide and distributed over an area of about 24 × 24 mm. Examination of pollen slides was performed using an optical microscope (400× or 1000×, as appropriate). The percentage of representation for each type of pollen was calculated by counting at least 500 pollen grains per sample. The pollen grains were classified as pollen type, as genus or a single species, when it was possible. Following the recommendations suggested by [13], the pollen frequency classes were determined.

2.4. Physico-Chemical Analysis and Color Determination

Quality parameters including honey freshness (HMF and diastase content), moisture, electrical conductivity, pH and free acidity were determined in duplicate and following methodologies proposed by the International Honey Commission [14].

2.4.1. Honey Freshness

HMF content was determined through the White spectrophotometric method. The absorbance of a honey solution was measured at 284 and 336 nm with a UV-visible spectrophotometer (Thermo Scientific Helios Gamma, Chorley, UK) against a blank. The determination of diastase activity was based on the quantity of starch converted by a honey solution and the absorbance of the resulting blue color. It was determined spectrophotometrically at 660 nm using a UV-visible spectrophotometer (Thermo Scientific Helios Gamma, Chorley, UK) at different times to an end point below 0.235. Diastase activity was expressed in diastase index (DI) Gothe Scale.

2.4.2. Other Quality Parameters

Water content was determined with a Carl-Zeiss Jena refractometer by measuring the refractive indices at 20 °C. The percentage of water was calculated using the CHATAWAY table. Electrical conductivity was performed at 20 °C in a 20% (w/v) honey solution (dry matter basis) in CO_2-free deionized distilled water using a portable conductivity meter (Knick Portamess® 913 Conductivity, Beuckestr, Berlin, Germany). The values were expressed in mS/cm. pH was measured by a pHmeter (WTW inoLab pH 750) on a solution containing 10 g of honey dissolved in 75 mL of distilled water; the same solution was titrated for free acidity with 0.1 M sodium hydroxide (NaOH) solution up to a pH of 8.3. The results were expressed in meq/kg.

2.4.3. Color of Honey

Color determination was carried out using a HANNA Honey colorimeter (HANNA C221 Honey Color Analyzer, Rhode Island, RI, USA), previously calibrated with glycerin (Glycerol HANNA instruments, Rhode Island, RI, USA), which gives the values in millimeters Pfund.

2.4.4. Polyphenol and Flavonoid Content

The total phenolic content (TPC) was determined using an adapted Folin-Ciocalteu method [15]. Briefly, a honey solution for each sample was prepared (0.1 g/mL) and mixed with 10 mL of distilled water, 1 mL of Folin–Ciocalteu reagent and 4 mL of 7.5% sodium carbonate (Na_2CO_3) solution up to a final volume of 25 mL. After incubation at room temperature in dark for 1 h, the absorbance of the solution was measured by spectrophotometry at 765 nm. Gallic acid (GA) was chosen as a standard, using various concentrated solutions (0.01–0.50 mg/mL), and then a calibration curve was obtained. The results were expressed as gallic acid equivalents in mg/100 g of honey.

Total flavonoid content (TFC) was measured using an adaptation of the Dowd method [16]. The same first stock honey solution (0.1 g/mL) prepared for the determination of phenol content was used. This solution was dissolved until a concentration of 0.33 g/mL and 0.5 mL of 5% aluminum chloride ($AlCl_3$) solution was added prior to incubation in dark for 30 min. The reaction yields a yellow color and their absorbance was determined spectrophotometrically at 425 nm after incubation in dark for 30 min. The TFC was calculated using a calibration curve with different quercetin solutions (0.002 to 0.01 mg/mL) and the results were expressed as mg equivalent of quercetin per 100 g of honey.

2.4.5. Mineral Composition Analysis

Mineral content of honey was quantified by inductively coupled plasma mass spectrometry (ICP-MS) and by Atomic Absorption spectrometry (AAS) using a Spectrometer

Varian SpectraAA-600 (Agilent Technologies, Santa Clara, CA, USA). First, the samples were warmed and sonicated to facilitate the homogenization of honey, then 5 mL of HNO_3-H_2O_2 9:2 were added to 0.5 g aliquots of the homogenized honey and afterwards were digested in a microwave oven (CEM MARSX press model) [17]. ASS was used to determine Na, K, Ca, Mg and Fe and ICP-MS for quantifying Mn, Cu, Zn, Cd, Pb and P. The results were expressed as mg per 100 g of honey.

2.5. Data Analysis

Sensorial data was analyzed using the XLSTAT Sensory tool by Addinsoft (Paris, France). This package let to obtain the characterization of the samples considering the attributes perceived by the 8 tasters. The procedure for sample characterization and panel evaluation was used. Significant statistical differences were set as p-value < 0.05. The radar charts for monofloral samples were done using the factorial values obtained in the characterization of products.

The package Analyzing data of XLSTAT was used as an exploratory statistical tool to reduce dimensionality of the multivariate data and to visualize them graphically, with minimal loss of information. PCA analysis was applied to identify groups of samples according to their botanical origin. This multivariate analysis allowed to summarize the information includes in the variables studied into a small number of principal components or factors providing a simplified interpretation of data variance through mathematical methods.

3. Results and Discussion

3.1. Sensorial Profile of Samples

Organoleptic properties are the first attributes that consumer can observed. These comprise visual properties as state, color, smell and aroma perceptions and taste. The descriptors considered in this work, to describe honey samples, had different discriminant power (Figure 1). The most useful were visual color, the different families of smell (animal, chemical, floral, fruity and vegetal), the persistence of smell, aroma perceptions (mainly fruity, vegetal, animal and floral families) and the saltiness for taste descriptors. The rest of the descriptors have been poorly detected so presented non-significant values.

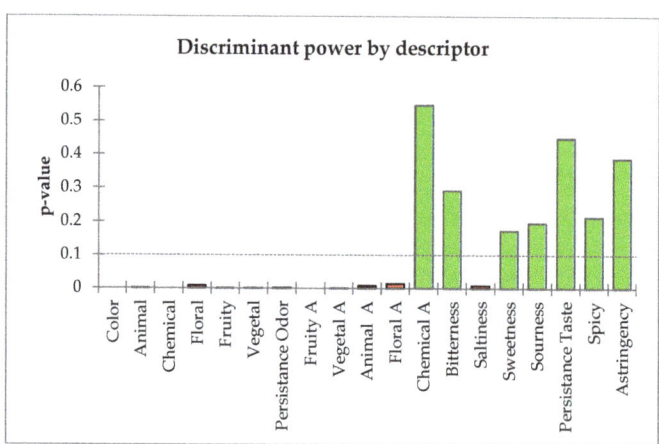

Figure 1. Sensorial descriptors used and their discriminant power.

All the samples were crystallized when the sensorial analysis was performed. Hence the scale used for color descriptors was from white to brown. The clearest samples were classified like straw color (two samples), 13 had gold color, five orange color and 10 brown color. The size of crystals was very fine or fine in 14 samples, medium in 15 samples while only one sample had large crystals. The tasters described the intensity of the smell as low being the predominant the vegetal in 21 samples, fruity in three and floral in two samples; however, secondary perceptions were mainly fruity, floral and in six samples, animal. The persistence of the odor was generally low or extremely low. Regarding the taste, besides the sweetness, tasters detected a degree of sourness and saltiness in most of the samples while bitterness was less common. When the aroma was considered, the most common perceptions were fruity, floral and in lesser extent vegetal. The most usual descriptors for vegetal smell or aroma were leaves, wood, mint, resin or cinnamon, for fruity were fresh fruit, tropical fruit, peach, dry grapes or fig, for floral were rose, lavender and violet and for animal leather and wax (except two samples in which tasters made associations with urine). Astringency was detected in a few samples. Finally, it should be mentioned that 15 samples had an intense smell to smoke (degraded family), that could be caused during the honey harvest. To protect quality beekeepers should improve honey management techniques.

3.2. Pollen Spectra and Main Botanical Origin of Samples

Ninety-six different pollen types were identified in the pollen spectrum of the honey samples including 77 nectariferous taxa and 19 taxa which do not produce nectar. Samples present a great diversity of pollen types, the mean value was 26, but 12 samples had more than 30 different pollen types. The botanical families Fabaceae, Asteraceae, Apiaceae and Lamiaceae provide the 36.0% of the pollen types, standing out the diversity of Fabaceae plants which were 15.3% of identified types. Within this family, *Genista* and *Hedysarum* pollen types were the most frequent pollen grains. The first was present in 96.7% of the samples and the second one in 83.3% of them.

Genista was dominant pollen in three samples and the maximum value reached was 73.9% whereas *Hedysarum* was dominant pollen in two samples and had a maximum value of 63.2%. For *Asteraceae* plants, *Galactites* t. was found in 70.0% but the values in the pollen spectra were lower than 7%, while *Aster* t., identified in 50.0% of honeys, had a maximum value of 39.7%. The most frequent Lamiaceae pollen was *Stachys* t., found in 70% of samples with a maximum value of 9.7%. Other common pollen type was *Thymus* t. (46.7% of samples) but values were always below 3%. Finally, Apiaceae plants mainly as *Foeniculum vulgare* t. and *Pimpinella anisum* t. were found in 73.3% and 50.0% honeys, respectively (Table 3).

Table 3. Frequency classes of the main pollen types.

Family	Pollen Type	Max. (%)	D (≥45%)	A (45–15%)	R (15–3%)	I (3–1%)	Present [1]
Amaryllidaceae	*Allium* t.	5.7	-	-	1	-	11
Anacardiaceae	*Pistacia*	30.0	-	2	7	4	22
Apiaceae	*Daucus carota* t.	3.9	-	-	2	2	7
Apiaceae	*Foeniculum vulgare* t.	9.0	-	-	5	6	22
Apiaceae	*Pimpinella anisum* t.	6.0	-	-	3	4	15
Asparagaceae	*Asparagus acutifolius*	21.0	-	1	-	-	1
Asteraceae	*Aster* t.	39.7	-	1	3	5	15
Asteraceae	*Galactites* t.	6.9	-	-	3	3	21
Boraginaceae	*Echium*	4.4	-	-	2	6	17
Capparaceae	*Capparis spinosa*	12.7	-	-	1	-	4
Cistaceae	*Cistus* t.	11.7	-	-	1	6	19
Cyperaceae	*Carex*	5.0	-	-	1	2	6
Ericaceace	*Erica*	72.8	4	7	7	4	25
Fabaceae	*Astragalus*	22.1	-	1	-	-	1
Fabaceae	*Ceratonia siliqua*	7.3	-	-	1	1	2

Table 3. Cont.

Family	Pollen Type	Max. (%)	D (≥45%)	A (45–15%)	R (15–3%)	I (3–1%)	Present [1]
Fabaceae	Genista t	73.9	3	15	8	1	29
Fabaceae	Hedysarum	63.2	1	-	7	6	20
Fabaceae	Lotus t	7.3	-	-	4	5	21
Fabaceae	Spartium junceum t.	16.8	-	1	5	1	8
Fabaceae	Trifolium repens t.	17.9	-	1	10	4	26
Fagaceae	Castanea	17.1	-	1	1	-	2
Fagaceae	Quercus	7.6	-	-	2	2	12
Lamiaceae	Stachys t.	9.7	-	-	6	7	21
Lythraceae	Punica granatum	10.5	-	-	1	3	5
Myrtaceae	Eucalyptus	41.9	-	2	4	4	16
Myrtaceae	Myrtus communis	25.9	-	3	12	4	24
Oleaceae	Olea europaea	6.1	-	-	1	1	4
Rosaceae	Crataegus t.	7.1	-	-	4	2	8
Rosaceae	Prunus t.	4.9	-	-	2	1	12
Rosaceae	Rubus	58.7	1	1	3	7	21
Salicaceae	Salix	11.6	-	-	4	1	11
Tamaricaceae	Tamarix	5.7	-	-	1	1	4

[1] Max.: maximum value reached by the pollen type in samples, D: number of samples in which the pollen type is dominant (percentage ≥ 45%), A: number of samples where the pollen type is between 45% and 15%, R: number of samples where the pollen type is between 15% and 3%, I: number of samples where the pollen type is between 3% and 1%, present: number of samples where the pollen type was identified. t.: pollen type common for different plant genera.

Despite the frequency of the mentioned botanical families in the honey samples, some other plants highlighted in the pollen spectra of honeys. The first was *Erica* genus, represented by the species *E. arborea* and *E. multiflora*. This pollen type was found in 70.0% of samples and had a maximum value of 72.8%, being dominant pollen in four samples and secondary pollen in seven samples. In addition, Myrtaceae is well represented in the area due to the presence of *Eucalyptus*, used to reforestation, and *Myrtus* that grows in Mediterranean area commonly below 600 m, both are present as secondary pollen in samples. *Pistacia* is other representative Mediterranean taxa, was present in 73.3% of honeys frequently with values under 15% but was secondary pollen in two samples. The area of study had an important rainy period that contributes to facilitate the introduction of some forest species in mountain areas as *Castanea*. The genus was only found in two samples, but its high bee value merit is to be considered. In the case of Rosaceae plants, the stand out, *Rubus*, presented in 70% of samples as being in one dominant pollen. Another mention should be made to the presence of *Asparagus* pollen, with a value of 21.0% in one sample.

The honey samples had medium pollen content in quantitative terms. Values ranged from 1117 to 23,196 grains/g, with an average of 6710 grains/g. Most of the samples (66.7%) were classified in class II of Maurizio and 23.3% in class III. Only 10% of samples had low pollen content (Class I). This is according to the use of centrifugation to extract honeys from the combs.

3.3. Quality Evaluation of the Honey Samples

Overall, most of the honey samples showed acceptable quality parameters that were in accordance with the international legislation on honey quality [18].

The evaluation of the degree of freshness gave good results for most of the samples. HMF values were below the limit of 40 mg/kg, having a mean value of 9.0 mg/kg. The lowest value was 1.3 mg/kg and the highest value was 31.8 mg/kg. Regarding diastase content, samples had extremely low values with a mean value of 11.3 DI. The minimum diastase level was 4.8 DI and the maximum was 29.0 DI. Ten honey samples had a diastase activity below 8 DI (the minimum legal limit in international standards without considerations about HMF content).

One important parameter for honey stability is water content. It varied from 16.4% to 19.8%, with a mean value of 18.1%. Most of the samples were collected during July and August and only one in April, but there were no differences in water content in relation to the date of harvest. Electrical conductivity was in general medium with a mean value of 0.60 mS/cm, a minimum of 0.29 mS/cm and a maximum value of 1.35 mS/cm. For pH, the values oscillated from 3.5 to 4.4 with a mean value of 3.9. The values of electrical conductivity correspond mainly to blossom honeys; however, six samples presented values over 0.80 mS/cm. Regarding to free acidity, all samples had values below the maximum permitted in European legislation which is 50 meq/kg [18], ranging from 14.1 meq/kg to 46.5 meq/kg with a mean value of 32.4 meq/kg. According to Pfund scale, the color of samples varied from extra light amber (37 mm), to dark amber (135 mm) with a mean value of 77 mm. In addition to the differentiation between blossom honeys and honeydew honeys, the EC of honey is strongly related to its organic acids, ash content and proteins [19]. Therefore, the higher the content of the latter, the higher the obtained conductivity. A high content of organic acid and salts increases the free acidity present in honey.

3.3.1. Mineral Content

Considering the average value of samples, the most abundant minerals were K followed by P, Na, Ca and Mg. The average value of K was 135.1 mg/100 g varying from a minimum of 43.1 mg/100 g to a maximum of 394.5 mg/100 g. P varied from 17.6 mg/100 g to 36.6 mg/100 g and had an average of 25.8 mg/100 g and Na, averaged 11.0 mg/100 g and ranged from 3.5 to 24.9 mg/100 g. Ca and Mg, had similar mean values of 7.5 mg/100 g and 5.5 mg/100 g, respectively, whereas Fe, Mn, Cu, Zn, were found with the lowest values ranging from the maximum value of 1.2 mg/100 g for Fe in some samples to values lower than 0.2 mg/100 g. Cd and Pb were always under the limit of detection. Normally, darker honeys had higher mineral content so heather honeys, chestnut honeys and honeydew honeys have the main values [20]. It has been proved by [21] that the botanical origin of honey has an impact on its elemental composition.

3.3.2. Total Phenolic and Flavonoid Content

The phenol content (TPC) and flavonoid content (TFC) are other interested parameters in honeys. TPC values ranged from 41.8 mg GAE/100 g to 128.3 mg GAE/100 g, with a mean value of 70.6 mg GAE/100 g, while TFC were in accordance with TPC and ranged from 2.3 mg QE/100 to 9.7 mg QE/100 g, with a mean value of 4.9 mg QE/100 g.

These results were included within the interval found by [22] on honey samples from the same region and close to those reported by [23] on honeys from Algeria. Furthermore, TPC values were included in the range reported by [24] on Italian honey. Regarding TFC it is possible to find a great variation in values depending on the methodology used, but these values are higher than other reported from Moroccan honey [25] and Czech honey [26].

Both parameters are related with botanical origin of honey and hence with color, mineral content and electrical conductivity [20]. Previous studies have demonstrated a strong relationship between the phenolic profile of different honey types and their antioxidant capacities [20,27,28]. Therefore, further study of biological activities of honey produced in Babors Kabylia, such as the antioxidant power of honey samples, would be very interesting and necessary to show the correlation between the botanical origin and the content of phenolic compounds.

The most critical values for the quality of these samples correspond to water content and diastase index. All the analyzed samples presented a water content below the limit established by international legislation. However, values were higher than those reported for honeys from semiarid areas of Algeria [29]. Indeed, honeys produced in dry regions should have lower moisture content than honeys produced in humid regions. Similar values were reported by [7,30] on honeys produced in a nearby region, as well as those produced in the Mediterranean coast [6], coinciding with the humid area of Algeria. In any cases, the studied samples had relatively high-water content that should be considered due to their relationship with stability and conservation of honey. Furthermore, high quality honeys should contain low HMF content and high Diastase index. Both parameters are thermosensitive and increase with aging and prolonged heating of honey [31]. Diastase content was frequently lower than the stablished limit in international legislation; however, HMF content does not exceed the limit. Similar values to those found in these sample was reported by [31] for Algerian honeys but higher values were also found [7]. It is essential to remark that water content and diastase index are dependent on several factors, including botanical and geographical origin, environmental and seasonal conditions, the degree of maturity and the beekeepers' handling during the honey harvest [32]. Studies on the common practices of honey management by beekeepers are necessary to determine the degree of influence on honey quality. In any case, the lack of professionalism combined with the quasi-total absence of training sessions related to beekeeping, hive maintenance, honey resources in the region and harvesting management, in addition to the propagation of Varroa disease and the absence of a legislative framework for quality recognition, can lead to problems linked to the quality of Algerian honey. In this context, to promote competitiveness with imported honey in the conventional market, it is recommended to promote knowledge of the characteristics of local honey and the participation of all stakeholders who can take actions to maintain the quality of the product.

3.4. Multivariate Analysis Applied to the Interpretation of Samples Characteristics and Typification of the Honey Samples

Data on the sensorial analysis, the main pollen types in the pollen spectra of samples and some physico-chemical parameters were used for PCA analysis. In total, 34 standardized variables were introduced to create the covariance matrix. The five first factors explained 92.0% of the variance of the data, and the first two 77.6%. The first principal component (PC1) represented 62.3% of the variance and had the main correlation coefficients with the parameters of visual color (0.843), Pfund color (0.968), TPC (0.976), TFC (0.959), and *Erica* (0.856). The second one (15.2% of the variance) had the highest positive correlation coefficient with *Genista* (0.936) and negative with *Hedysarum* (−0.733). Some of the variables are highly correlated and appeared together in the biplot (Figure 2a). At the right, sensorial perceptions such as vegetal odor and aroma, saltiness, bitterness and sourness taste and persistence of taste and smell, pollen types such as *Erica* or *Pistacia* and electrical conductivity, color, TPC, TFC and PH. At the left, are animal smell and sweetness taste together the pollen types *Eucalyptus*, *Myrtus*, *Rubus*, *Aster* and *Hedysarum*.

In the biplot of the two first components, the cases are dispersed but some similar samples appeared close (Figure 2b). At the right, it can be seen a group of samples with reference H (heather samples) and one Hd (honeydew sample). Up are the samples with the highest content in *Genista* pollen type and down, at the left, are the samples with the highest content of *Hedysarum*, *Aster* and the sample with *Rubus* as dominant pollen. In the center, there are a group of samples classified as polyfloral honeys (P).

Analyzing the results, main taxa for honey production in this hotspot for biodiversity were *Erica*, *Hedysarum*, *Genista* plants and in lesser extent *Myrtus*, *Eucalyptus* and some Asteraceae plants. The importance of these plants for honey production was mentioned before by [4,33,34].

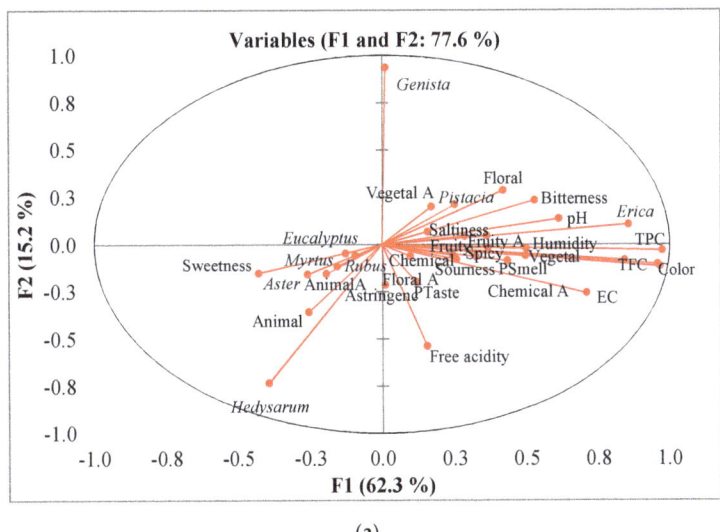

(a)

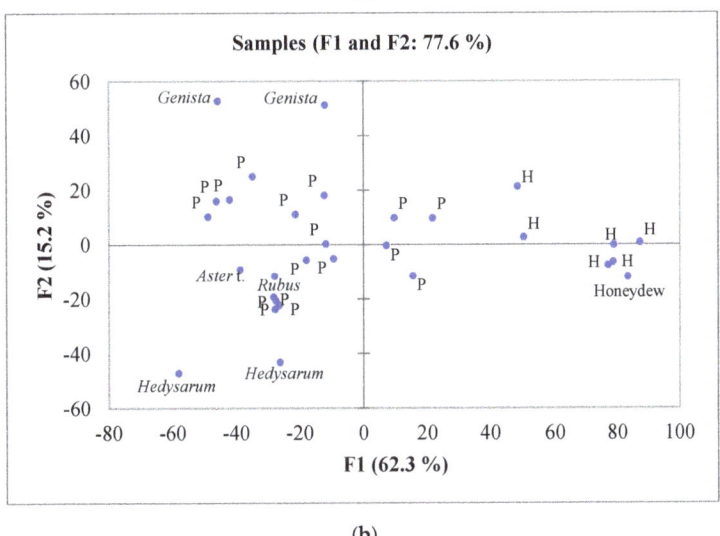

(b)

Figure 2. Principal Component Analysis (PCA). (**a**) Loading biplot of the variables included in the analysis, (**b**) Score biplot of the samples regarding component 1 and 2; P: Polyfloral, H: Heather.

The shrubs occupy a large surface in deforested areas and slopes of the mountains. These Mediterranean maquis has a great interest for apiculture, being the most representative species *Erica arborea* developing mainly on nutrient-poor acidic soils and *Erica multiflora* growing in calcareous soils. Both have been found in samples and give recognized and high valued unifloral honeys. Other plants such as *Ceratonia siliqua* L., *Pistacia*, *Calicotome*, *Spartium junceum* L., and *Genista*, and some Lamiaceae like *Thymus* or *Lavandula*, *Phillyrea latifolia* L. or different species of *Cistus* and *Quercus* form plant communities with *Erica*, so their pollen appears in honeys samples. It is worth mentioning the controversy on the apicultural value of Genisteae plants; while some species are considered that do not produce nectar (such as *Calicotome spinosa*), others are considered good nectariferous source for honeys such as *Retama sphaerocarpa* [35] or *Spartocytisus supranubius* (L.f.) Christ ex G.

Kunkel [36]. In the case of some *Genista* species, *Chamaecytisus* and *Retama* nectar secretion was reported [37]. Furthermore, through the pollen spectra of samples, the great diversity in pollen types can be observed, and further, the high number of pollen types by sample in accordance with biodiversity of the area. In this sense, melissopalynology is a key analysis with which to identify geographical origin of honeys and to determine melliferous species in the area where honey has been collected [38].

Mainly, the apiaries were in altitudes lower than 600 m above sea level and near to villages. In these areas, shrubs and spontaneous plants are most common; therefore, they predominate in the pollen spectra of honeys. Spontaneous species have high melliferous yield; among them, standing out are the Fabaceae plants such as *Hedysarum*, which is represented in the area for some species [39], many species of *Trifolium*, *Lotus* and other as *Securigera atlantica* Boiss and Reut., *Scorpiurus muricatus* L., *Tetragonolobus purpureus* Moench, *Astragalus echinatus* Murray. and *Medicago polymorpha* L. The importance of Fabaceae for honey production in the region was mentioned previously, either to produce monofloral honeys and/or to participate importantly in the production of polyfloral honeys [4,33,40–42]. Their importance also appears in studies carried out on honeys from the Mediterranean regions of Algeria [2,7,34,43].

Other important herbaceous plants for honey yield were Lamiaceae and Asteraceae; both are the most visited after Fabaceae [44]. The best represented pollen type in samples belonging to the family Lamiaceae was *Stachys*, which includes plants like *Stachys ocymastrum* or *Phlomis bovei* Noë and many other plants of this family; these are probably exclusives of the area. Something similar occurs with Asteraceae plants; the best represented pollen type was *Aster* t. with a pollen grain morphology similar for plants such as *Aster*, *Helichrysum stoechas* (L.) Moench, *Dittrichia viscosa* (L.) Greuter or *Phagnalon saxatile* (L.) *Cass.*, frequently occurring plants in the area. Other pollen types of Asteraceae were identified in samples in lesser extents, but due to the possible contribution of these plants to honey production and the fact that they normally appear underrepresented in the pollen spectra of honeys [35], a deep investigation about vegetation and plant distribution in the area could help in the identification of honeybee resources.

3.5. Characteristics of the Different Honey Types

The interpretation of the data allowed us to classify 12 samples, such as unifloral six heather honeys, two sulla (*Hedysarum*) honeys, two *Genista* honeys, one blackberry (*Rubus*) honey, one Asteraceae honey), one like honeydew honey and the rest, as polyfloral.

Table 4 shows the mean and standard deviation values for physicochemical and main pollen parameters. Polyfloral honeys showed the highest variation between samples in most of the variables as correspond to the heterogeneity of the group. A great variation in pollen content within polyfloral samples was noted, ranging from class I, II and III, although some heather honeys, *Hedysarum*, Asteraceae and honeydew were found to be in class III. Regarding color, polyfloral honeys were mainly light amber, as found in *Genista*, *Hedysarum*, Asteraceae and *Rubus* honeys. However, honeydew and heather honeys were the darkest honeys. The honeys presented similar water content that ranged from 16.8 to 19.6%. However, in terms of electrical conductivity and pH, honeydew honey presented the highest values. At the other extremes were *Genista* honey with the lowest electrical conductivity and *Rubus* honey with the lowest pH.

Table 4. Values (mean and standard deviation) of the studied variables considering the botanical origin of samples.

	Asteraceae (n = 1)	Rubus (n = 1)	Heather (n = 6)	Sulla (n = 2)	Genista (n = 2)	Honeydew (n = 1)	Polyfloral (n = 17)
Main Pollen	*Aster t*	*Rubus*	*Erica*	*Hedysarum*	*Genista t*		
	39.8	58.7	54.2 ± 16.9	58.8 ± 6.2	72.0 ± 2.6	18	27 ± 6
N. Pollen Types	31	19	19 ± 4	19 ± 6	20 ± 2		
PK (pollen/g)	13686	5292	7248 ± 4192	11110 ± 11338	4174 ± 2982	14465	5118 ± 5440
Maurizio classes	III	II	II, III	III	II	III	I, II, III
Humidity (%)	16.8	19.2	18.4 ± 0.6	18.1 ± 2.4	17.9 ± 0.7	19.6	17.9 ± 1.0
EC (mS/cm)	0.57	0.82	0.75 ± 0.2	0.54 ± 0.3	0.33 ± 0.0	1.35	0.55 ± 0.2
pH	3.7	3.5	4.0 ± 0.2	3.6 ± 0.2	3.9 ± 0.0	4.4	3.8 ± 0.1
Free acidity	33.5	35.5	34.7 ± 10.2	39.0 ± 2.8	15.5 ± 0.7	38.5	32.8 ± 8.2
	52	58	122 ± 7	53 ± 23	53 ± 22	135	66 ± 17
Color (mm Pfund)	Light Amber	Light Amber	Dark Amber	Light Amber	Light Amber	Dark Amber	Light Amber
Diastase Index	8.5	7.6	7.5 ± 2.2	14.8 ± 2.8	8.6 ± 2.4	16.9	12.5 ± 5.9
HMF (mg/100 g)	10.3	10.5	4.9 ± 0.8	7.0 ± 5.2	4.1 ± 0.5	5.4	11.5 ± 7.9
Polyphenols (mg/100 g)	49.6	53.9	109.8 ± 16.4	52.6 ± 7.8	52.6 ± 11.2	128.3	60.5 ± 15.8
Flavonoids (mg/100 g)	3.3	3.6	7.9 ± 1.3	3.8 ± 1.6	3.2 ± 1.3	8.7	4.1 ± 1.3
Na (mg/100 g)	9.4	11.9	12.0 ± 4.6	9.1 ± 5.1	6.9 ± 1.9	9.8	11.5 ± 5.9
K (mg/100 g)	122.4	170.5	193 ± 60.5	105.5 ± 71.4	66.9 ± 33.6	394.4	109.5 ± 55.7
P (mg/100 g)	27.6	23.9	25.7 ± 1.8	26.6 ± 4.8	20.0 ± 3.4	36.1	25.8 ± 4.1
Ca (mg/100 g)	6.2	4.0	12.9 ± 4.8	6.2 ± 3.2	2.9 ± 0.7	14.7	6.1 ± 1.6
Mg (mg/100 g)	2.1	2.9	10.7 ± 3.5	2.7 ± 1.7	2.9 ± 1.8	11.2	4.3 ± 2.6
Fe (mg/100 g)	0.2	0.1	0.3 ± 0.2	0.3 ± 0.1	0.15 ± 0.1	1.3	0.15 ± 0.1
Mn (mg/100 g)	<0.2	<0.2	<0.35	<0.2	<0.2	0.4	<0.2
Cu (mg/100 g)	<0.2	<0.2	<0.2	<0.2	<0.2	<0.2	<0.2
Zn (mg/100 g)	0.1	0.1	0.1 ± 0.1	0.1 ± 0.0	0.1 ± 0.0	0.1	0.1 ± 0.1
Cd (mg/100 g)	<0.1	<0.1	<0.1	<0.1	<0.1	<0.1	<0.1
Pb (mg/100 g)	<0.1	<0.1	<0.1	<0.1	<0.1	<0.1	<0.1

Although *Genista* honeys had the lowest free acidity, all honey types had similar values. The same occurred with the HMF content. Honeydew honey had the highest average value of diastase index, while the lowest level was observed in heather honeys. This difference could be explained by the fact that the concentration of diastase depends on several factors, including handling conditions, and could not be a reliable indicator of the honey's origin. It could therefore be explained by the fact that the honeys may have been heated or improperly stored. The TPC and TFC also varied depending on the type of honey. Honeydew honey followed by heather honeys had the higher values of these compounds. Finally, the mineral content was higher in honeydew honey. Secondly, the high content of K, Ca, Mg and Fe in heather honeys were noteworthy. However, followed by Honeydew honey, the P content was higher in Asteraceae and *Hedysarum* honeys. *Rubus* honey type showed a light amber color but also a low TPC and TFC, although this type of honey showed a wide variation in color and varied from extra light amber to dark amber [45]. It is therefore necessary to study more *Rubus* honeys from this region to verify this finding.

The sensorial profile of the different honey types can be seen in Figure 3. The six samples of heather honey presented a common pattern with dark color (brown), when crystallized, predominant vegetal smell (leaves and resin), some floral features and fruity (dry and tropical fruit). The persistence of the smell was low whereas the persistence of the taste was medium, and samples presented saltiness, bitterness and sourness in similar intensity. The aroma was mainly fruity and vegetal. The *Genista* samples had similar pattern in smell and aroma but different color, one had straw color and the other orange color (both were crystallized). The smell was mainly floral and in lesser extent fruity and aroma was mainly fruity. The persistence of smell was low, and the persistence of the taste was medium, the intensity of the sweetness was high and presented some saltiness and sourness notes. Sulla honeys were the most different regarding their sensorial profile, one was darker than the other and the smell and the aroma of one of them was animal (leather) while the other had floral and fruity smell. The persistence of the smell and the taste were similar in both cases. Asteraceae samples presented orange color, fruity smell and aroma with low intensity and persistence for smell, aroma and taste. Regarding the blackberry honey, highlighted the floral and fruity smell and the fruity aroma as well as the predominance of sweetness in taste. Finally, the honeydew sample had brown color, vegetal odor, fruity aroma and the highest persistence of smell.

The published scientific information about sensory profile of honey is scarce [12,35,46,47] and no information is available about sensorial properties of honey produced in Algerian country. The heather honeys were the most known and have similar sensorial profile to those previously described [35,48]. Regarding the sulla honeys were mainly described with floral smell but these samples were different in the smell having one of them fruity smell and the other animal smell, both with very low intensity [49]. Finally, the sensorial properties of Genista honeys are poorly described in literature, but its sensorial properties are near to other honey obtained from Fabaceae plants as *Spartocytisus supranubius* [36]. Other honey types are poorly known regarding their sensorial properties.

This is the first study focuses on the sensory properties of honey and their relationships with the botanical origin and the physicochemical properties of honeys from Babors Kabylia Region. Considering that this area is a hotspot for biodiversity and have a high potential for honey production, further studies could contribute to increase knowledge in the characteristics of the honey, the types that can be obtained and the most relevant plant species for honey yield and definitely contribute to the valorization of local beekeeping.

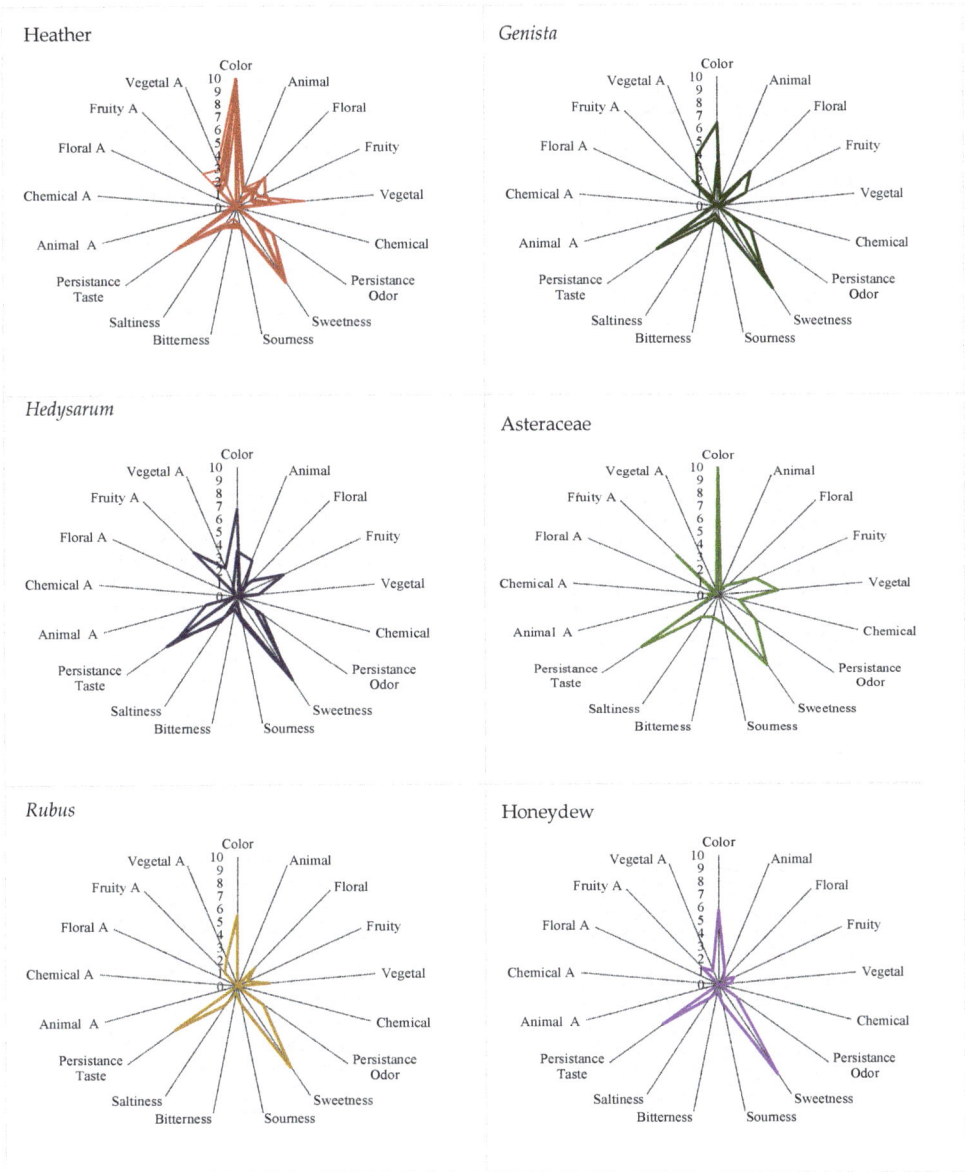

Figure 3. Chart diagrams for sensorial perceptions of monofloral honeys.

4. Conclusions

Honey samples collected from Babors Kabylia, which is an important biogeographical zone with great beekeeping potential, were analyzed to characterize the honey of this region using physico-chemical, melissopalynological and sensory parameters. The richness of the region in terms of melliferous plants has manifested through the production of a variety of monofloral honeys, including heather, *Genista*, sulla, Asteraceae and blackberry besides to the polyfloral honeys. Samples had a diverse sensory profile in terms of smell/flavor (vegetal, fruity and floral) highly appreciated by consumers. However, it is necessary to mention that some quality parameters related to hive management have been affected. It is therefore recommended to enhance knowledge on beekeeping management and to promote professional training for beekeepers to better manage the hive and improve both the quality and quantity of honey.

Author Contributions: Conceptualization, A.G. and F.B.; methodology, A.G., M.S.R.-F. and O.E.; formal analysis, A.G. and R.N.; data curation, M.S.R.-F. and O.E.; writing—original draft preparation, A.G. and L.H.; writing—review and editing, M.C.S.; supervision, F.B. All authors have read and agreed to the published version of the manuscript.

Funding: This research received no external funding.

Informed Consent Statement: Informed consent was obtained from all subjects involved in the study.

Acknowledgments: The authors express their gratitude to the beekeepers of the region. They would also like to thank the staff of the Apiculture Laboratory of the Animal Production Research Division of the National Institute for Agronomic Research—Algeria (INRAA) for their collaboration.

Conflicts of Interest: The authors declare no conflict of interest.

References

1. Hussein, M.H. A review of beekeeping in Arab countries. *Bee World* **2000**, *81*, 56–71. [CrossRef]
2. Boutabia, L.; Telailia, S.; Chefrour, A. Spectre pollinique de miels d'abeille (*Apis mellifera* L.) de la région d'El Tarf (Nord-Est algérien). *Livest. Res. Rural. Dev.* **2016**, *28*, 1–8.
3. Véla, E.; Benhouhou, S. Évaluation d'un nouveau point chaud de biodiversité végétale dans le Bassin méditerranéen (Afrique du Nord). *Comptes Rendus Biol.* **2007**, *330*, 589–605. [CrossRef]
4. Ouchemoukh, S.; Louaileche, H.; Schweitzer, P. Physicochemical characteristics and pollen spectrum of some Algerian honeys. *Food Control* **2007**, *18*, 52–58. [CrossRef]
5. Haderbache, L.; Mouna, B.; Arezki, M. Ziziphus lotus and Euphorbia bupleuroides Algerian honeys. *World Appl. Sci. J.* **2013**, *24*, 1536–1543.
6. Otmani, I.; Abdennour, C.; Dridi, A.; Kahalerras, L.; Salem, A.H. Characteristics of the bitter and sweet honey from Algeria Mediterranean coast. *Vet. World* **2019**, *12*, 551–557. [CrossRef]
7. Homrani, M.; Escuredo, O.; Rodríguez-Flores, M.S.; Fatiha, D.; Mohammed, B.; Homrani, A.; Seijo, M.C. Botanical Origin, Pollen Profile, and Physicochemical Properties of Algerian Honey from Different Bioclimatic Areas. *Foods* **2020**, *9*, 938. [CrossRef] [PubMed]
8. Tomczyk, M.; Tarapatskyy, M.; Dżugan, M. The influence of geographical origin on honey composition studied by Polish and Slovak honeys. *Cze. J. Food Sci.* **2019**, *37*, 232–238. [CrossRef]
9. di Marco, G.; Manfredini, A.; Leonardi, D.; Canuti, L.; Impei, S.; Gismondi, A.; Canini, A. Geographical, botanical and chemical profile of monofloral Italian honeys as food quality guarantee and territory brand. *Plant Biosyst. Int. J. Deal. All Asp. Plant Biol.* **2017**, *151*, 450–463. [CrossRef]
10. Gharzouli, R. Flore et Végétation de la Kabylie des Babors Etude Floristique et Phytosociologique des Groupements Forestiers et Post-Forestier des Djbels Takoucht, Adrar Ou-Mellal, Tababort et Babor. Ph.D. Thesis, Ferhat Abbas University, Sétif, Algeria, 2007.
11. Rana, S.; Mishra, M.; Yadav, D.; Subramani, S.K.; Katare, C.; Prasad, G. Medicinal uses of honey: A review on its benefits to human health. *Prog. Nutr.* **2018**, *20*, 5–14.
12. Marcazzan, G.L.; Caretta, C.M.; Marchese, C.M.; Piana, M.L. A review of methods for honey sensory analysis. *J. Apic. Res.* **2018**, *57*, 75–87. [CrossRef]
13. Louveaux, J.; Maurizio, A.; Vorwohl, G. Methods of Melissopalynology. *Bee World* **1978**, *59*, 139–157. [CrossRef]
14. Bogdanov, S.; Martin, P.; Lüllmann, C. Harmonized methods of the European honey commission. International Honey Commission. *Apidologie* **1997**, 1–59.
15. Singleton, V.L.; Rossi, J.A. Colorimetry of total phenolics with phosphomolybdic-phosphotungstic acid reagents. *Am. J. Enol. Vitic.* **1965**, *16*, 144–158.
16. Grand, A.A.; Vennat, B.; Pourrat, A.; Legret, P. Standardization of propolis extract and identification of principal constituents. *J. Pharm. Belg.* **1994**, *49*, 462–468.
17. Caroli, S.; Forte, G.; Iamiceli, A.L.; Galoppi, B. Determination of essential and potentially toxic trace elements in honey by inductively coupled plasma-based techniques. *Talanta* **1999**, *50*, 327–336. [CrossRef]
18. The Council of the European Union. Council Directive 2001/110/EC of 20 December 2001 relating to honey. *Off. J. Eur. Communities* **2002**, *L10*, 47–52.
19. Karabagias, I.K.; Badeka, A.V.; Kontakos, S.; Karabournioti, S.; Kontominas, M.G. Botanical discrimination of Greek unifloral honeys with physico-chemical and chemometric analyses. *Food Chem.* **2014**, *165*, 181–190. [CrossRef]
20. Escuredo, O.; Míguez, M.; González, M.F.; Seijo, M.C. Nutritional value and antioxidant activity of honeys produced in a European Atlantic area. *Food Chem.* **2013**, *138*, 851–856. [CrossRef]

21. Bilandžić, N.; Gajger, I.T.; Kosanoviʹc, M.; Calopek, B.; Sedak, M.; Kolanoviʹc, B.S.; Varenina, I.; Luburiʹc, D.B.; Varga, I.; Đokiʹc, M. Essential and toxic element concentrations in monofloral honeys from southern Croatia. *Food Chem.* **2017**, *234*, 245–253. [CrossRef]
22. Tafinine, Z.M.; Ouchemoukh, S.; Tamendjari, A. Antioxidant activity of some Algerian honey and propolis. *Ind. Crops Prod.* **2016**, *88*, 85–90. [CrossRef]
23. Dahmani, K.; Houdeib, J.B.; Zouambi, A.; Bendeddouche, B.; Fernández-Muiño, M.; Osés, S.M.; Sancho, M.T. Quality Attributes of Local and Imported Honeys Commercialized in Algeria. *J. Apic. Sci.* **2020**, *1*, 251–262. [CrossRef]
24. di Marco, G.; Gismondi, A.; Panzanella, L.; Canuti, L.; Impei, S.; Leonardi, D.; Canini, A. Botanical influence on phenolic profile and antioxidant level of Italian honeys. *J. Food Sci. Technol.* **2018**, *55*, 4042–4050. [CrossRef] [PubMed]
25. Bouyahya, A.; Abrini, J.; Touys, A.E.; Lagrouh, F.; Dakka, N.; Bakri, Y. Analyse phytochimique et évaluation de l'activité antioxydante des échantillons du miel marocain. *Phytothérapie* **2018**, *16*, S220–S224. [CrossRef]
26. Lachman, J.; Hejtmankova, A.; Sýkora, J.; Karban, J.; Orsak, M.; Rygerova, B. Contents of major phenolic and flavonoid antioxidants in selected Czech honey. *Czech. J. Food Sci.* **2010**, *28*, 412–426. [CrossRef]
27. Dong, R.; Zheng, Y.N.; Xu, B.J. Phenolic Profiles and Antioxidant Capacities of Chinese Unifloral Honeys from Different Botanical and Geographical Sources. *Food Bioprocess Technol.* **2013**, *6*, 762–770. [CrossRef]
28. Cheung, Y.; Meenu, M.; Yu, X.; Xu, B. Phenolic acids and flavonoids profiles of commercial honey from different floral sources and geographic sources. *Int. J. Food Prop.* **2019**, *22*, 290–308. [CrossRef]
29. Haouam, L.; Tahar, A.; Dailly, H.; Lahrichi, A.; Chaqroune, A.; Abdennour, C. Physicochemical properties and major elements contents of Algerian honeys from semi-arid regions. *Emir. J. Food Agric.* **2016**, *28*, 107–115. [CrossRef]
30. Tafinine, Z.M.; Ouchemoukh, S.; Bachir Bey, M.; Louaileche, H.; Tamendjari, A. Effect of storage on hydroxymethylfurfural (HMF) and color of some Algerian honey. *Int. Food Res. J.* **2018**, *25*, 1044–1050.
31. Haouam, L.; Dailly, H.; Bruneau, E.; Tahar, A. The quality of honeys influenced by the traditional heating method. *J. Microbiol. Biotechnol. Food Sci.* **2019**, *8*, 1276–1280.
32. Flores, M.S.R.; Escuredo, O.; Seijo, M.C. Assessment of physicochemical and antioxidant characteristics of Quercus pyrenaica honeydew honeys. *Food Chem.* **2015**, *166*, 101–106. [CrossRef] [PubMed]
33. Ouchemoukh, S.; Ouchemoukh, N.A.; Romero, M.G.; Aboud, F.; Giuseppe, A.; Gutierrez, A.F.; Carretero, A.S. Characterisation of phenolic compounds in Algerian honeys by RP-HPLC coupled to electrospray time-of-flight mass spectrometry. *LWT Food Sci. Technol.* **2017**, *85*, 460–469. [CrossRef]
34. Zerrouk, S.; Seijo, M.C.; Boughediri, L.; Escuredo, O.; Rodríguez-Flores, M.S. Palynological characterisation of Algerian honeys according to their geographical and botanical origin. *Grana* **2014**, *53*, 147–158. [CrossRef]
35. Oddo, L.P.P.; Piana, L.; Bogdanov, S.; Bentabol, A.; Gotsiou, P.; Kerkvliet, J.; Martin, P.; Morlot, M.; Ortiz Valbuena, A.; Ruoff, K.; et al. Botanical species giving unifloral honey in Europe. *Apidologie* **2004**, *35*, S82–S93. [CrossRef]
36. Bonvehi, J.S.; Manzanares, A.B.; Vilar, J.M.S. Quality evaluation of broom honey (*Spartocytisus supranubius* L.) produced in Tenerife (The Canary Islands). *J. Sci. Food Agric.* **2004**, *84*, 1097–1104. [CrossRef]
37. Galloni, M.; Cristofolini, G. Floral rewards and pollination in Cytiseae (Fabaceae). *Plant Syst. Evol.* **2003**, *238*, 127–137. [CrossRef]
38. Dimou, M.; Thrasyvoulou, A. Pollen analysis of honeybee rectum as a method to record the bee pollen flora of an area. *Apidologie* **2009**, *40*, 124–133. [CrossRef]
39. Berrekia, R.A.; Abdelguerfi, A.; Bounaga, N.; Guittonneau, G.G. Répartition des espèces spontanées du genre *Hedysarum* selon certains facteurs du milieu en Algérie. *Fourrages* **1991**, *126*, 187–207.
40. Chefrour, E.; Draiaia, R.; Tahar, A.; Kaki, Y.A.; Bennadja, S.; Battesti, M.J. Physicochemical characteristics and pollen spectrum of some north-east Algerian honeys. *Afr. J. Food Agric. Nutr. Dev.* **2009**, *9*, 1276–1293. [CrossRef]
41. Makhloufi, C.; Kerkvliet, J.; Schweitzer, P. Characterisation of some monofloral Algerian honeys by pollen analysis. *Grana* **2015**, *54*, 156–166. [CrossRef]
42. Nair, S.; Meddah, B.; Aoues, A. Melissopalynological Characterization of North Algerian Honeys. *Foods* **2013**, *2*, 83–89. [CrossRef] [PubMed]
43. Hamel, T.; Boulemtafes, A. Plants foraged by bees in the Edough peninsula (Northeast Algeria). *Livest. Res. Rural Dev.* **2017**, *29*, 1–13.
44. Hamel, T.; Bellili, M.; Hamza, A.M.; Boulemtafes, A. Nouvelle contribution à l'étude de la flore mellifère et caractérisation pollinique de miels de la Numidie (Nord-Est algérien). *Livest. Res. Rural Dev.* **2019**, *31*, 1–24.
45. Escuredo, O.; Silva, L.R.; Valentão, P.; Seijo, M.C.; Andrade, P.B. Assessing Rubus honey value: Pollen and phenolic compounds content and antibacterial capacity. *Food Chem.* **2012**, *130*, 671–678. [CrossRef]
46. Piana, M.L.; Oddo, L.P.; Bentabol, A.; Bruneau, E.; Bogdanov, S.; Declerck, C.G. Sensory analysis applied to honey: State of the art. *Apidologie* **2004**, *35*, S26–S37. [CrossRef]
47. Araujo, D.; Cacho, P.R.P.; Serrano, S.; Palomares, R.D.; Soldevilla, H.G. Sensory Profile and Physico-Chemical Properties of Artisanal Honey from Zulia, Venezuela. *Foods* **2020**, *9*, 339. [CrossRef]
48. Flores, M.S.R.; Falcão, S.I.; Escuredo, O.; Seijo, M.C.; Vilas-Boas, M. Description of the volatile fraction of Erica honey from the northwest of the Iberian Peninsula. *Food Chem.* **2021**, *336*, 127758. [CrossRef]
49. Floris, I.; Satta, A.; Ruiu, L. Honeys of Sardinia (Italy). *J. Apic. Res.* **2007**, *46*, 198–209. [CrossRef]

Article

Demanding New Honey Qualitative Standard Based on Antibacterial Activity

Marcela Bucekova [†], Veronika Bugarova [†], Jana Godocikova and Juraj Majtan *

Laboratory of Apidology and Apitherapy, Department of Genetics, Institute of Molecular Biology, Slovak Academy of Sciences, Dubravska cesta 21, 845 51 Bratislava, Slovakia; marcela.bucekova@gmail.com (M.B.); veronika.bugarova@savba.sk (V.B.); jana.godocikova@savba.sk (J.G.)
* Correspondence: juraj.majtan@savba.sk; Tel.: +421-903869413
† These authors contributed equally to this work.

Received: 14 August 2020; Accepted: 8 September 2020; Published: 9 September 2020

Abstract: Honey is a functional food with health-beneficial properties and it is already used as a medical device in wound care management. Whether ingested orally or applied topically, honey must fulfill the requirements of international standards based on physicochemical characteristics. However, there is an urgent need for some additional standards reflecting biological properties. The aim of the study was to evaluate the antibacterial activity of 36 commercial honey samples purchased from supermarkets and local food shops and compare their efficacy to that of three honey samples from local beekeepers and three types of medical-grade honey. Furthermore, the hydrogen peroxide (H_2O_2) content and protein profile were assessed in all honey samples. Analysis of the antibacterial activity of commercial honeys revealed that 44% of tested samples exhibited low antibacterial activity, identical to the activity of artificial honey (sugars only). There was a significant correlation between the overall antibacterial activity and H_2O_2 content of honey samples. However, in some cases, honey samples exhibited high antibacterial activity while generating low levels of H_2O_2 and vice versa. Honey samples from local beekeepers showed superior antibacterial activity compared to medical-grade honeys. The antibacterial activity of honey can be easily altered by adulteration, thermal treatment or prolonged storage, and therefore it fulfils strict criteria to be suitable new additional quality standard.

Keywords: hydrogen peroxide; functional food; bacterial pathogen; commercial honey; quality standard

1. Introduction

Honey, a traditional sweetener, is considered as a functional food, and several recent clinical studies have proved its health-beneficial properties such as improving lipid profile [1], reducing postoperative pain [2] and inflammation [3] and modulating of hypertension [4]. Besides its oral consumption, honey has successfully been used topically in the treatment of a broad spectrum of surgical and chronic wounds [5,6] and mucositis [7] as well as herpes simplex labialis [8]. Whether applied topically or ingested orally, honey must fulfil all the requirements of international standards and possesses proven biological activity. In addition, honey has to be sterilised by gamma radiation when used in wound care management. In the case of medical usage, only honey of high quality and with guaranteed biological activity should be a part of honey-based medical products.

Several quality standards have been recognised and listed as 'Current international honey standards' which are specified in the European Honey Directive (2002) [9] as well as in the Codex Alimentarius Standard for Honey (2001) [10]. Honey samples meeting all strictly defined composition criteria, including mainly moisture, sucrose and hydroxymethylfurfural (HMF) content, are recommended for human consumption and can be placed on the market.

On the other hand, the range of particular current honey standards is too wide and, in most cases, adulterated and/or heat-processed honey is still within the range for tested criteria and therefore classified as honey with proved quality. Furthermore, most importantly, none of the above-mentioned criteria are related to the biological activity of honey.

One of the most important and well-described aspects of honey's biological activity is its antibacterial activity that is mediated via multiple mechanisms of action such as osmotic pressure, low pH value and water activity, and disruption of bacterial cell membranes due to the presence of antibacterial peptide defensin-1.

Bee defensin-1, a regular but quantitatively variable antibacterial component of honey [11], is mainly effective against Gram-positive bacteria [12–14]. Furthermore, defensin-1 possesses antibiofilm activity against established multi-species biofilm [15] and exhibits wound healing properties [16].

After dilution, the antibacterial activity of honey is mainly mediated via enzymatically generated hydrogen peroxide (H_2O_2) in the diluted honey [17], excluding manuka honey where H_2O_2 is not accumulated [18]. This activity can be negatively affected by uncontrolled thermal processing or prolonged storage [19]. The major antibacterial compound in manuka honey is methylglyoxal (MGO) [20,21] that is primarily able to inhibit the growth of Gram-positive bacteria [22,23]. On the other hand, MGO in manuka honey seems to be ineffective against *Pseudomonas aeruginosa* [24]. Similar to that of different blossom honeys, the total antibacterial action of manuka honey is based on multiple mechanisms of action rather than single components. Therefore, the antibacterial potential of honey could be a suitable new additional international quality standard.

Our recent study aimed to determine the antibacterial efficacy of 233 different blossom honey samples collected in Slovakia [25]. Linden honeys showed the greatest antibacterial efficacy followed by sunflower, multi-floral and acacia honey samples. The lowest antibacterial activity was assessed in rapeseed honey samples.

The goal of the study was to (i) evaluate the antibacterial potential of commercial honeys purchased from supermarkets ($n = 19$) and local shops ($n = 17$) in Slovakia, (ii) characterise the protein profile of honey samples and (iii) determine the overall H_2O_2 content in honey samples. Moreover, antibacterial activity together with the capability to generate H_2O_2 was assessed in three honey samples from local beekeepers and in three medical-grade honey samples in order to compare the overall antibacterial potential of different honeys.

2. Materials and Methods

2.1. Honey Samples

A total of 36 commercial samples of honey purchased from supermarkets in 2017 ($n = 19$) or local food shops in 2019 ($n = 17$) were evaluated. Most of the samples, particularly from those from local food shops, declared the botanical and geographical origin on the label (Table 1).

Three honey samples (acacia, linden and honeydew) from local beekeepers and three different commercial products based on medical-grade honeys: Vivamel® (Toasama, Domzale, Slovenia), Revamil® (Bfactory, Rhenen, Netherlands) and Activon Tube® (Advancis Medical, Nottingham, UK) were also evaluated. All samples were immediately stored at 4 °C in the dark.

As a negative control artificial honey was prepared by dissolving 39 g D-fructose, 31 g D-glucose, 8 g maltose, 3 g sucrose, and 19 g distilled water, as described elsewhere [26].

Table 1. Commercial honey samples tested in the study.

	No. of Honey Samples	Type of Honey	Geographical Origin of Honey
Supermarkets	1	multi-floral	Unknown *
	2	acacia	Slovakia
	3	honeydew	Slovakia
	4	linden	Unknown *
	5	multi-floral	Unknown *
	6	honeydew	Unknown *
	7	multi-floral	Unknown *
	8	multi-floral	Unknown *
	9	forest	Unknown *
	10	linden	Unknown *
	11	multi-floral	Unknown *
	12	acacia	Unknown *
	13	forest	Unknown *
	14	multi-floral	Unknown *
	15	multi-floral	Unknown *
	16	rapeseed	Slovakia
	17	multi-floral	Slovakia
	18	acacia	Slovakia
	19	honeydew	Slovakia
Local food shops	20	acacia	Slovakia
	21	multi-floral	Slovakia
	22	forest	Slovakia
	23	linden	Slovakia
	24	honeydew	Slovakia
	25	rapeseed	Slovakia
	26	multi-floral	Slovakia
	27	acacia	Slovakia
	28	honeydew	Slovakia
	29	honeydew	Slovakia
	30	multi-floral	Slovakia
	31	acacia	Slovakia
	32	multi-floral	Slovakia
	33	linden	Slovakia
	34	multi-floral	Slovakia
	35	multi-floral	Slovakia
	36	multi-floral	Slovakia

* honey originated from European union/non-European union countries.

2.2. Microorganisms

The tested bacterial isolates *Pseudomonas aeruginosa* CCM1960 and *Staphylococcus aureus* CCM4223 were acquired from the Department of Medical Microbiology at the Slovak Medical University in Bratislava, Slovakia.

2.3. Determining the Antibacterial Activity

The honey samples were subjected to an antibacterial minimum inhibitory concentration (MIC) assay to determine the antibacterial activity against *P. aeruginosa* and *S. aureus* following the method of Bucekova et al. [27]. Bacteria were cultured in Mueller-Hinton broth (MHB) at 37 °C overnight. Bacterial culture was suspended in phosphate-buffered saline (PBS), with a pH of 7.2, and the turbidity of the suspension was adjusted to 10^8 colony-forming units (CFU)/mL and diluted with MHB medium (pH 7.3 ± 0.1) to a final concentration of 10^6 CFU/mL. The final volume in each well of sterile 96-well polystyrene U-shaped plates (Sarstedt, Germany) was 100 µL, consisting of 90 µL of sterile MHB medium or diluted honey sample and 10 µL of bacterial suspension. After 18 h incubation at 37 °C and 1250 rpm, the inhibition of bacterial growth was determined by monitoring the optical density at

490 nm using a Synergy HT microplate reader (BioTek Instruments, Winooski, VT, USA). The final MIC values correspond to the lowest concentrations of honey that completely inhibited bacterial growth. All the tests were performed in triplicate and repeated three times.

Each honey sample dilution was prepared from a 50% honey solution (w/w in MHB medium) by further dilution with the MHB medium, resulting in final concentrations of 40%, 35%, 30%, 25%, 20%, 18%, 16%, 14%, 12%, 10%, 8%, 6% and 4%.

2.4. Determining the H_2O_2 Content

The H_2O_2 content in the honey samples was determined using a Megazyme GOX assay kit (Megazyme International Ireland Ltd., Bray, Ireland) based on the release of H_2O_2 after glucose oxidase catalysis of the oxidation of β-D-glucose to D-glucono-δ-lactone. As a standard, 9.8–312.5 µM diluted H_2O_2 was used. Honey solutions (40% w/w in 0.1 M potassium phosphate buffer, pH 7.0) were prepared and immediately measured. Each honey sample and H_2O_2 standard was tested in duplicate in a 96-well microplate. The absorbance of the reaction was measured at 510 nm using a Synergy HT microplate reader (BioTek Instruments, Winooski, VT, USA).

2.5. Determining the Protein Profile of Honey Samples

For protein determination, 15 µL aliquots of diluted honey samples (50% w/w in distilled water) were loaded on 12% SDS-PAGE gels and separated using a Mini-Protean II electrophoresis cell (Bio-Rad, Hercules, CA, USA). Protein content was assessed after gel staining with Coomassie Brilliant Blue R-250 (Sigma-Aldrich, Darmstadt, Germany) or Serva Blue (Serva, Heidelberg, Germany).

2.6. Statistical Analysis

The Pearson correlation test was used to analyse the correlation between the antibacterial activity and the H_2O_2 content in the honey. The data are expressed as mean values with the standard deviation (SD). Data with p values smaller than 0.05 were considered statistically significant. GraphPad Prism was used to perform all the statistical analyses (GraphPad Software Inc., La Jolla, CA, USA).

3. Results

3.1. Antibacterial Activity of Commercial Honey Samples

Antibacterial activity was determined in commercial ($n = 36$), local beekeeper ($n = 3$) and medical-grade ($n = 3$) honey samples as well as in artificial honey. The activity of the commercial honey samples against bacteria was expressed as an MIC value (Figure 1). Commercial honey samples from local food shops (Nos. 20–36) exhibited a greater antibacterial effect compared to samples from supermarkets (Nos. 1–19). The MIC values of about 50% of supermarket honey samples against *S. aureus* were identical to those of artificial honey, compared to 25% of honey samples from local food shops. The average MIC value of supermarket and local food shop honey samples against *S. aureus* was 28.6% and 19.2%, respectively. The highest antibacterial activity was exhibited by honey sample No. 29 with an MIC value of 4% against *S. aureus*.

Artificial honey at a concentration of 25% was able to inhibit the bacterial growth of *P. aeruginosa*. Due to the high susceptibility of *P. aeruginosa* to the sugar content in honey samples, this bacterium is not a suitable model for testing the antibacterial activity of honey.

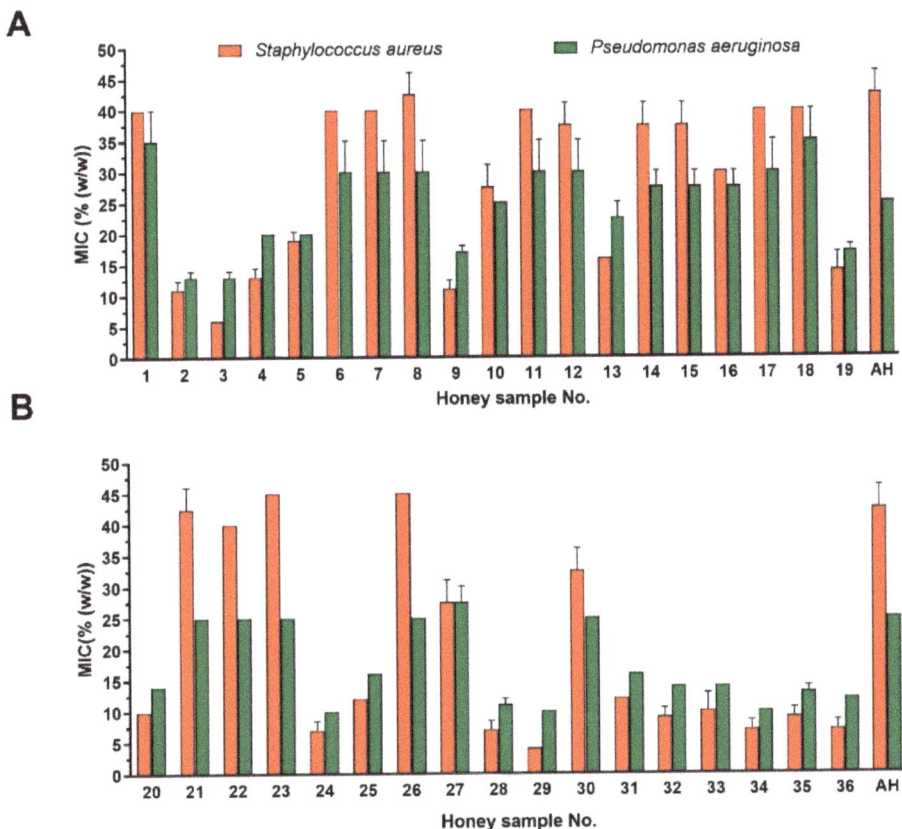

Figure 1. Antibacterial activity of the commercial honey samples ($n = 36$) purchased in Slovakia. The antibacterial activity was evaluated against two bacterial pathogens by a minimum inhibitory concentration (MIC) assay. (**A**) Honey samples purchased from supermarket ($n = 19$) (**B**) Honey samples purchased from local food shop ($n = 17$). The data are expressed as the mean values with the standard deviation (SD). AH—artificial honey.

3.2. H_2O_2 Content of Commercial Honey Samples

H_2O_2, a major antibacterial compound in honey, was determined in 36 commercial honey samples (Figure 2). The average H_2O_2 content in supermarket and local food shop honey samples was 256 and 565 µM, respectively. Honey samples purchased in local food shops were more potent in H_2O_2 production (Figure 2B). Three samples from supermarkets (sample Nos. 10, 11 and 12) accumulated very low levels of H_2O_2, at concentration below 50 µM (Figure 2A).

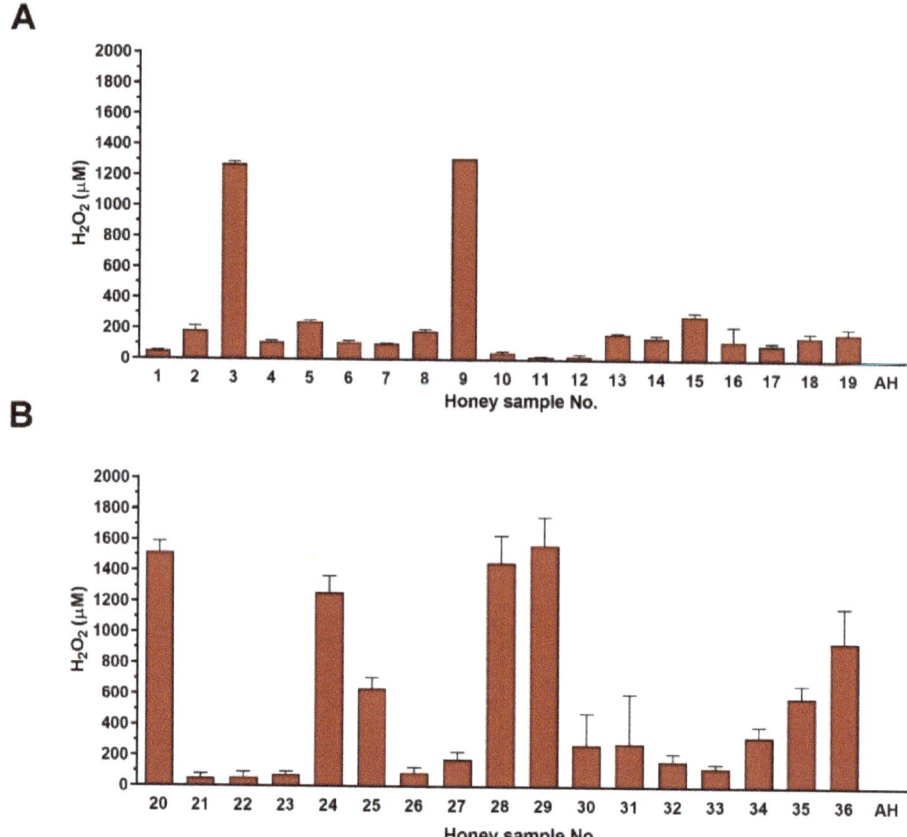

Figure 2. Hydrogen peroxide (H_2O_2) content in commercial honey samples ($n = 36$). (**A**) supermarket honey samples ($n = 19$), (**B**) local food shop honey samples ($n = 17$). Determination of the H_2O_2 content was carried out in diluted honey samples (40% (w/w)) after 24 h of incubation at 37 °C. The data are expressed as the mean values with the standard deviation (SD). AH—artificial honey.

The correlation analysis revealed a significant correlation between the H_2O_2 values and the antibacterial activity of both supermarket and local shop honey samples against *S. aureus* ($r = -0.582$, $p < 0.01$; $r = -0.648$, $p < 0.01$) as well as *P. aeruginosa* ($r = -0.584$, $p < 0.001$; $r = -0.682$, $p < 0.01$) (Figure 3).

Despite a significant correlation between the content of H_2O_2 and the antibacterial activity of all commercial honeys was revealed, some of tested honey samples (for example Nos. 19 and 33) did not show any correlation among MICs and H_2O_2 content.

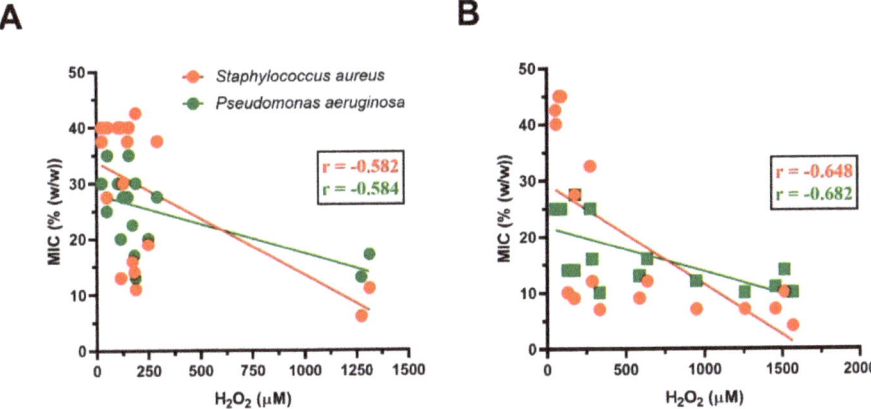

Figure 3. The relationship between the content of H_2O_2 and the antibacterial activity of commercial honeys: (**A**) supermarket honey samples ($n = 19$), (**B**) local food shop honey samples ($n = 17$) against two bacterial isolates. A Pearson correlation test was used for the correlation analysis.

3.3. Antibacterial Activity and H_2O_2 Content of Local and Medical-Grade Honey Samples

Samples of three different types of honey (honeydew, linden and acacia) from local beekeepers and three medical-grade honeys were used for the determination of antibacterial activity. The antibacterial activity of these samples is shown in Figure 4A. All honeys from local beekeepers exhibited high antibacterial activity, with average MIC values ranging from 7% to 12% depending on the bacterium. The antibacterial effect of local honey samples was even higher than that of medical-grade honeys where MIC values were in the range from 9% to 22.5%. Among the medical-grade honeys tested, the lowest antibacterial activity was documented for Revamil®. The local honeys, honeydew and acacia honey samples in particular, produced a high level of H_2O_2 (Figure 4B). Interestingly, the level was significantly higher than in medical-grade honeys. Activon Tube®, a manuka honey based-wound care product, did not accumulate H_2O_2 due to inactivated glucose oxidase enzyme. Revamil®, a blossom honey-based product, produced a low level of H_2O_2 and thus showed moderate antibacterial activity.

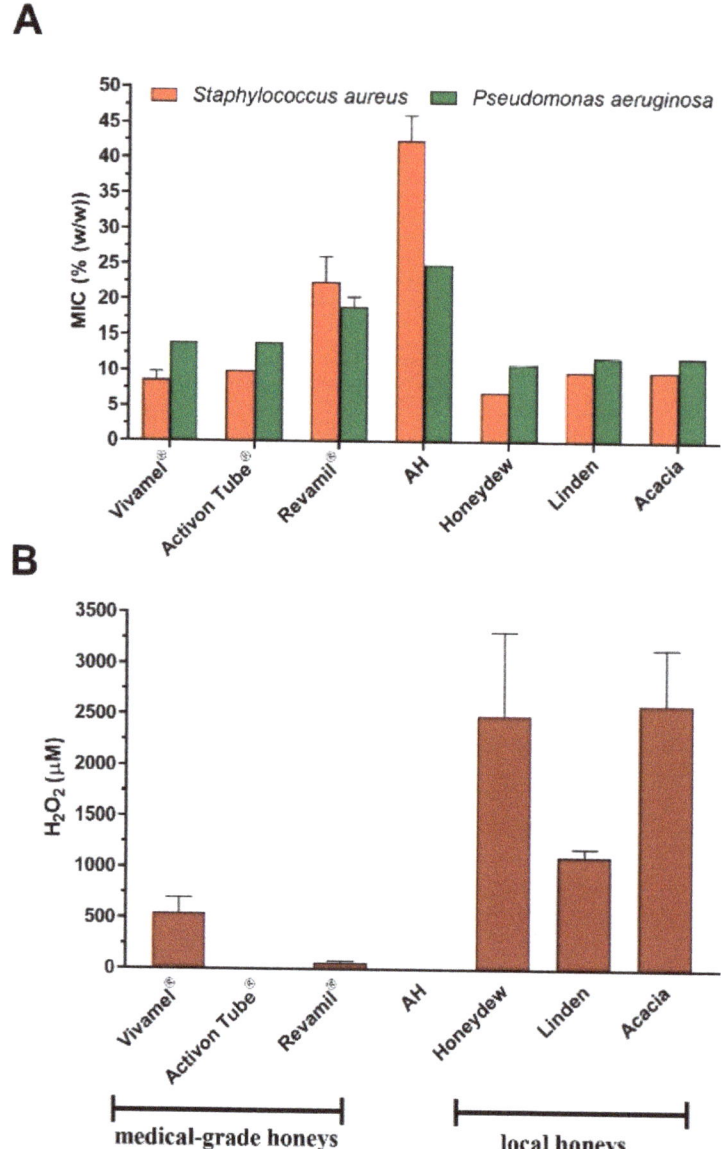

Figure 4. Antibacterial activity and the content of H_2O_2 in the honey samples from local beekeepers ($n = 3$) and medical-grade honey ($n = 3$). (**A**) The antibacterial activity was evaluated against two bacterial pathogens by a minimum inhibitory concentration (MIC) assay. (**B**) Determination of the H_2O_2 content was carried out in diluted honey samples (40% (w/w) after 24 h of incubation at 37 °C. The data are expressed as mean values with the standard deviation (SD).

4. Discussion

Natural functional foods including honey with diverse biological activity have become very popular among consumers and a plethora of studies have discovered a number of bioactive ingredients that can provide a beneficial effect on human health. However, rising demand for honey and natural healthy foods in general causes the limited availability and high price of honey and thus honey has become the object of adulteration. International honey standards need to be updated since nowadays more adulterated honeys meet currently defined limits in term of quality. Furthermore, industrial processing of honey such as liquefaction or thermal treatment causes an increase in HMF, the major indicator of honey freshness and overheating during processing, but to values below the maximum permissible level of 40 mg/kg of honey [28–31]. Similarly, another standard, diastase activity (DN), also changes after thermal processing but, in most cases, the minimum permissible level of 8 is maintained [29,32]. As mentioned before, international honey standards do not take into account the biological effects of honey, although some of the effects including the antibacterial effect are well described and have proved to be sensitive parameters. Antibacterial activity is also the sole criterion for selecting honey for medical usage.

In this study, we demonstrated that the antibacterial activity of commercial honeys purchased in supermarkets and local food shops is not uniform. Unfortunately, the antibacterial activity of more than 40% of the tested commercial honeys was identical to that of artificial honey. According to protein profile analysis of all honey samples, only three samples (Nos. 10, 11 and 12) showed a very low protein content (Figure S1). Missing crucial protein components, including bee-derived glucose oxidase, in these three samples resulted in negligible accumulation of H_2O_2. The typical protein profile was recognised in the rest of the honey samples, showing major royal jelly protein 1 (MRJP1) as the dominant honey protein (Figures S1 and S2) with molecular weight of 55 kDa.

The MRJP1 protein is multifunctional and has a nutritional function in larval jelly [33] and a presumed function in the bee brain that is associated with learning ability [34]. It also acts as a precursor protein of short antimicrobial peptide jelleines [35], however, its direct antibacterial activity is questioned [36,37].

Based on present and previous results [17], H_2O_2 content is not a suitable parameter for determining honey quality, even though a statistically significant correlation between antibacterial activity and H_2O_2 content was calculated. In some cases, honey samples exhibited high antibacterial activity while generating low levels of H_2O_2 and vice versa. Similarly, analysis of Polish honey samples revealed that some samples with the highest antibacterial activity were characterised by low levels of H_2O_2 [38]. In addition, we have recently showed that linden Slovak honey samples, which exhibited the strongest antibacterial effect, showed weak or no correlation between the antibacterial activity and H_2O_2 content [25]. Therefore, we assume that some certain types of honey (e.g., linden honey) may contain either non-peroxide antibacterial factors or specific minor components which act synergistically with other components to increase honey antibacterial potential.

Therefore, the antibacterial activity of honey depends on the presence of H_2O_2 but its concentration can vary from honey to honey and the presence of some additional minor constituents such as polyphenols is essential for a pronounced antibacterial effect. Honey as a highly complex product contains various known and yet unknown compounds of botanic origin that might alter H_2O_2 levels. In addition, the interaction of H_2O_2 with some particular polyphenols and/or other minor constituents can result in augmentation of honey antibacterial activity [17]. A very recent study by Brudzinski (2020) discussed the role of the colloidal structure of honey in H_2O_2 production, suggesting a relationship between the concentration of macromolecules, the propensity to form colloidal assemblies, H_2O_2 production and antibacterial effect [39].

In this study, three different types of medical-grade honey-based wound care products were evaluated. All products consist of 100% honey and no other additives were added. In contrast to natural honey, medical-grade honey undergoes a process of sterilisation, usually by gamma radiation. Gamma radiation of honey eliminates vegetative microbial cells as well as microbial spores [40–42] without affecting the overall antibacterial activity of honey [40,42–45]. However, it is likely that the whole process of medical-grade honey manufacture, including tubing, packing, sterilising and storing processes, results in a loss of some antibacterial activity. Taking into account the entire manufacturing process, it is necessary to monitor the antibacterial activity of processed honey at each manufacturing step.

In the case of manuka honey, the Unique Manuka Factor (UMF) grading system has been established, reflecting the concentration of MGO, a major antibacterial compound found in manuka honey. A very recent study [46] demonstrated that UMF grade, as an indicator of the antibacterial potential of manuka honey, surprisingly, does not correlate with the antibacterial efficacy of manuka honey, supposedly due to changes in MGO content over time. Therefore, the UMF grading system seems to be unreliable and there is a need to develop a more suitable system. Furthermore, UMF grading is solely based on the agar well diffusion assay using *S. aureus* as a model bacterium [47]. Nowadays, this is considered to have relatively low sensitivity and the obtained results need to be confirmed/updated by a more sensitive broth microdilution assay, a preferred method for determining the antibacterial effect of complex natural products. Lastly, a few recent studies have raised the issue of the dominant role of MGO in the antibacterial effect of manuka honey [24,46].

Last but not least, determination of honey antibacterial activity could give beekeepers a competitive advantage due to the high demand for biologically active honey. Detrimental changes in the antibacterial effect of honey induced by prolonged storage or uncontrolled heating at high temperature may also negatively affect other biological effects of honey such as antioxidant, anti-inflammatory and wound healing activity. Many of these types of activity are mediated by polyphenolic compounds which are a light/heat sensitive group of biologically active compounds [48]. Therefore, it is likely that honey with low antibacterial activity, identical to artificial honey, loses most of its health-beneficial properties and acts solely as a sweetener.

5. Conclusions

In conclusion, analysis of the antibacterial activity of commercial honeys purchased in Slovakia revealed that more than 40% of the samples tested exhibited low antibacterial activity, identical to that of artificial honey (sugars only). There was a significant correlation between the overall antibacterial activity and H_2O_2 content of honey samples. However, in some cases, honey samples exhibited high antibacterial activity while generating low levels of H_2O_2 and vice versa. Honey samples from local beekeepers showed superior antibacterial activity compared to medical-grade honeys which are recommended for the treatment of infected wounds. The antibacterial activity of honey, as one of its well-described biological properties, should be used as an additional quality standard reflecting its biological properties. Further studies are needed to characterise the effect of long-term storage as well as heat processing on antibacterial activity of honey in order to improve this new potential quality standard.

Supplementary Materials: The following are available online at http://www.mdpi.com/2304-8158/9/9/1263/s1, Figure S1: Protein profile of commercial honey samples (n = 19) from supermarkets and three samples from local beekeepers, Figure S2: Protein profile of commercial honey samples (n = 17) from local food shops.

Author Contributions: Conceptualization, J.M.; Methodology, M.B., V.B., J.G.; Formal Analysis, J.G., V.B., M.B.; Investigation, J.M., M.B.; Resources, J.M.; Writing—Original Draft Preparation, J.M.; Writing—Review & Editing, J.M., M.B., V.B., J.G.; Visualization, J.M., Funding Acquisition, J.M. All authors have read and agreed to the published version of the manuscript.

Funding: This research was supported by the Scientific Grant Agency of the Ministry of Education of the Slovak Republic and the Slovak Academy of Sciences VEGA 2/0004/18.

Conflicts of Interest: The authors declare no conflict of interest.

References

1. Rasad, H.; Entezari, M.H.; Ghadiri, E.; Mahaki, B.; Pahlavani, N. The effect of honey consumption compared with sucrose on lipid profile in young healthy subjects (randomized clinical trial). *Clin. Nutr.* **2018**, *26*, 8–12.
2. Geißler, K.; Schulze, M.; Inhestern, J.; Meißner, W.; Guntinas-Lichius, O. The effect of adjuvant oral application of honey in the management of postoperative pain after tonsillectomy in adults: A pilot study. *PLoS ONE* **2020**, *15*, e0228481.
3. Ghazali, W.S.; Romli, A.C.; Mohamed, M. Effects of honey supplementation on inflammatory markers among chronic smokers: A randomized controlled trial. *BMC Complement. Altern. Med.* **2017**, *17*, 175.
4. Ab Wahab, S.Z.; Hussain, N.H.N.; Zakaria, R.; Kadir, A.A.; Mohamed, N.; Tohit, N.M.; Norhayati, M.N.; Hassan, I.I. Long-term effects of honey on cardiovascular parameters and anthropometric measurements of postmenopausal women. *Complement. Ther. Med.* **2018**, *41*, 154–160.
5. Goharshenasan, P.; Amini, S.; Atria, A.; Abtahi, H.; Khorasani, G. Topical application of honey on surgical wounds: A randomized clinical trial. *Complement. Med. Res.* **2016**, *23*, 12–15.
6. Fox, C. Honey as a dressing for chronic wounds in adults. *Br. J. Community Nurs.* **2002**, *7*, 530–534.
7. Ramsay, E.I.; Rao, S.; Madathil, L.; Hegde, S.K.; Baliga-Rao, M.P.; George, T.; Baliga, M.S. Honey in oral health and care: A mini review. *J. Oral Biosci.* **2019**, *61*, 32–36.
8. Semprini, A.; Singer, J.; Braithwaite, I.; Shortt, N.; Thayabaran, D.; McConnell, M.; Weatherall, M.; Beasley, R. Kanuka honey versus aciclovir for the topical treatment of herpes simplex labialis: A randomised controlled trial. *BMJ Open* **2019**, *9*, e026201.
9. EU Council. Council Directive 2001/110/CE Concerning Honey. *Off. J. Eur. Communities* **2002**, *L10*, 47–52.
10. Codex Alimentarius Commission. *Codex Standards for Honey. Codex Stan 12-1981, Rev. 1 (1987), Rev. 2 (2001)*; Codex Alimentarius Commission: Rome, Italy, 2001.
11. Valachova, I.; Bucekova, M.; Majtan, J. Quantification of bee-derived defensin-1 in honey by competitive enzyme-linked immunosorbent assay, a new approach in honey quality control. *Czech J. Food Sci.* **2016**, *34*, 233–243.
12. Fujiwara, S.; Imai, J.; Fujiwara, M.; Yaeshima, T.; Kawashima, T.; Kobayashi, K. A potent antibacterial protein in royal jelly. Purification and determination of the primary structure of royalisin. *J. Biol. Chem.* **1990**, *265*, 11333–11337.
13. Bachanova, K.; Klaudiny, J.; Kopernicky, J.; Simuth, J. Identification of honeybee peptide active against *Paenibacillus larvae* larvae through bacterial growth-inhibition assay on polyacrylamide gel. *Apidologie* **2002**, *33*, 259–269.
14. Shen, L.; Liu, D.; Li, M.; Jin, F.; Din, M.; Parnell, L.D.; Lai, C.Q. Mechanism of action of recombinant acc-royalisin from royal jelly of Asian honeybee against gram-positive bacteria. *PLoS ONE* **2012**, *7*, e47194.
15. Sojka, M.; Valachova, I.; Bucekova, M.; Majtan, J. Antibiofilm efficacy of honey and bee-derived defensin-1 on multispecies wound biofilm. *J. Med. Microbiol.* **2016**, *65*, 337–344.
16. Bucekova, M.; Sojka, M.; Valachova, I.; Martinotti, S.; Ranzato, E.; Szep, Z.; Majtan, V.; Klaudiny, J.; Majtan, J. Bee-derived antibacterial peptide, defensin-1, promotes wound re-epithelialisation in vitro and in vivo. *Sci. Rep.* **2017**, *7*, 7340.
17. Bucekova, M.; Jardekova, L.; Juricova, V.; Bugarova, V.; Di Marco, G.; Gismondi, A.; Leonardi, D.; Farkasovska, J.; Godocikova, J.; Laho, M.; et al. Antibacterial activity of different blossom honeys: New findings. *Molecules* **2019**, *24*, E1573.
18. Majtan, J.; Bohova, J.; Prochazka, E.; Klaudiny, J. Methylglyoxal may affect hydrogen peroxide accumulation in manuka honey through the inhibition of glucose oxidase. *J. Med. Food* **2014**, *17*, 290–293.
19. Bucekova, M.; Juricova, V.; Monton, E.; Martinotti, S.; Ranzato, E.; Majtan, J. Microwave processing of honey negatively affects honey antibacterial activity by inactivation of bee-derived glucose oxidase and defensin-1. *Food Chem.* **2018**, *240*, 1131–1136.
20. Mavric, E.; Wittmann, S.; Barth, G.; Henle, T. Identification and quantification of methylglyoxal as the dominant antibacterial constituent of Manuka (*Leptospermum scoparium*) honeys from New Zealand. *Mol. Nutr. Food Res.* **2008**, *52*, 483–489.
21. Adams, C.J.; Manley-Harris, M.; Molan, P.C. The origin of methylglyoxal in New Zealand manuka (*Leptospermum scoparium*) honey. *Carbohydr. Res.* **2009**, *344*, 1050–1053.

22. Cooper, R.A.; Molan, P.C.; Harding, K.G. The sensitivity to honey of Gram-positive cocci of clinical significance isolated from wounds. *J. Appl. Microbiol.* **2002**, *93*, 857–863.
23. French, V.M.; Cooper, R.A.; Molan, P.C. The antibacterial activity of honey against coagulase-negative staphylococci. *J. Antimicrob. Chemother.* **2005**, *56*, 228–231.
24. Bouzo, D.; Cokcetin, N.N.; Li, L.; Ballerin, G.; Bottomley, A.L.; Lazenby, J.; Whitchurch, C.B.; Paulsen, I.T.; Hassan, K.A.; Harry, E.J. Characterising the mechanism of action of an ancient antimicrobial, manuka honey, against *Pseudomonas aeruginosa* using modern transcriptomics. *mSystems* **2020**, *5*, e00106-20.
25. Farkasovska, J.; Bugarova, V.; Godocikova, J.; Majtan, V.; Majtan, J. The role of hydrogen peroxide in the antibacterial activity of different floral honeys. *Eur. Food Res. Technol.* **2019**, *245*, 2739–2744.
26. Majtan, J.; Majtan, V. Is manuka honey the best type of honey for wound care? *J. Hosp. Infect.* **2010**, *73*, 305–306.
27. Bucekova, M.; Valachova, I.; Kohutova, L.; Prochazka, E.; Klaudiny, J.; Majtan, J. Honeybee glucose oxidase —Its expression in honeybee workers and comparative analyses of its content and H_2O_2-mediated antibacterial activity in natural honeys. *Naturwissenschaften* **2014**, *101*, 661–670.
28. Bartakova, K.; Drackova, M.; Borkovcova, I.; Vorlova, L. Impact of microwave heating on hydroxymethylfurfural content in Czech honeys. *Czech J. Food Sci.* **2011**, *29*, 328–336.
29. Hebbar, H.U.; Nandini, K.E.; Lakshmi, M.C.; Subramanian, R. Microwave and infrared heat processing of honey and its quality. *Food Sci. Technol. Res.* **2003**, *9*, 49–53.
30. Kędzierska-Matysek, M.; Florek, M.; Wolanciuk, A.; Skałecki, P.; Litwińczuk, A. Characterisation of viscosity, colour, 5-hydroxymethylfurfural content and diastase activity in raw rape honey (*Brassica napus*) at different temperatures. *J. Food Sci. Technol.* **2016**, *53*, 2092–2098.
31. Al-Ghamdi, A.; Mohammed, S.E.A.; Ansari, M.J.; Adgaba, N. Comparison of physicochemical properties and effects of heating regimes on stored *Apis mellifera* and *Apis florea* honey. *Saudi J. Biol. Sci.* **2019**, *26*, 845–848.
32. Kowalski, S.; Lukasiewicz, M.; Bednarz, S.; Panus, M. Diastase number changes during thermal and microwave processing of honey. *Czech J. Food Sci.* **2012**, *30*, 21–26.
33. Schmitzova, J.; Klaudiny, J.; Albert, S.; Schroder, W.; Schrockengost, W.; Hanes, J.; Judova, J.; Simuth, J. A family of major royal jelly proteins of the honeybee *Apis mellifera* L. *Cell. Mol. Life Sci.* **1998**, *54*, 1020–1030.
34. Hojo, M.; Kagami, T.; Sasaki, T.; Nakamura, J.; Sasaki, M. Reduced expression of major royal jelly protein 1 gene in the mushroom bodies of worker honeybees with reduced learning ability. *Apidologie* **2010**, *41*, 194–202.
35. Fontana, R.; Mendes, M.A.; de Souza Monson, B.; Konno, K.; César, L.M.M.; Malaspina, O. Jelleines: A family of antimicrobial peptides from the royal jelly of honeybees (*Apis mellifera*). *Peptides* **2004**, *25*, 919–928.
36. Brudzynski, K.; Sjaarda, C.; Lannigan, R. MRJP1-containing glycoproteins isolated from honey, a novel antibacterial drug candidate with broad spectrum activity against multi-drug resistant clinical isolates. *Front. Microbiol.* **2015**, *6*, 711.
37. Bucekova, M.; Majtan, J. The MRJP1 honey glycoprotein does not contribute to the overall antibacterial activity of natural honey. *Eur. Food Res. Technol.* **2016**, *242*, 625–629.
38. Grecka, K.; Kuś, P.M.; Worobo, R.W.; Szweda, P. Study of the anti-staphylococcal potential of honeys produced in Northern Poland. *Molecules* **2018**, *23*, 260.
39. Brudzynski, K. A current perspective on hydrogen peroxide production in honey. A review. *Food Chem.* **2020**. [CrossRef]
40. Postmes, T.; van den Bogaard, A.E.; Hazen, M. The sterilization of honey with cobalt 60 gamma radiation: A study of honey spiked with spores of *Clostridium botulinum* and *Bacillus subtilis*. *Experientia* **1995**, *51*, 986–989.
41. Midgal, W.; Owczarczyk, H.B.; Kedzia, B.; Holderna-Kedzia, E.; Madajczyk, D. Microbial decontamination of natural honey by irradiation. *Radiat. Phys. Chem.* **2000**, *57*, 285–288.
42. Saxena, S.; Gautam, S.; Sharma, A. Microbial decontamination of honey of Indian origin using gamma radiation and its biochemical and organoleptic properties. *J. Food Sci.* **2010**, *75*, M19–M27.
43. Molan, P.C.; Allen, K.L. The effect of gamma-irradiation on the antibacterial activity of honey. *J. Pharm. Pharmacol.* **1996**, *48*, 1206–1209.
44. Sabet Jalali, F.S.; Ehsani, A.; Tajik, H.; Ashtari, S. In vitro assessment of efficacy of gamma irradiation on the antimicrobial activity of iranian honey. *J. Anim. Vet. Adv.* **2007**, *6*, 996–999.

45. Horniackova, M.; Bucekova, M.; Valachova, I.; Majtan, J. Effect of gamma radiation on the antibacterial and antibiofilm activity of honeydew honey. *Eur. Food Res. Technol.* **2017**, *243*, 81–88.
46. Girma, A.; Seo, W.; She, R.C. Antibacterial activity of varying UMF-graded Manuka honeys. *PLoS ONE* **2019**, *14*, e0224495.
47. Allen, K.L.; Molan, P.C.; Reid, G.M. A survey of the antibacterial activity of some New Zealand honeys. *J. Pharm. Pharmacol.* **1991**, *43*, 817–822.
48. Barnes, J.S.; Foss, F.W.; Schug, K.A. Thermally accelerated oxidative degradation of quercetin using continuous flow kinetic electrospray-ion trap-time of flight mass spectrometry. *J. Am. Soc. Mass Spectrom.* **2013**, *24*, 1513–1522.

© 2020 by the authors. Licensee MDPI, Basel, Switzerland. This article is an open access article distributed under the terms and conditions of the Creative Commons Attribution (CC BY) license (http://creativecommons.org/licenses/by/4.0/).

Article

Antioxidant and Anti-Inflammatory Activities of Safflower (*Carthamus tinctorius* L.) Honey Extract

Li-Ping Sun [1], Feng-Feng Shi [1], Wen-Wen Zhang [1], Zhi-Hao Zhang [1,2] and Kai Wang [1,*]

[1] Institute of Apicultural Research, Chinese Academy of Agricultural Sciences, Beijing 100093, China; sunliping01@caas.cn (L.-P.S.); shi9914@126.com (F.-F.S.); zhangwen_w@126.com (W.-W.Z.); talent-zzh@163.com (Z.-H.Z.)
[2] College of Animal Science (College of Bee Science), Fujian Agriculture and Forestry University, Fuzhou 350002, China
* Correspondence: wangkai@caas.cn

Received: 21 May 2020; Accepted: 29 July 2020; Published: 2 August 2020

Abstract: Safflower honey is a unique type of monofloral honey collected from the nectar of *Carthamus tinctorius* L. in the *Apis mellifera* colonies of northwestern China. Scant information is available regarding its chemical composition and biological activities. Here, for the first time, we investigated this honey's chemical composition and evaluated its in vitro antioxidant and anti-inflammatory activities. Basic physicochemical parameters of the safflower honey samples in comparison to established quality standards suggested that safflower honeys presented a good level of quality. The in vitro antioxidant tests showed that extract from *Carthamus tinctorius* L. honey (ECH) effectively scavenged DPPH and ABTS$^+$ free radicals. In lipopolysaccharides (LPS) activated murine macrophages inflammatory model, ECH treatment to the cells inhibited the release of nitric oxide and down-regulated the expressions of inflammatory-relating genes (iNOS, IL-1β, TNF-α and MCP-1). The expressions of the antioxidant genes TXNRD, HO-1, and NQO-1, were significantly boosted in a concentration-dependent manner. ECH decreased the phosphorylation of IκBα and inhibited the nuclear entry of the NF-κB-p65 protein, in LPS-stimulated Raw 264.7 cells, accompany with the increased expressions of Nrf-2 and HO-1, suggesting that ECH achieved the anti-inflammatory effects by inhibiting NF-κB signal transduction and boosting the antioxidant system via activating Nrf-2/HO-1 signaling. These results, taken together, indicated that safflower honey has great potential into developing as a high-quality agriproduct.

Keywords: safflower (*Carthamus tinctorius* L.) honey; chemical analysis; anti-inflammatory; antioxidant; NF-κB; Nrf-2

1. Introduction

Honey used widely as a food and medicine [1]. It is often studied due to its biological activities, among them the anti-inflammatory, anti-bacterial, and anti-oxidation properties are the most published [2–5]. Others include the resistance to bromobenzene-induced liver injury in mice [6] and potential treatment for colitis [7]. However, there are some special honeys, which come from particular floral sources, for example, Manuka, which can play an anti-ulcer role through antioxidant and anti-inflammatory effects [4,8], or Gelam honey, which was able to effectively inhibit airway inflammation in an ovalbumin-induced allergic asthma mouse model [9]. There are significant differences in the antioxidant activities among different honeys [10]. Safflower (*Carthamus tinctorius* L.) honey is brewed by honeybees collecting safflower nectar, which is a local speciality honey type of Xinjiang Providence, China. This honey is mainly produced in the Balruk Mountains of Xinjiang (82°12′–83°30′ longitude in the East and 45°24′–46°3′ latitude in the north), and contains a variety of natural sweeteners with bioactive components [11–13], which is the largest safflower planting base in China. Our recent preliminary results suggested that the safflower honey has several superiorities, such as comprehensive and rich nutrients, and a large number of bioactive substances [14]. Nevertheless, detailed information is still unavailable regarding its chemical composition and biological activities, which limits the application and development of this special agriproducts.

Recent decades, several studies reported on the anti-inflammatory and antioxidant activities of honey, which are mainly attribute to its abundant phenolic and flavonoid contents [15,16]. Honey contains rich phenolic acids and flavonoids, which contribute to the antioxidant, antimicrobial, anti-inflammatory, anti-proliferation, anti-cancer, and anti-metastasis effects [17]. Studies showed that these compounds in honey were able to inhibit the pro-inflammatory activity of nitric oxide synthase (iNOS) and had anti-inflammatory effects [17]. Bangladesh honey samples are rich in phenolic acids and flavonoids with high antioxidant potential [18]. Honeys from specific floral sources or collected by different bee species, given their special and rare compounds, such as certain phenolic acids and flavonoids, for instance, in stingless honey, can show anti-inflammatory and antioxidant effects [19]. Data published with Camellia honey presented an antioxidant effect closely related to its phenolic content [20]. We therefore inferred that the safflower honey also some activities, such as the anti-inflammatory and antioxidant properties. In our previous studies, we evaluated the anti-inflammatory activities by the bee products using the bacterial lipopolysaccharides (LPS) challenged murine macrophage model (RAW 264.7 cells), in which model a number of typical inflammatory responses can be mimicked in vitro, including the releases of inflammatory mediators, accompanying with the oxidative stress to the cells. LPS induced RAW 264.7 cells were shown to be with an increase in nitric oxide (NO) release [21,22], which is mediated via the activation of the inflammation related-nuclear factor kappa-B (NF-κB) signaling pathway [23], and thereafter induced oxidative stress. NO has a significant correlation with oxidative stress [24], and its release depends on the expression of inducible NO synthase (iNOS). LPS-activated Raw 267.4 cells also lead to the rapid phosphorylation and degradation of IκBα [25]. The phosphorylation of IκB-α or IκB-β can activate the NF-κB signaling pathway and promote NF-κB-p65 protein in the nucleus. However, the NF-κB-p65 protein can mediate the synthesis/releases of tumor necrosis factor alpha (TNF-α), interleukin-1β (IL-1β), monocyte chemoattractant protein 1 (MCP-1), which further regulate the transcription of other inflammatory mediators [26–28]. In addition, the NF-E2-related factor 2 (Nrf-2) signaling activation regulates the expression of a series of downstream antioxidant factors (4-nitroquinoline-N-oxide (NQO), HO-1 as well as the thioredoxin reductase(TXNRD)) [29,30]. Activation of these internal anti-oxidant enzymes alleviated the cells against oxidative stress [31].

In the current study, *Carthamus tinctorius* L. safflower honey were collected to carry out a screening of physical and chemical indicators and the extract from *Carthamus tinctorius* L. safflower honey (ECH) were obtained for preliminary characterizations on major main phenolic acids and flavonoid components based on the high-performance liquid chromatography–quadrupole-time of flight mass spectrometry (HPLC-QTOF-MS). Furthermore, we tested the in vitro free radical scavenging activities by ECH and its anti-inflammatory and antioxidant potentials were evaluated in in LPS-activated Raw 264.7 murine macrophages.

2. Materials and Methods

2.1. Chemicals and Reagents

Primary antibodies including IκB-α, p-IκBα, and NF-κB-p65 were purchased from Cell Signaling Technology (Danvers, MA, USA). Nrf-2, HO-1, and β-actin were purchased from Cambridge Bio (Boston, MA, USA). The NanoDrop 2000 Ultra Micro Spectrophotometer was purchased from Thermo Fisher Scientific (Pittsburgh, PA, USA). Gallic acid, sodium nitrite, the Prime Script TM RT Master Mix kit and TB Green® Premix Ex Taq TM were purchased from Biotech Co. Ltd. (Shanghai, China). Protein loading buffer, 20% SDS, electroporation solution and Tris-HCl buffer were purchased from Solarbio (Beijing, China). BCIP/NBT alkaline phosphatase developer and 40% Acr-Bis were purchased from Beyotime Biotechnology (Shanghai, China). The CCK-8 kit was purchased from Dojindo

Laboratories (Kumamoto, Japan). Amberlite XAD-2 resin was purchased from Sigma-Aldrich Trading Co. Ltd. (Shanghai, China). Other chemicals, including the LPS (*Escherichia coli* 0127: B8) and alkaline phosphatase-conjugated secondary antibody (anti-rabbit IgG) and the standards applied in the chemical analysis were purchased from Sigma (St. Louis, MO, USA).

2.2. Safflower Honey Samples and Physical Deteriminations

Safflower honey samples were obtained from three *Apis mellifera* L. colonies during the safflower (*C. tinctorius* L.) flower season in 2018. Each honey from the single hive was considered as one sample, 500 g each. The apiaries were located in the Yumin County, Tacheng City, China, which were belongs to bases of Jiangsu Rigao Bee Products Co., Ltd. (Xuyi, China). We collected the capped comb honeys in the day, and transfer them to the refrigerator at 4 °C, afterwards the honey samples were taken to the lab and storage at −20 °C until usage. The moisture, free acid, amylase, hydroxymethylfurfural, fructose, glucose, sucrose, and ash contents of the Xinjiang safflower honey samples were determined by the method specified in CODEX STAN 12-1981. Melissopalynology was applied to analyze the botanical origin of the pollens in the safflower honeys, following pervious published method [32]. The typical pollen grain in a safflower honey sample was shown in the Figure S1.

2.3. Extraction on the Safflower Honey

The preparation of the extract from *C. tinctorius* safflower honey (ECH) refers to the method of Mu et al. [33] and was improved. XAD-2 resin was soaked in 95% ethanol for 24 h, washed two to three times with ultrapure water until there was no ethanol, and then placed for standby. We weighed 800 g of the Xinjiang safflower honey sample, mixed it with hydrochloric acid solution (adjusted using hydrochloric acid pH = 2 with ultrapure water) at 1:5 (w/v), and used an ultrasonic instrument for 30 min until it dissolved. The activated XAD-2 resin was weighed to 800 g, treated with ultrasonic to make it free of air, then mixed with Xinjiang safflower honey aqueous solution, mixed evenly for 1 h, and left standing overnight. We discarded the supernatant, added XAD-2 resin into the glass column, washed with 2 volumes of hydrochloric acid water (pH = 2) and 3 volumes of ultra-pure water, and then eluted with 8 volumes of ethanol to collect the eluate, evaporated the ethanolic extract to a solid residue with a rotary evaporator with vacuum, and then dissolved the residue in 15 mL of pure water. This aqueous solution was extracted with 20 mL of ethyl acetate. We collected the ethyl acetate layer, repeated this four times, for each extraction for 20 min. The collected ethyl acetate layer was blown dry with nitrogen to obtain the resulting ECH, and placed in a −20 °C refrigerator for storage. We dissolved the ECH with an appropriate amount of ethanol for the further usages.

2.4. Preliminary Analysis of ECH Phenolic Flavonoids by HPLC-QTOF-MS

The HPLC-QTOF-MS analysis method of ECH was established by our laboratory [34]. Chromatographic conditions involved using a proshell 120 EC-C18 column (100 × 2.1 mm, particle size 2.7 μm), typical parameters of chromatographic were as follows: column temperature, 30 °C; injection volume, 5 μL; flow rate, 0.2 mL/min. The elution procedure is shown in Table 1.

Table 1. Mobile phase elution procedures.

Time (min)	Phase A% (0.1% Formic Acid)	Phase B% (100% Acetonitrile)
0	90	10
0–15	75	25
15–20	70	30
20–30	65	35
30–35	30	70
35–40	30	70
40–42	90	10
42–50	90	10

Mass spectrometry conditions were conducted with an electrospray (ESI) ion source in negative ion mode The typical parameters of the mass spectrometer were as follows: drying gas temperature, 350 °C; drying gas flow rate, 6 L/min; sprayer pressure, 35 psi; capillary voltage 3500 V; atomizing gas temperature, 350 °C; and atomizing gas flow rate, 9 L/min. Qualitative and quantitative analysis were carried out by accuracy mass and extracted ion chromatography (EIC). Polyphenolic compounds ion chromatograms were extracted by Mass Hunter Qualitative Analysis software (Agilent Technologies) for ECH.

2.5. In Vitro Free Radical Scavenging Ability Determination Experiment

2.5.1. DPPH Free Radical Scavenging Experiment

For the determination of the DPPH· clearance rate, we referred to WU et al.'s method [35], and made certain adjustments. We placed 100 µL each of DPPH working solution and ECH into a 1.5 mL centrifuge tube, shook and mixed, and react at room temperature in the dark for 30 min. We took 100 µL of the reaction liquid to a 96-well plate (100 µL/well), and measured the absorbance at 517 nm. For A1, the same method was used to determine the absorbance when adding 100 µL of 95% ethanol solution instead of DPPH working solution. The absorbance of the blank group (100 µL DPPH solution and 100 µL of 95% ethanol solution) was recorded as A0. The calculation formula of the clearance rate is:

$$\text{clearance rate}\% = 1 - \frac{A1 - A2}{A0} \times 100$$

The removal ability of the sample is expressed by IC_{50}.

2.5.2. ABTS$^+$ Free Radical Scavenging Experiment

For the determination of the ABTS$^+$ clearance rate, we referred to YANG et al. [36] and made certain adjustments. The ABTS solution was generated by the reaction of 15 mL 7 mM ATBS solution and 246 µL 140 mM potassium persulfate aqueous solution in the dark for 16 h. When used, it was diluted with methanol to the absorbance of 0.70 ± 0.02 at 734 nm. We placed 250 µL ABTS methanol working solution and 150 µL ECH (to dissolve) in a 1.5 mL centrifuge tube, shook and mixed, avoided light for the reaction for 10 min. We took 150 µL of the reaction liquid to a 96-well plate (150 µL/well), and measured the absorbance value, recorded as A1. The same method was used to determine the absorbance when adding 250 µL of methanol instead of ABTS methanol working solution. The absorbance of the blank group (250 µL ABTS solution and 150 µL (ECH solvent) solution) was recorded as A0, with parallel values. The calculation formula of the clearance rate is:

$$\text{clearance rate}\% = 1 - \frac{A1 - A2}{A0} \times 100$$

The removal ability of the sample is expressed by IC_{50}.

2.6. Cell Experiment

Raw 264.7 cells were a gift from Professor Hu Fuliang, college of Animal Science, Zhejiang University. Raw 264.7 cells were cultured with DMEM high glucose medium containing 10% heat-inactivated fetal bovine serum and double antibodies (streptomycin (100 µg/mL) and penicillin (100 IU/mL)) at 37 °C Incubation was in a 5% CO_2 incubator [37], to maintain the stable growth of the cells.

2.6.1. ECH Measurement on Cell Relative Survival Rate

Raw 264.7 cells in the logarithmic growth phase were inoculated into 96-well plates with 1×10^5 cells/mL, 100 µL per well, incubated overnight in a 5% CO_2 incubator at 37 °C, until the cell adherence reached 70–80%. The cells were incubated with 2.5 µg/mL, 5 µg/mL, 10 µg/mL, 15 µg/mL, or 20 µg/mL ECH for 24 h, and the culture medium was aspirated. The cells were washed with new cell culture medium for two to three times, and a blank control group was set. We added 10 µL CCK-8 reagent and cell culture medium into each well, incubated for 3 h, and then the absorbance at 450 nm was determined [38,39].

2.6.2. Determination of the Nitric Oxide (NO) Concentration in LPS-Induced Cells Treatment with ECH

The NO concentration was measured with a Griess reagent. We referred to the method of Lee et al. and adjusted [40]. Raw 264.7 cells in the logarithmic growth phase were inoculated into 24-well plates with 1×10^5 cells/mL, 500 µL per well, incubated in a 37 °C, 5% CO_2 incubator, until the cell adherence reached 70–80%, and we set the DXMS positive control group, blank control group, LPS treatment control group, and experimental group. To the experimental group, we successively added 2.5 µg/mL and 5 µg/mL ECH for 1 h, followed by LPS (1 µg/mL) for 24 h. The cell culture fluid was collected, centrifuged at 5000 r/min for 10 min, and the supernatant was collected. We mixed 100 µL of each sample with an equal volume of Griess reagent and added these samples to a 96-well plate, and incubated at room temperature for 10 min. We measured the absorbance at 540 nm.

2.6.3. ECH Detection of the mRNA Expression Related to Inflammation and Oxidation in LPS-Induced Cells

Raw 264.7 cells in the logarithmic growth phase were inoculated into 24-well plates with 1×10^5 cells/mL, 500 µL per well, incubated in a 37 °C, 5% CO_2 incubator, until the cell adherence reached 70–80%, and we set the DXMS positive control group, blank control group, LPS treatment control group, and experimental group. To the experimental group, we successively added 2.5 µg/mL and 5 µg/mL ECH and incubated for 1 h, and then induced with LPS (1 µg/mL) for 6 h. The cells were collected, and then the total cell RNA was extracted using a CarryHelix RNA extraction kit, and its concentration and purity were determined with a NanoDrop 2000 ultramicro spectrophotometer. Using 1 µg of extracted total RNA as a template, the PrimeScript® TM RT Master Mix reverse transcription kit was used to perform reverse transcription of the cDNA, and the product was placed in a −20 °C refrigerator for use. Real-time quantitative PCR was performed with a TB Green® Premix Ex TaqTM kit. The total volume of the reaction system was 10 µL: TB Green® Premix Ex TaqTM 5.0 µL, cDNA template 0.2 µL, RNase Free dH_2O 4.4 µL, upstream and downstream primers 0.2 µL each, related to use. The primer sequences are shown in Table 2.

Table 2. The real-time PCR-related primer sequences [25,34,41].

Gene	Upstream Primer Sequence	Downstream Primer Sequence
iNOS	5′-TTTCCAGAAGCAGAATGTGACC-3′	5′-AACACCACTTTCACCAAGACTC-3′
IL-1β	5′-CCAACAAGTGATATTCTCCATGAG-3′	5′-ACTCTGCAGACTCAAACTCCA-3′
TNF-α	5′-CCACGCTCTTCTGTCTACTG-3′	5′-ACTTGGTGGTTTGCTACGAC-3′
MCP-1	5′-AAGAAGCTGTAGTTTTTGTCACCA-3′	5′-TGAAGACCTTAGGGCAGATGC-3′
HO-1	5′-ACATTGAGCTGTTTGAGGAG-3′	5′-TACATGGCATAAATTCCCACTG-3′
TXNRD	5′-AGGATTTCTGGCTGGTATCG-3′	5′-CTCGCTGTTTGTGGATTGAG-3′
NQO	5′-TTCAACCCCATCATTTCC-3′	5′-TCAGGCGTCCTTCCTTATA-3′

2.6.4. ECH Detection of Related Protein Expression in LPS-Induced Cells

Raw 264.7 cells in the logarithmic growth phase were inoculated in 6-well plates with 1×10^5 cells/mL, 1 mL per well, and incubated in a 37 °C, 5% CO_2 incubator, so that the cell adherence reached 90%, and we set the DXMS positive control group, blank control group, LPS treatment control group, and experimental group. To the experimental group we successively added 2.5 µg/mL and 5 µg/mL ECH and incubated for 1 h, and then induced with LPS (1 µg/mL) for 0.5 h. The collected cells were washed twice with PBS, and inhibited with NP-40 protein lysate Cellular protein was extracted from the reagent, and the concentration of the extracted protein was measured by BCA. The protein samples of each group of cells were detected by immunoblotting. The total sample loading was 20 µg of total protein, with β-actin as a reference, through 12% SDS-PAGE gel electrophoresis, transfer, hybridization, alkaline phosphatase color development and other steps to obtain the hybridization bands of IκBα, P-IκBα, Nrf-2 and HO-1 [42].

2.6.5. Effect of ECH on LPS-Induced Nuclear Localization of p65 (NF-κB)

Raw 264.7 cells in the logarithmic growth phase were inoculated into the slide confocal small dish with 1×10^5 cells/mL, incubated in a 37 °C, 5% CO_2 incubator overnight, and we set up the DXMS positive control group, blank control group, LPS treatment control group, and experimental group. To the experimental group, we added 5 µg/mL of ECH and incubated for 1 h. Then, we induced with LPS (1 µg/mL) for 0.5 h. A methanol-acetone mixture (1:1, *v/v*) was used as the fixing solution for 30 min. Permeabilization was performed with 0.5% Triton X-100 for 30 min, blocked with 10% serum blocking solution for 30 min at room temperature. We added primary antibody (NF-κB-p65) (1:50 dilution) and secondary antibody goat anti-rabbit (IgG) (1:500 dilution) and incubated for 1 h, with DAPI (4′,6-diamidino-2-phenylindole) (1:2000 dilution) stained nuclear processing coverslips, observed with a laser confocal scanning microscope, and analyzed the results [42].

2.7. Statistics Analysis

All experimental data were obtained from at least three repeated experiments. Data are presented as the Mean ± SD, and the statistical significance between two groups was determined using Student's *t*-test, and a *p* value of <0.05 was considered as statistically significant.

3. Results

3.1. Physical and Chemical Analysis on the Safflower Honey

All data regarding the physical and chemical indicators of Xinjiang safflower honey (see Table 3), using the relevant methods in the EU standard, revealed better results than the standard values cited. Among them, the moisture content was 18.2%, which meets the EU standard for honey (not more than 20%). The acidity (1 mol/L sodium hydroxide titration) was 25.0 mL/kg, which is lower than the EU standard limit of 50 mL/kg. Hydroxymethylfurfural, sucrose, and ash were not detected; the three were far below the relevant EU standard limits. The fructose content was 36.9%, the glucose content was 25.2%, and the total of the two was 62.1%, higher than the EU standard (not less than 60%); the amylase value was 21.1 mL/(g·h), which is much higher than the EU standard (not less than 8), which is twice the EU requirements.

Table 3. The physical and chemical indicators of safflower honey.

Physical and Chemical Indicators	Result	Standard Limited	Standard Method
Moisture (%)	18.2 ± 0.25	≤20	2001/110/EC
Acidity (mL/kg)	25.0 ± 0.43	≤50	2001/110/EC
Amylase value (mL/(g·h))	21.1 ± 0.36	≥8	2001/110/EC
Hydroxymethylfurfural (mg/kg)	ND	≤40	2001/110/EC
Fructose (%)	36.9 ± 0.26	≥60	2001/110/EC
Glucose (%)	25.2 ± 0.14		
Sucrose (%)	ND	≤5	2001/110/EC
Ash (%)	ND	≤0.1	2001/110/EC

Note: ND means not detected.

3.2. Preliminary Analysis of Phenolic Flavonoids in ECH by HPLC-QTOF-MS

HPLC-QTOF-MS detection of ECH was carried out with 24 phenolic flavonoid standards. As shown in Table 4, eight phenolic acids were detected, with a total content of 7.197 mg/kg honey. Vanillic acid and p-hydroxybenzoic acid were the highest phenolic acids, with 3.196 mg/kg honey and 1.524 mg/kg honey, respectively. Protocatechualdehyde was not detected. Seven flavonoids were detected, with a total content of 7.633 mg/kg honey, with quercetin and myricetin being the highest, with 3.196 mg/kg honey and 1.524 mg/kg honey, respectively. Rutin, quercetin-3-O-glucoside, kaempferol, naringenin and pinobanksin were low in content. Morin, luteolin, diosmetin, pinocembrin, galanin, caffeic acid phenethyl ester, chrysin and kaempferol-3-O-glucoside were not detected. As far as the content of safflower honey phenolic acid flavonoids is concerned, the phenolic acid components were characterized by vanillic acid and p-hydroxybenzoic acid, and the flavonoids were characterized by quercetin and myricetin. The diversity and content of phenolic acid and flavonoids play a positive role in the antioxidant and anti-inflammatory activities of safflower honey.

Table 4. High-performance liquid chromatography combined with a quadrupole time-of-flight mass spectrometry (HPLC-QTOF/MS) analysis of phenolic flavonoids in safflower.

Compound Name	Standard Curve Equation	R^2	Concentration Ranges (ng/mL)	RT (min)	[M − H]⁻	Content (mg/kg)
gallic acid	y = 573.42x − 32178.21	0.9983	50–2000	1.765	168.7	0.636
Protocatechuic acid	y = 925.79x + 37631.28	0.9938	50–2000	2.501	153	0.115
protocatechualdehyde	y = 338.47x + 18654.55	0.9991	50–1000	3.589	136.9	ND
p-hydroxybenzoic acid	y = 437.73x + 2112.43	0.9975	50–2000	3.963	137	1.524
vanillic acid	y = 17.86x − 166.86	0.9973	50–2000	4.865	167	3.196
caffeic acid	y = 1377.37x + 6941.68	0.9984	50–2000	5.046	179	0.041
ferulic acid	y = 108.21x − 1692.97	0.9989	50–2000	10.924	193.1	0.248
cinnamic acid	y = 12.75x − 100.06	0.9983	50–2000	18.060	146.9	0.559
rutin	y = 1017.39x − 35010.60	0.9993	50–2000	11.689	609.1	0.388
quercetin-3-O-glucoside	y = 2782.17x − 74249.70	0.9993	50–2000	11.906	462.8	0.300
kaempferol-3-O-glucoside	y = 2350.39x + 111703.32	0.9952	50–2000	13.776	446.8	ND
myricetin	y = 0.91x − 77.39	0.9953	50–2000	15.033	317	1.021
morin	y = 611.76x + 3378.56	0.9962	50–2000	16.578	301	ND
luteolin	y = 1103.83x + 98837.97	0.9961	50–2000	18.06	285	ND
quercetin	y = 3032.78x − 1438777.72	0.998	100–2000	18.153	301	5.342
kaempferol	y = 10.83x − 72.24	0.9987	50–2000	20.439	271.1	0.126
naringenin	y = 14.58x + 410.22	0.9969	50–2000	20.439	271.1	0.072
pinobanksin	y = 71.63x + 450.56	0.995	50–2000	20.709	271.1	0.384
diosmetin	y = 4225.03x + 392274.94	0.9916	50–2000	21.102	298.9	ND
chrysin	y = 498.99x + 27824.49	0.9989	50–2000	25.205	253.1	ND
pinocembrin	y = 340.00x + 29581.47	0.9902	50–2000	25.739	255.1	ND
galanin	y = 120.24x + 9565.86	0.9952	50–2000	25.907	269	ND
caffeic acid phenethyl ester	y = 15593.82x + 3956352.11	0.9976	100–2000	26.155	283.1	ND

Note: ND means not detected.

3.3. In Vitro Antioxidant Free Radical Scavenging Capacity of ECH

DPPH and ABTS⁺ free radical scavenging experiments are commonly used to evaluate natural antioxidants [15]. It can be seen from Table 5 that the concentration of ECH inhibiting 50% DPPH free radical was 68.23 ± 0.40 µg/mL, and the concentration of ECH inhibiting 50% ABTS⁺ free radical was 81.88 ± 0.54 µg/mL.

Table 5. In vitro ECH (extract from *Carthamus tinctorius* L. honey) free radical scavenging activity.

	IC_{50} (µg/mL)
DPPH	68.23 ± 0.40
ABTS+	81.88 ± 0.54

3.4. In Vitro Antioxidant, Anti-Inflammatory Activies by ECH

3.4.1. Effects of ECH on Raw 264.7 Cell Survival

By treating Raw 264.7 cells with ECH in concentrations from 2.5 to 20 µg/mL and using the CCK-8 reagent to detect cell viability, we found the most suitable ECH concentration to ensure that adding ECH concentration to Raw 264.7 cells produced no obvious toxic effects (see Figure 1). Compared with the control group, there was no significant difference in the growth of Raw 264.7 cells when the ECH concentration were 2.5 µg/mL and 5 µg/mL. When the ECH concentration was 10 µg/mL, the Raw 264.7 cells growth was highly significantly inhibited. When the ECH concentrations were 15 µg/mL and 20 µg/mL, the inhibitory effect on the growth of Raw 264.7 cells was highly significant. Therefore, the safe concentration of ECH available in this experiment was 2.5–5 µg/mL.

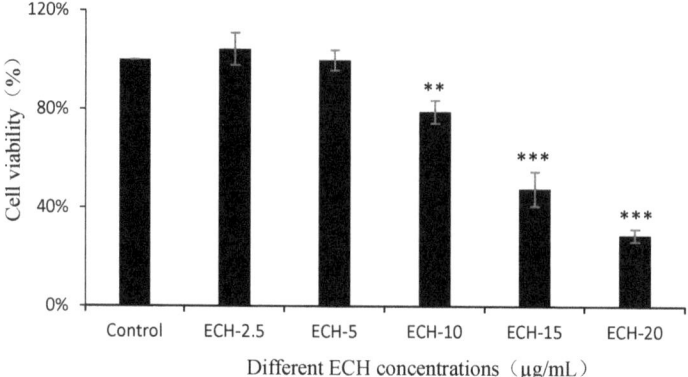

Figure 1. The effects of various concentrations of ECH on RAW 264.7 cell viability. Raw 264.7 cells were pretreated with ECH (2.5 to 20 µg/mL) or not for 24 h, cell viability was tested using the CCK-8 method. ** ($p < 0.01$) and *** ($p < 0.001$) indicates significant difference compared with the control group.

3.4.2. Effect of ECH on LPS-Induced Nitric Oxidase (NO) Release in RAW 264.7 Cells

A typical Griess method was used to determine the amount of NO released by different treatment cells (see Figure 2). Dexamethasone (DXMS) treatment at 100 µg/mL (referred as DXMS-100) to the LPS-activated cells was used as the positive control group. Compared with the Control group, the concentrations of NO releases in Raw 264.7 cells was increased about nine times following LPS treatment, indicating the model of Raw 264.7 cells inflammation model was successfully established. After co-treatment with ECH and LPS, the NO release decreased by about two-thirds compared with the LPS group, and the decreasing effects even better than the DXMS-100 group. Therefore, ECH was shown to effectively reduce the amount of NO released during macrophage inflammation induced by LPS.

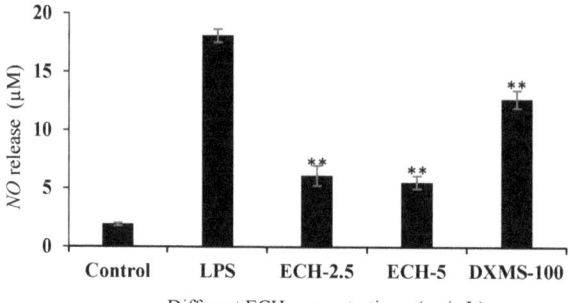

Figure 2. The effects of ECH on the nitric oxide (NO) release in LPS-activated macrophages. RAW 264.7 cells were pretreated with/without indicated concentrations of ECH or dexamethasone (100 µg/mL, positive control) for 1 h then stimulated with LPS (1 µg/mL) for 24 h. NO concentrations in the cell culture medium were measured using the Griess method. ** ($p < 0.01$) indicates significant difference compared with the LPS group.

3.4.3. Effect of ECH on LPS-Induced Inflammation and Oxidation-Related Gene Expression in RAW 264.7 Cells

Next, RT-qPCR was applied to detect the expression levels of inflammation and oxidation-related genes in LPS-induced Raw 264.7 cells (see Figure 3). After LPS treatment, the expression of NQO decreased significantly, the expression of iNOS, IL-1β, TNF-α and MCP-1 increased significantly. Compared with cells treated with LPS alone, the expressions of iNOS, IL-1β, TNF-α and MCP-1 in ECH-treated groups were decreased, and the inhibitory effect was stronger with the increase of the ECH concentration. For HO-1, TXNRD, and NQO, ECH significantly increased these antioxidant-related gens expressions.

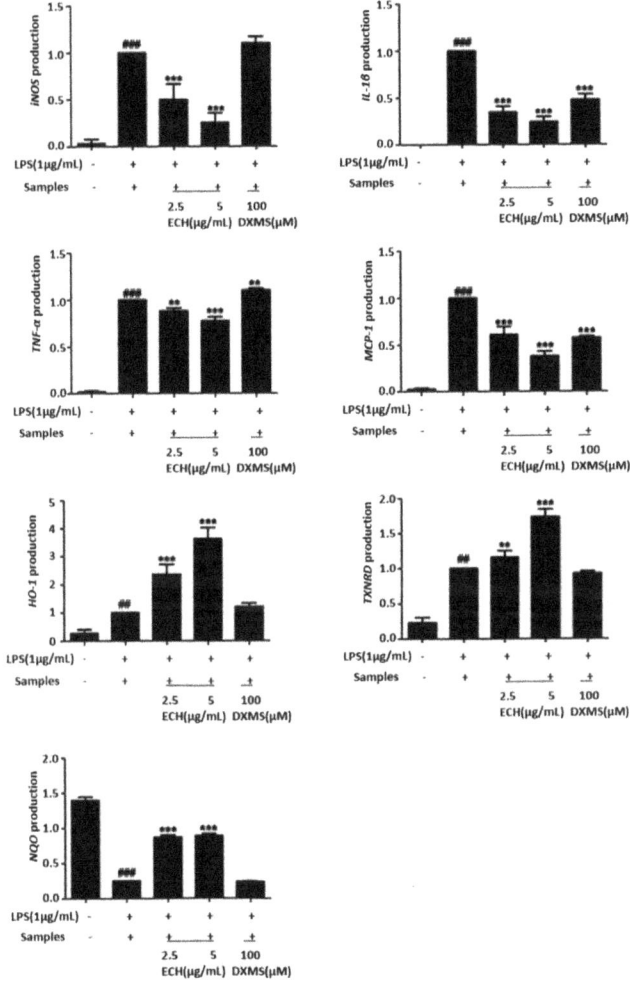

Figure 3. Effects of ECH on the expressions of antioxidant and inflammatory genes in LPS stimulated RAW 264.7 cells. RAW 264.7 cells were pretreated with/without indicated concentrations of ECH or dexamethasone (100 μg/mL, positive control) for 1 h then stimulated with LPS (1 μg/mL) for 6 h. mRNA expression in RAW 264.7 cells were measured using RT-qPCR. ** ($p < 0.01$) and *** ($p < 0.001$) indicates significant difference compared with the LPS group. ## ($p < 0.01$) and ### ($p < 0.001$), compared with the normal control group.

3.4.4. Effect of ECH on the Expressions of Inflammation and Anti-Oxidant Signaling Related Proteins in LPS-Activated RAW 264.7 Cells

The expression of inflammatory and anti-oxidation-related proteins in LPS-induced Raw 264.7 cells was detected by western blot (Figure 4). The expression level of P-IκBα in the LPS group was significantly higher than that of IκBα, however, after ECH treatment, the expression of P-IκBα in the LPS induced group was significantly reduced compared to the LPS treatment group only. IκBα phosphorylation was inhibited, and 5 µg/mL was found with a more potent inhibitive effects thant the low dosage group (2.5 µg/mL). The expression levels of Nrf-2 and HO-1 increased significantly in cells induced by LPS after ECH pretreatment. When the concentration of ECH increased, the expression levels of Nrf-2 and HO-1 increased more obviously.

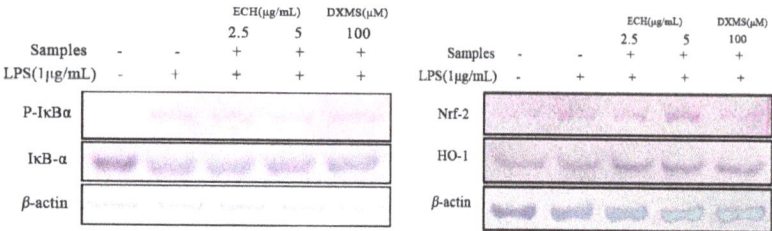

Figure 4. Effect of ECH on the expressions of inflammation and anti-oxidant signaling related proteins in LPS-induced activated RAW 264.7 cells RAW 264.7 cells were pretreated or not with indicated concentrations of ECH for 1 h then were activated with LPS (1 µg/mL) for 30 min (left) or 6 h (right). Whole cell lysates were analyzed by Western blotting analysis using specific antibodies.

3.4.5. Effect of ECH on the Nuclear Localization of NF-κB-p65 Induced by LPS

A laser confocal scanning microscope was used explore the effect of 5 µg/mL ECH on the nuclear localization of NF-κB-p65 in LPS-induced inflammatory cells (see Figure 5). Compared with the control group, NF-κB-p65 entry into the nucleus was significantly increased during LPS treatment alone, indicating with the activation of NF-κB. After 5 µg/mL ECH incubation and LPS treatment, NF-κB-p65 entry into the nucleus was significantly inhibited; DXMS-100 group treatment of NF-κB-p65 into the nucleus was also significantly inhibited.

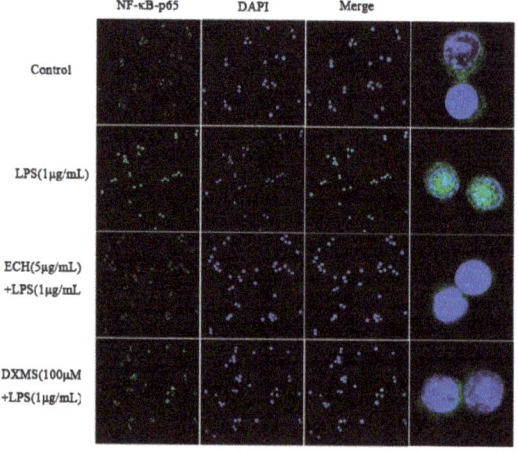

Figure 5. The inhibited effects of ECH on the transport from the cytoplasm to the nucleus of NF-κB-p65 proteins.

4. Discussion

Using the methods recommended in the European Community Directives, we found that all the physical and chemical indexes results of Xinjiang safflower honey samples were in line with the requirements. Hydroxymethylfurfural is a substance produced by honey during storage or heating. It is an important criterion for evaluating the quality of honey [43,44]. As the honey storage time becomes longer or the heating temperature rises higher, the more its content increase. Amylase is an important indicator of freshness in honey [1]. Honey enzymes will inactivate (or the value will drop) when honey is heated to a temperature higher than 40 °C. The amylase value measured in our experiment was 21.1 ± 0.3 mL/g honey, which is much higher than the 8 mL/g honey of EU standard (2001/110/EC). We did not detect hydroxymethylfurfural nor sucrose in the safflower honey samples. The moisture complied with EU standard 2001/110/EC to not be higher than 20 g/100 g. The safflower honey results for moisture were better than carob honey from Morocco as studied by Redouan et al. [45], The carob honey moisture was 19.5 g/100 g. Therefore, the Xinjiang safflower honey samples were fresh and mature honey.

As polyphenols endowed honey with distinct color, taste, flavor and biological activities. Polyphenol analysis is therefore considered as an important tool for determining the quality of honey. A total number of 14 polyphenolic compounds were preliminary characterized using the HPLC-Q-TOF/MS analysis, and the sum of each component content was 14.83 mg/kg honey. Solid phase extraction (SPE) using the Amberlite XAD-2 resin is one of the most widely used extraction methods for isolating individual phenolic and flavonoid compounds in honey [46]. In the present study, we didn't measure the phenolic compounds recovery, which is a major limitation. Nevertheless, the honey was reported with a mean recovery of 43.7% (ranging from 6.7%, gallic acid, to 65.2%, naringin) [47]. In addition, the discrepancies of the extraction recoveries have been found among different types of honey, which might be attributed to sugars as well as some other matrix constitutions. Among them, eight phenolic acid components, and the vanillic acid and p-hydroxybenzoic acid contents were higher, which were 3.196 mg/kg honey and 1.524 mg/kg honey, respectively, accounting for 65.6% of the total phenolic acid content. Protocatechuic acid, gallic acid, caffeic acid, ferulic acid, and other active components were also detected. In addition, seven flavonoid components were detected, among which quercetin and myricetin were higher, at 5.342 mg/kg honey and 1.021 mg/kg honey, respectively, accounting for 83.4% of the total flavonoid content. Rutin, quercetin-3-O-glucoside, kaempferol and naringenin were lower. Protocatechualdehyde, kaempferol-3-O-glucoside, morin, luteolin, diosmetin, chrysin, pinocembrin, galanin, and CAPE, were not detected. We propose the first reason is that different honey plants, different regions, and other environments have a great impact on the polyphenol composition. Secondly, our polyphenol extraction method has a low extraction rate of flavonoids. In a previous study, Bangladesh honey samples (Mustard flower honey, Kalijira honey, Padma flower honey, and Teel/sesame honey) had higher levels of caffeic acid and benzoic acid in phenolic acid substances, while gallic acid, kaempferol, myricetin, and naringenin were also detected [18]. Sundarban, Bangladesh, which is a multifloral honey collected from the largest mangrove forest in the world, showed various phenolic compounds, including gallic acid, vanillic acid, pyrogallol, trans-cinnamic acid, and other phenolic substances, as well as naringin, rutin, and quercetin flavonoids. Among all these phenolic compounds, trans-cinnamic acid and naringin had the highest contents [48]. There are more than 30 kinds of polyphenols in honey. Naringenin and hesperetin typically only appear in specific honeys, such as orange honey [17].

The DPPH· and ABTS$^+$ free radical scavenging capacity experiments used the IC$_{50}$ (semi-inhibitory rate, i.e., the concentration of the analyte required to scavenge 50% of free radicals) as the criterion for evaluating the antioxidant activity. The literature showed that due to the rich polyphenol content, honey has a very significant antioxidant effect. Five types of honey in Croatia showed high DPPH· free radical scavenging activity (black locust honey 125.48 mg/mL, lime honey 42.77 mg/mL, sage honey 25.04 mg/mL, chestnut honey 16.02 mg/mL, and honeydew honey 8.69 mg/mL [49]), In eight kinds of carob honey from different areas of Morocco, the DPPH free radical IC$_{50}$ values ranged from

12.54 mg/mL to 23.52 mg/mL [45] and seven kinds of Turkish single flower honey (Heather, Oak, Chestnut, Pine, Astragalus, Acacia, Lavender) demonstrated, ABTS$^+$·free radical IC$_{50}$ values from 0.06 g/mL–3.68 g/mL [50]. In our experiment, the DPPH and ABTS$^+$·free radical IC$_{50}$ values of ECH were 68.23 ± 0.40 µg/mL and 81.88 ± 0.54 µg/mL, respectively, and they had high antioxidant activity.

In this experiment, ECH can significantly reduce the mRNA levels of the related inflammatory factors iNOS, IL-1β, TNF-α and MCP-1, while up-regulating the expression of the antioxidant genes HO-1, TXRND and NQO. Due to the expression of iNOS when suppressed, the amount of NO released is reduced accordingly. Through the detection of IκBα phosphorylation and the NF-κB signaling pathway, we found that ECH was likely to inhibit the phosphorylation of IκBα to produce P-IκBα, thereby inhibiting the activation of the NF-κB signaling pathway and reducing inflammation. The Nrf-2/HO-1 signaling pathway demonstrated a protective effect against oxidative stress. Nrf-2 can regulate the gene expression of HO-1 at the transcription level, while HO-1 can inhibit the production of inflammatory factors in LPS-stimulated Raw 264.7 cells and protect the cells under oxidative stress [27]. We measured the expression of Nrf-2 and HO-1, which increased in line with the improvement of the ECH concentration. The 2.5 µg/mL and 5 µg/mL ECH anti-inflammatory and anti-oxidant effects were superior to the positive control group Dexamethasone, indicating that ECH has good anti-inflammatory and anti-oxidant effects.

ECH is rich in phenolic acids and flavonoids and also has good anti-oxidation and anti-inflammatory effects. Thus, we speculated that the existence of these compounds provides a material basis for the antioxidant and anti-inflammatory activities of safflower honey extract. Vanillic acid can play an anti-inflammatory and analgesic role through anti-oxidation and inhibiting the production of pro-inflammatory cytokines associated with NF-κB [51]. Protocatechuic acid and p-hydroxybenzoic acid showed significant antioxidant activity in thyme honey [52]. The O$_3$-H$_{15}$ bond of gallic acid is easy to break to scavenge free radicals, providing antioxidant effects [53]. Gallic acid can also block LPS-induced activation of TLR4/NF-κB (Toll-like receptor 4/nuclear factor-κB) to inhibit the inflammation of Raw 264.7 cells [54]. Caffeic acid reduces LPS-induced neuroinflammation through the regulation of cytokine networks, down-regulates NF-κB-dependent pro-inflammatory genes, and reduces oxidative stress [55]. Ferulic acid can exert anti-inflammatory activity by inhibiting nitroso-oxidative stress and pro-inflammatory cytokine production [56]. We found that the greatest common denominator of these substances is that they all contain different amounts of hydroxyl groups; thus, we infer that the vast majority of the substances involved in antioxidant activities are all related to these hydroxyl groups. It was reported that quercetin can scavenge active oxygen, inhibit damage caused by oxidative stress, and prevent the development of TNF-α and the secretion of iNOS and IL-1β in Raw 264.7 cells induced by lipopolysaccharide (LPS) [57]. Myricetin exerted anti-inflammatory effects by inhibiting NF-κB-p65 in the NF-κB pathway [58]. Rutin [59] and naringenin [60] alleviate oxidative stress and exert anti-inflammatory effects by regulating mechanisms related to the Nrf-2/HO-1 signaling pathway. Quercetin-3-O-glucoside can inhibit the expression of TNF-α, IL-1β and IL-6 and the activation of NF-κB, and simultaneously up-regulate the expression of Nrf-2 and HO-1, and inhibit the inflammation and oxidation to resist the induction of cisplatin of acute kidney injury in mice [61]. Kaempferol can up-regulate the expression of iNOS in rat articular chondrocytes stimulated by IL-1β, and can inhibit the degradation of IκBα and the activation of NF-κB in rat articular chondrocytes stimulated by IL-1β [62]. Based on this, we propose that these polyphenolic compounds in ECH are the main contributors to the anti-oxidation and anti-inflammatory effects of safflower honey.

In the present study, we must explain the limitations of our experiment. First, as the identified polyphenols might had multiple isoforms and exist in the honey in a very complex form, their quantifications in our honey samples might have some bias. Second, phenolic compounds recovery from XAD-2 resin could to be estimated which can partly counteract the strong matrix effects in the honey. Third, in this study we only preliminary detected several polyphenolic compounds in our safflower honey samples, we still did not propose a "marker" compound in our honey samples, which is with great importance for the standardization and quality control on this special honey type.

Nevertheless, our research group are keeping working on the isolation and identification on the marker compound of the safflower honey, which is helpful to discriminate from other commercial honey types. In such work accurate recovery rate will be examined on the single maker compound.

5. Conclusions

In summary, we found that the main physical and chemical indicators of safflower monofloral honey samples, a local specialty honey type of Xinjiang Providence, China were in-line with the requirements of EU or Codex standards. We also preliminary characterized 14 polyphenolic compounds in the extracts from *Carthamus tinctorius* L. honey (ECH). ECH showed a strong ability to scavenge DPPH·and ABTS$^+$·free radicals in vitro. ECH also inhibited the phosphorylation of $I\kappa B\alpha$ to activate the NF-κB signaling pathway, promoted the expression of antioxidant genes, thereby inhibiting inflammatory process induced by LPS. Our research provides basic information into development on this special honey in the future.

Supplementary Materials: The following are available online at http://www.mdpi.com/2304-8158/9/8/1039/s1. Figure S1: Micrograph of safflower pollen from a safflower honey sample.

Author Contributions: Conceptualization, L.-P.S. and K.W.; methodology, L.-P.S. and F.-F.S.; validation, L.-P.S., F.-F.S. and W.-W.Z.; formal analysis, F.-F.S.; investigation, L.-P.S.; resources, L.-P.S.; data curation, F.-F.S.; writing—original draft preparation, F.-F.S.; writing—review and editing, L.-P.S., F.-F.S., Z.-H.Z., W.-W.Z. and K.W.; visualization, F.-F.S. and Z.-H.Z.; supervision, L.-P.S. and K.W.; project administration, K.W. All authors have read and agreed to the published version of the manuscript.

Funding: This research was funded by the National Natural Science Foundation of China (Grant No. 31702287, 31602016) and the Agricultural Science and Technology Innovation Program (CAAS-ASTIP-2019-IAR).

Acknowledgments: The authors gratefully acknowledge Fuliang Hu from the School of Animal Science, Zhejiang University, for him provide the Raw 267.4 cells during the experiment.

Conflicts of Interest: The authors declare no conflict of interest.

References

1. Ma, T.; Wang, Q.; Cheng, N.; Cao, W. Effects of thermal processing on honey quality. *Food Ferment. Ind.* **2019**, *45*, 245–249. [CrossRef]
2. Mandal, M.D.; Mandal, S. Honey: Its medicinal property and antibacterial activity. *Asian Pac. J. Trop. Biomed.* **2011**, *1*, 154–160. [CrossRef]
3. Israili, Z.H. Antimicrobial Properties of Honey. *Am. J. Ther.* **2014**, *21*, 304–323. [CrossRef] [PubMed]
4. Almasaudi, S.B.; El-Shitany, N.A.; Abbas, A.T.; Abdel-dayem, U.A.; Ali, S.S.; Al Jaouni, S.S.; Steve, H. Antioxidant, Anti-inflammatory, and Antiulcer Potential of Manuka Honey against Gastric Ulcer in Rats. *Oxidative Med. Cell. Longev.* **2016**, *2016*, 1–10. [CrossRef] [PubMed]
5. Kassim, M.; Achoui, M.; Mustafa, M.R.; Mohd, M.A.; Yusoff, K.M. Ellagic acid, phenolic acids, and flavonoids in Malaysian honey extracts demonstrate in vitro anti-inflammatory activity. *Nutr. Res.* **2010**, *30*, 650–659. [CrossRef]
6. Zhao, H.; Cheng, N.; He, L.; Peng, G.; Cao, W. Hepatoprotective Effects of the Honey of Apis cerana Fabricius on Bromobenzene-Induced Liver Damage in Mice. *J. Food Sci.* **2018**, *83*, 509–516. [CrossRef]
7. Nooh, H.Z.; Nour-Eldien, N.M. The dual anti-inflammatory and antioxidant activities of natural honey promote cell proliferation and neural regeneration in a rat model of colitis. *Acta Histochem.* **2016**, *118*, 588–595. [CrossRef]
8. Almasaudi, S.B.; Abbas, A.T.; Al-Handi, R.R.; El-Shitany, N.A.; Abdel-dayem, U.A.; Ali, S.S.; Saleh, R.M.; Al Jaouni, S.K.; Kamal, M.A.; Harakeh, S.M. Manuka Honey Exerts Antioxidant and Anti-Inflammatory Activities That Promote Healing of Acetic Acid-Induced Gastric Ulcer in Rats. *Evid. Based Complement. Altern. Med.* **2017**, *2017*, 1–12. [CrossRef]
9. Shamshuddin, N.S.S.; Zohdi, R.M. Gelam honey attenuates ovalbumin-induced airway inflammation in a mice model of allergic asthma. *J. Tradit. Complement. Med.* **2016**, *8*, 39–45. [CrossRef]
10. Gül, A.; Pehlivan, T. Antioxidant Activities of some Monofloral Honey Types Produced Across Turkey. *Saudi J. Biol. Sci.* **2018**, *25*, 1056–1065. [CrossRef]

11. Guo-Tai, W.U.; Yu-Peng, W.U.; Xiao-Fei, H.E.; Wang, X.F.; Niu, T.H.; Ren, Y. Progresses on Chemistry, Pharmacology and Application of Medihoney. *J. Bee* **2017**, *1*, 3–6.
12. Dżugan, M.; Sowa, P.; Kwaśniewska, M.; Wesołowska, M.; Czernicka, M. Physicochemical Parameters and Antioxidant Activity of Bee Honey Enriched with Herbs. *Plant Foods Hum. Nutr.* **2017**, *72*, 74–81. [CrossRef] [PubMed]
13. Khan, R.U.; Naz, S.; Abudabos, A.M. Towards a better understanding of the therapeutic applications and corresponding mechanisms of action of honey. *Environ. Sci. Pollut. Res.* **2017**, *24*, 27755–27766. [CrossRef] [PubMed]
14. Sun, L.P.; Zuo-Lin, Y.I.; Jin, X.L.; Chao, J.I.; Zhang, Z.Y. Chemical analysis on the safflower(Carthamus tinctorius)honey collected from Xinjiang. *Sci. Technol. Food Ind.* **2017**, *38*, 281–285.
15. Yuan, W.; Ge, J.; Meng, W.; Ni, C.; Wei, C.; Jing, Z. A review of antioxidant activity in honey. *Food Ferment. Ind.* **2014**, *40*, 111–114.
16. Erejuwa, O.O.; Sulaiman, S.A.; Ab Wahab, M.S. Honey: A novel antioxidant. *Molecules* **2012**, *17*, 4400–4423. [CrossRef]
17. Samarghandian, S.; Farkhondeh, T.; Samini, F. Honey and Health: A Review of Recent Clinical Research. *Pharmacogn. Res.* **2017**, *9*, 121–127. [CrossRef]
18. Moniruzzaman, M.; Yung An, C.; Rao, P.V.; Hawlader, M.N.I.; Azlan, S.A.B.M.; Sulaiman, S.A.; Gan, S.H. Identification of phenolic acids and flavonoids in monofloral honey from Bangladesh by high performance liquid chromatography: Determination of antioxidant capacity. *Biomed. Res. Int.* **2014**, *2014*, 737490. [CrossRef]
19. Ruiz-Ruiz, J.C.; Matus-Basto, A.J.; Acereto-Escoffié, P.; Segura-Campos, M.R. Antioxidant and anti-inflammatory activities of phenolic compounds isolated from Melipona beecheii honey. *Food Agric. Immunol.* **2017**, *28*, 1424–1437. [CrossRef]
20. Anand, S.; Pang, E.; Livanos, G.; Mantri, N. Characterization of Physico-Chemical Properties and Antioxidant Capacities of Bioactive Honey Produced from Australian Grown Agastache rugosa and its Correlation with Colour and Poly-Phenol Content. *Molecules* **2018**, *37*, 108. [CrossRef]
21. Ying, X.; Yu, K.; Chen, X.; Chen, H.; Hong, J.; Cheng, S.; Peng, L. Piperine inhibits LPS induced expression of inflammatory mediators in RAW 264.7 cells. *Cellular Immunol.* **2013**, *285*, 49–54. [CrossRef] [PubMed]
22. Choi, W.-S.; Shin, P.-G.; Lee, J.-H.; Kim, G.-D. The regulatory effect of veratric acid on NO production in LPS-stimulated RAW264.7 macrophage cells. *Cellular Immunol.* **2012**, *280*, 164–170. [CrossRef] [PubMed]
23. Zhai, X.-T.; Zhang, Z.-Y.; Jiang, C.-H.; Chen, J.-Q.; Ye, J.-Q.; Jia, X.-B.; Yang, Y.; Ni, Q.; Wang, S.-X.; Song, J.; et al. Nauclea officinalis inhibits inflammation in LPS-mediated RAW 264.7 macrophages by suppressing the NF-κB signaling pathway. *J. Ethnopharmacol.* **2016**, *183*, 159–165. [CrossRef] [PubMed]
24. Assis, P.O.A.D.; Guerra, G.C.B.; Araújo, D.F.D.S.; Araújo, R.F.D.; Queiroga, R.d.C.R.d.E. Intestinal anti-inflammatory activity of goat milk and goat yoghurt in the acetic acid model of rat colitis. *Int. Dairy J.* **2016**, *56*, 45–54. [CrossRef]
25. Wang, K.; Ping, S.; Huang, S.; Hu, L. Molecular Mechanisms Underlying the In Vitro Anti-Inflammatory Effects of a Flavonoid-Rich Ethanol Extract from Chinese Propolis (Poplar Type). *Evid. Based Complement. Altern. Med.* **2013**, *2013*. [CrossRef]
26. Pamukcu, B.; Lip, G.Y.H.; Shantsila, E. The nuclear factor–kappa B pathway in atherosclerosis: A potential therapeutic target for atherothrombotic vascular disease. *Thromb. Res.* **2011**, *128*, 117–123. [CrossRef]
27. Roy, A.; Park, H.-J.; Jung, H.A.; Choi, J.S. Estragole Exhibits Anti-inflammatory Activity with the Regulation of NF-κB and Nrf-2 Signaling Pathways in LPS-induced RAW 264.7 cells. *Nat. Prod. Sci.* **2018**, *24*, 13–20. [CrossRef]
28. Liu, J.; Tang, J.; Zuo, Y.; Yu, Y.; Luo, P.; Yao, X.; Dong, Y.; Wang, P.; Liu, L.; Zhou, H. Stauntoside B inhibits macrophage activation by inhibiting NF-κB and ERK MAPK signalling. *Pharmacol. Res.* **2016**, *111*, 303–315. [CrossRef]
29. Kaspar, J.W.; Niture, S.K.; Jaiswal, A.K. Nrf2:INrf2 (Keap1) signaling in oxidative stress. *Free Radic. Biol. Med.* **2009**, *47*, 1304–1309. [CrossRef]
30. Song, D.; Lu, Z.; Wang, F.; Wang, Y. 041 Biogenic nano-selenium particles effectively attenuate oxidative stress–induced intestinal epithelial barrier injury by activating the Nrf2 antioxidant pathway. *J. Anim. Sci.* **2017**, *95*, 20–21. [CrossRef]

31. Hou, X.; Xua, X.; Li, H.; He, S.; Wana, C.; Yina, P.; Liu, M.; Liu, F.; Xu, J. Punicalagin Induces Nrf2/HO-1 expression via upregulation of PI3K/AKT pathway and inhibits LPS-induced oxidative stress in RAW264.7 Macrophages. *Mediat. Inflamm.* **2015**, *2015*, 1–11. [CrossRef] [PubMed]
32. Kadri, S.M.; Zaluski, R.; Pereira Lima, G.P.; Mazzafera, P.; de Oliveira Orsi, R. Characterization of Coffea arabica monofloral honey from Espírito Santo, Brazil. *Food Chem.* **2016**, *203*, 252–257. [CrossRef] [PubMed]
33. Xue-Feng, M.U.; Sun, L.P.; Xiang, X.U.; Pang, J.; Wei, H.E.; Huang, L.; Shen, X.F. Optimization of Purification Process for DPPH Free Radical Scavenging Components from Acidic Aqueous Extract from Chinese Date Honey by Macroporous Resin Adsorption. *Food Sci.* **2011**, *32*, 98–102.
34. Guo, N.; Zhao, L.; Zhao, Y.; Li, Q.; Xue, X.; Wu, L.; Gomez Escalada, M.; Wang, K.; Peng, W. Comparison of the Chemical Composition and Biological Activity of Mature and Immature Honey: An HPLC/QTOF/MS-Based Metabolomic Approach. *J. Agric. Food Chem.* **2020**, *68*, 4062–4071. [CrossRef]
35. Wu, H.C.; Chen, H.-M.; Shiau, C.-Y. Free amino acids and peptides as related to antioxidant properties in protein hydrolysates of mackerel (Scomber austriasicus). *Food Res. Int.* **2003**, *36*, 949–957. [CrossRef]
36. Yang, H.; Dong, Y.; Du, H.; Shi, H.; Li, X. Antioxidant Compounds from Propolis Collected in Anhui, China. *Molecules* **2011**, *16*, 3444–3455. [CrossRef]
37. Gutierrez, R.M.P.; Hoyo-Vadillo, C. Anti-inflammatory Potential of Petiveria alliacea on Activated RAW264.7 Murine Macrophages. *Pharmacogn. Mag.* **2017**, *13*, 174. [CrossRef]
38. Li, Y.; Meng, T.; Hao, N.; Tao, H.; Zou, S.; Li, M.; Ming, P.; Ding, H.; Dong, J.; Feng, S. Immune regulation mechanism of Astragaloside IV on RAW264.7 cells through activating the NF-κB/MAPK signaling pathway. *Int. Immunopharmacol.* **2017**, *49*, 38–49. [CrossRef]
39. Cui, C.; Lu, H.; Hui, Q.; Lu, S.; Liu, Y.; Ahmad, W.; Wang, Y.; Hu, P.; Liu, X.; Cai, Y.; et al. A preliminary investigation of the toxic effects of Benzylpenicilloic acid. *Food Chem. Toxicol.* **2018**, *111*, 567–577. [CrossRef]
40. Ah, L.H.; Ram, S.B.; Ryeong, K.H.; Eun, K.J.; Bin, Y.W.; Ju, P.J.; Lim, L.M.; Young, C.J.; Seob, L.H.; Youn, H.D. Butanol extracts of Asparagus cochinchinensis fermented with Weissella cibaria inhibit iNOS-mediated COX-2 induction pathway and inflammatory cytokines in LPS-stimulated RAW264.7 macrophage cells. *Exp. Ther. Med.* **2017**, *14*, 4986–4994.
41. Wang, K.; Hu, L.; Jin, X.-L.; Ma, Q.-X.; Marcucci, M.C.; Netto, A.A.L.; Sawaya, A.C.H.F.; Huang, S.; Ren, W.-K.; Conlon, M.A.; et al. Polyphenol-rich propolis extracts from China and Brazil exert anti-inflammatory effects by modulating ubiquitination of TRAF6 during the activation of NF-κB. *J. Funct. Foods* **2015**, *19*, 464–478. [CrossRef]
42. Wang, B.; Chang, H.; Su, S.; Sun, L.; Wang, K. Antioxidative and Anti-inflammatory Activities of Ethanol Extract of Geopropolis from Stingless Bees. *Sci. Agric. Sin.* **2019**, *52*, 939–948.
43. Özkök, D.; Silici, S. Effects of honey HMF on enzyme activities and serum biochemical parameters of Wistar rats. *Environ. Sci. Pollut. Res.* **2016**, *23*, 20186–20193. [CrossRef] [PubMed]
44. Shapla, U.M.; Solayman, M.; Alam, N.; Khalil, M.I.; Gan, S.H. 5-Hydroxymethylfurfural (HMF) levels in honey and other food products: Effects on bees and human health. *Chem. Cent. J.* **2018**, *12*, 35. [CrossRef]
45. El-Haskoury, R.; Kriaa, W.; Lyoussi, B.; Makni, M. Ceratonia siliqua honeys from Morocco: Physicochemical properties, mineral contents, and antioxidant activities. *J. Food Drug Anal.* **2018**, *26*, 67–73. [CrossRef] [PubMed]
46. Michalkiewicz, A.; Biesaga, M. Solid-phase extraction procedure for determination of phenolic acids and some flavonols in honey. *J. Chromatogr. A* **2008**, *1187*, 18–24. [CrossRef]
47. Yung An, C.; Hossain, M.M.; Alam, F.; Islam, M.A.; Khalil, M.I.; Alam, N.; Gan, S.H. Efficiency of Polyphenol Extraction from Artificial Honey Using C_{18} Cartridges and Amberlite®XAD-2 Resin: A Comparative Study. *J. Chem.* **2016**, *2016*, 8356739. [CrossRef]
48. Afroz, R.; Tanvir, E.M.; Paul, S.; Bhoumik, N.C.; Gan, S.H.; Khalil, M.I. DNA Damage Inhibition Properties of Sundarban Honey and its Phenolic Composition. *J. Food Biochem.* **2016**, *40*, 436–445. [CrossRef]
49. Flanjak, I.; Kenjerić, D.; Bubalo, D.; Primorac, L. Characterisation of selected Croatian honey types based on the combination of antioxidant capacity, quality parameters, and chemometrics. *Eur. Food Res. Technol.* **2016**, *242*, 467–475. [CrossRef]
50. Kaygusuz, H.; Tezcan, F.; Bedia Erim, F.; Yildiz, O.; Sahin, H.; Can, Z.; Kolayli, S. Characterization of Anatolian honeys based on minerals, bioactive components and principal component analysis. *LWT Food Sci. Technol.* **2016**, *68*, 273–279. [CrossRef]

51. Calixto-Campos, C.; Carvalho, T.T.; Hohmann, M.S.N.; Pinho-Ribeiro, F.A.; Fattori, V.; Manchope, M.F.; Zarpelon, A.C.; Baracat, M.M.; Georgetti, S.R.; Casagrande, R.; et al. Vanillic Acid Inhibits Inflammatory Pain by Inhibiting Neutrophil Recruitment, Oxidative Stress, Cytokine Production, and NFκB Activation in Mice. *J. Nat. Prod.* **2015**, *78*, 1799–1808. [CrossRef] [PubMed]
52. Spilioti, E.; Jaakkola, M.; Tolonen, T.; Lipponen, J.; Virtanen, V.; Chinou, I.; Kassi, E.; Karabournioti, S.; Moutsatsou, P. Phenolic acid composition, antiatherogenic and anticancer potential of honeys derived from various regions in Greece. *PLoS ONE* **2014**, *9*, e94860. [CrossRef] [PubMed]
53. Rajan, V.K.; Muraleedharan, K. A computational investigation on the structure, global parameters and antioxidant capacity of a polyphenol, Gallic acid. *Food Chem.* **2017**, *220*, 93–99. [CrossRef] [PubMed]
54. Huang, L.; Hou, L.; Xue, H.; Wang, C. Gallic acid inhibits inflammatory response of RAW264.7 macrophages by blocking the activation of TLR4/NF-κB induced by LPS. *Chin. J. Cell. Mol. Immunol* **2016**, *32*, 1610–1614.
55. Basu Mallik, S.; Mudgal, J.; Nampoothiri, M.; Hall, S.; Dukie, S.A.; Grant, G.; Rao, C.M.; Arora, D. Caffeic acid attenuates lipopolysaccharide-induced sickness behaviour and neuroinflammation in mice. *Neurosci. Lett.* **2016**, *632*, 218–223. [CrossRef] [PubMed]
56. Sadar, S.S.; Vyawahare, N.S.; Bodhankar, S.L. Ferulic acid ameliorates TNBS-induced ulcerative colitis through modulation of cytokines, oxidative stress, iNOs, COX-2, and apoptosis in laboratory rats. *EXCLI J.* **2016**, *15*, 482–499.
57. Batiha, G.E.-S.; Besbishy, A.M.; Ikram, M.; Mulla, Z.S.; El-Hack, M.E.A. The Pharmacological Activity, Biochemical Properties, and Pharmacokinetics of the Major Natural Polyphenolic Flavonoid: Quercetin. *Foods* **2020**, *9*, 374. [CrossRef]
58. Hou, W.; Hu, S.; Su, Z.; Wang, Q.; Meng, G.; Guo, T.; Zhang, J.; Gao, P. Myricetin attenuates LPS-induced inflammation in RAW 264.7 macrophages and mouse models. *Future Med. Chem.* **2018**, *10*, 2253–2264. [CrossRef]
59. Tian, R.; Yang, W.; Xue, Q.; Gao, L.; Huo, J.; Ren, D.; Chen, X. Rutin ameliorates diabetic neuropathy by lowering plasma glucose and decreasing oxidative stress via Nrf2 signaling pathway in rats. *Eur. J. Pharmacol.* **2016**, *771*, 84–92. [CrossRef]
60. De Oliveira, M.R.; Andrade, C.M.B.; Fürstenau, C.R. Naringenin Exerts Anti-inflammatory Effects in Paraquat-Treated SH-SY5Y Cells Through a Mechanism Associated with the Nrf2/HO-1 Axis. *Neurochem. Res.* **2018**, *43*, 894–903. [CrossRef]
61. Chao, C.-S.; Tsai, C.-S.; Chang, Y.-P.; Chen, J.-M.; Chin, H.-K.; Yang, S.-C. Hyperin inhibits nuclear factor kappa B and activates nuclear factor E2-related factor-2 signaling pathways in cisplatin-induced acute kidney injury in mice. *Int. Immunopharmacol.* **2016**, *40*, 517–523. [CrossRef] [PubMed]
62. Zhuang, Z.; Ye, G.; Huang, B. Kaempferol Alleviates the Interleukin-1β-Induced Inflammation in Rat Osteoarthritis Chondrocytes via Suppression of NF-κB. *Med. Sci. Monit.* **2017**, *23*, 3925–3931. [CrossRef] [PubMed]

© 2020 by the authors. Licensee MDPI, Basel, Switzerland. This article is an open access article distributed under the terms and conditions of the Creative Commons Attribution (CC BY) license (http://creativecommons.org/licenses/by/4.0/).

Article

Botanical Origin, Pollen Profile, and Physicochemical Properties of Algerian Honey from Different Bioclimatic Areas

Mounia Homrani [1], Olga Escuredo [2], María Shantal Rodríguez-Flores [2], Dalache Fatiha [1], Bouzouina Mohammed [3], Abdelkader Homrani [1] and M. Carmen Seijo [2,*]

1. Laboratory of Sciences and Technics of Animal Production (LSTPA), Abdelhamid Ibn Badis University (UMAB), 27000 Mostaganem, Algeria; mounia-homrani@hotmail.fr (M.H.); fdalache2@yahoo.fr (D.F.); abdelkader.homrani@univ-mosta.dz (A.H.)
2. Department of Vegetal Biology and Soil Sciences, Faculty of Sciences, University of Vigo, As Lagoas, 32004 Ourense, Spain; oescuredo@uvigo.es (O.E.); mariasharodriguez@uvigo.es (M.S.R.-F.)
3. Laboratory of Vegatal Protection, Abdelhamid Ibn Badis University (UMAB), 27000 Mostaganem, Algeria; lvp@univ-mosta.dz
* Correspondence: mcoello@uvigo.es

Received: 27 May 2020; Accepted: 9 July 2020; Published: 16 July 2020

Abstract: The palynological and physicochemical analysis of 62 honey samples produced in different biogeographical areas of Algeria was conducted. Results showed high variety in the botanical origin of samples and their physicochemical profile. Twenty-six samples were polyfloral honey, 30 were unifloral honey from different botanical sources such as *Eucalyptus*, *Citrus*, *Apiaceae*, *Punica*, *Erica*, *Rosmarinus*, *Eriobotrya*, or *Hedysarum*, and 6 were characterized as honeydew honey. Pollen analysis allowed the identification of 104 pollen types belonging to 51 botanical families, whereas the physicochemical profile showed important variations between samples. Multivariate techniques were used to compare the characteristics of samples from different biogeographical areas, showing significant differences between humid-area samples, located in the northeast of the country, and samples taken in semiarid, subhumid, and arid zones. Principal-component analysis (PCA) extracted nine components explaining 72% of data variance, being 30%, the sum of Component 1 and Component 2. The plot of both components showed samples grouped upon botanical and geographical origin. The results of this paper highlighted the great variability in honey production of Algeria, evidencing the importance of honey characterization to guarantee authenticity and to valorize local production.

Keywords: characterization; Algerian honey; botanical origin; biogeographical origin

1. Introduction

Beekeeping in Algeria is part of agricultural life. It is practiced in mountainous regions such as the Aures Mountains, Kabylie, and Dahra, in the coastal plains, and the valleys of the big wadis, but it is more intensive in the northern part of the country where the flora provides resources for honey throughout most of the year [1]. Beekeeping plays an important role in the rural areas of the Mediterranean region due to the suitable climatic conditions and high biodiversity. Highly valued honey supplies are a welcome income for farmers and hobby beekeepers [2].

The composition of this food, obtained by bees from nectar, honeydew, or both sugary resources, is highly dependent on honeybee activity and the biogeography of the area in which it was produced. So, physicochemical properties such as color, pH, electrical conductivity, sugar content, organic acids, ash content, polyphenols, flavonoids, and other phytochemicals differ with different environmental conditions and plant communities visited by honeybees [3].

The Mediterranean region of Algeria is an outstanding biogeographical crossroads that resulted from a complex history and highly heterogeneous environmental factors [4]. Algeria has significant plant biodiversity, with about 3150 species [5], of which more than 500 species are endemic and characteristic of the Mediterranean basin, the Atlas Mountains, the high steppe plateau, the large Saharan plateau, and the mountainous massifs of the Algerian Sahara [6]. The melliferous area in Algeria is estimated at 797,122 hectares with a predominance of forests and maquis, natural meadows, orchards with orange fruits, and other crops.

The main honey sources in the area are trees like *Eucalyptus* and some wild herbaceous plants, mainly from the *Apiaceae* (*Foeniculum, Daucus, Coriandrum, Eryngium, Pimpinella*), Asteraceae (*Galactites tomentosa, Centaurea, Echinops*), Brassicaceae (*Brassica* and *Sinapis*), and Fabaceae (*Hedysarum, Ononis natrix, Trifolium, Melilotus*) families [7–9]. Some plants, such as *Ziziphus lotus*, or grown plants, such as *Helianthus annuus* or *Citrus*, were also named as important plants sources for honey production [10]. In this context, the main honey types produced in the country are *Eucalyptus, Citrus*, forest (honeydew), jujube, sunflower, rosemary, and wild mustard. However, the production of other rare or less known honey types was mentioned in some papers [11,12].

The country is considered a traditional consumer of honey, but national production does not achieve self-sufficiency, so to cover the relevant needs, large quantities of honey are imported every year from countries such as China, India, and Saudi Arabia. The lack of national legislation and unknowns about Algerian honey characteristics have a negative impact on the scarce development of Algerian beekeeping. In the absence of specific legislation, there are also no requirements to assess the quality or geographical and botanical origin of honey in the market. Therefore, consumers and local producers are not protected against frauds, and honey of good quality is not valorized. Improving knowledge in the characteristics of local production is the main course of action for their valorization and to link agricultural products with the territory in which they were produced. These studies facilitate the recognition of specific properties with botanical origin and allow differentiating local production from imported honey. Currently, there is no designation of origin for local honey that would contribute to valorizing Algeria honey production.

Some authors studied the characteristics of honey produced in Algeria and found differing results as to the pollen content [8–10,13], physicochemical attributes and pollen profile of samples from different areas [7,14–16], their botanical origin [9,11,17–19], antibacterial and antioxidant activity [20], or specific compounds such as sugars and phenolic or volatile compounds [21,22]. However, honey production in Algeria remains poorly studied. On the basis of these facts, this paper aimed to provide scientific information on the botanical origin and physicochemical profile of honey samples from different bioclimatic areas of Algeria.

2. Material and Methods

2.1. Study-Area Bioclimatology

The territory of Algeria is composed of four principal structural units. Near the coast, in the north, there is a narrow region with several hills and plains named Tell. This region is densely populated and constitutes the main agricultural land in the country. It has a Mediterranean climate with mild winters and moderate rainfall. Mean annual temperatures are close to 22 °C in summer and 10 °C in winter. Rainfall is abundant along this coastal area, increasing the amount of annual precipitation from 400 to 670 mm in the west to near 1000 mm in the east. These weather conditions determined a subhumid-humid climate tendency.

In the central area, several mountain chains were determined as the high plateau area that separates the Mediterranean area from the Saharan Atlas and the desert. This is a highly diverse area dominated in the plains by steppelike morphology. This region is characterized by a semiarid-to-arid climate with irregular and low precipitation (200–400 mm); its bowl profile explains the presence of many salt lakes (called "chotts") collecting surface water. It continues to the south with the Saharan Atlas that delimits the Sahara desert. The Saharan Atlas is formed by three massifs that are the source of watercourses that supply the wells of oases along the northern edge of the desert; the massifs are Biskra, Laghouat, and Béchar. The Sahara is a windy and very arid area with a continental climate, high thermal amplitude, and extremely poor precipitation of no more than 130 mm.

2.2. Honey Samples

We collected 62 honey samples from different bioclimatic areas of Algeria (Table 1). Concretely, 34 samples were from the semiarid area, 15 were obtained in subhumid areas, 4 in humid areas, and 9 samples in the arid areas of the Sahara and Saharan Atlas (Figure 1). The samples were obtained directly from beekeepers during the 2015–2016 period and were stored refrigerated at 4 °C until analysis. All determinations were carried out after sample homogenization.

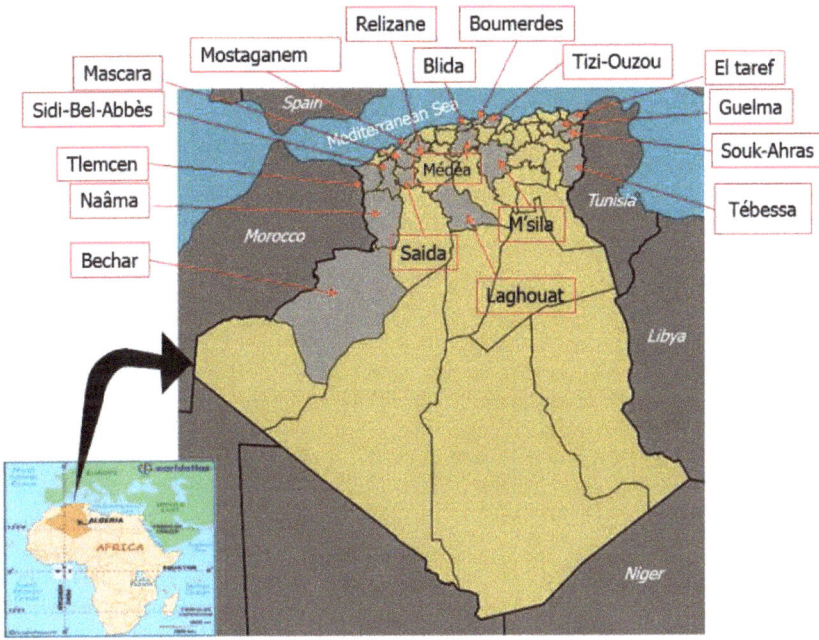

Figure 1. Geographical areas where the samples were taken.

Table 1. Geographical origin of samples, climate, and harvest period.

Localities	Geographical Situation	Climate	Harvest Period	No. Samples
Bechar	SW, Sahara	Arid	Summer	1
Laghouat	S, Sahara	Arid	Summer	1
M'Sila	E, High plateau	Arid	Summer	2
Naama	W, Sahara	Arid	Summer	3
Tebessá	E, High plateau	Arid	Spring, Summer	2
El Taref	NE, Tell	Humid	Summer	4
Boumerdes	N, Tell	Semiarid	Summer	1
Mascara	NW, Tell	Semiarid	Summer	2
Mostaganem	NW, Tell	Semiarid	Spring, Summer	15
Relizane	NW, Tell	Semiarid	Summer	2
Saida	NW, High plateau	Semiarid	Summer	3
Sidi Bel Abbés	NW, Tell	Semiarid	Summer, Winter	3
Souk-Ahras	NE, High plateau	Semiarid	Summer	1
Médea	N, High plateau	Semiarid	Summer	7
Blida	N, Tell	Subhumid	Spring	2
Guelmá	NE, Tell	Subhumid	Summer	5
Tizi-Ouzou	N, Tell	Subhumid	Summer	3
Tlemcen	NW, Tell	Subhumid	Summer	5

2.3. Melissopalynological Analysis

2.3.1. Quantitative Pollen Analysis

Ten grams of honey were weighed and dissolved in 20 mL of distilled water. The solution was centrifuged for 10 min at 4500 rpm, and the supernatant was removed until a final volume of 5 mL. The sediment was stirred, and then two aliquots of 10 µL were deposited over a slide to microscopically examine (Nikon UK Ltd., Surbiton, UK). All pollen in the aliquot was counted. The results are expressed as the number of pollen grains per gram of honey considering the average value of the two aliquots.

2.3.2. Qualitative Pollen Analysis

Ten grams of honey were weighed and fully dissolved in 20 mL of distilled water. The solution was centrifuged for 10 min at 4500 rpm, and the supernatant was drawn off. The sediment was again washed with distilled water, and another centrifugation in the same conditions was done. Then, two aliquots (100 µL) of the sediment were deposited into a slide over a heated surface until desiccation. Lastly, drops were covered with a 24 × 24 mm cover glass containing a drop of glicerogelatin with fuchsine.

Slide observation was carried out with a light microscope at 400× or 1000×, as appropriate to improve the identification of pollen types. For each honey sample, at least 500 pollen grains were counted. All pollen grains were classified as the level of pollen type due to difficulties in accessing information about the wild botanical species of apicultural interest in a large part of the territory.

The relative frequency, expressed as a percentage, of all identified pollen type was considered for the pollen spectra of the samples. The following frequency classes were used to establish the representation of pollen types: dominant pollen (≥45%), accompanying pollen (from ≥15% to <45%), important pollen (from ≥3% to <15%), minor pollen (from ≥1% to <3%) and present pollen (≤1%). However, for unifloral honeys, the percentage of the main pollen type was calculated excluding plants considered to be non-nectariferous species, namely, *Acacia, Buxus sempervirens, Casuarina, Cannabis, Chamaerops, Cistus, Cyperus, Olea europaea, Papaver rhoeas, Paronychia argentea, Pinus, Pistacia, Poaceae,* and *Quercus*.

2.3.3. Physicochemical Analysis

The quality parameters of moisture, electrical conductivity, pH, and hydroxymethylfurfural content were determined following the methodology of the International Honey Commission [23]. All determinations were made in duplicate. The results are expressed as the mean of the obtained values.

Moisture was determined with a Carl Zeiss Jena refractometer (Zeiss, Oberkochen, Germany) by measuring the refractive index at 20 °C. Moisture content was calculated using the Wedmore table, and results were expressed as percentages.

Electrical conductivity was measured at 20 °C in a 20% (w/v) honey solution (dry-matter basis) in CO_2-free deionized distilled water with an EUTECH instrument conductivity meter (Thermo Fisher Scientific, Massachusetts, USA), and results were expressed as mS/cm; pH was measured by a pH meter (WTW in Lab pH 750) in a solution containing 10 g of honey in 75 mL of distilled water.

Hydroxymethyfurfural (HMF) content was determined using the White spectrophotometric method. Briefly, 5 g of honey was dissolved in 25 mL of distilled water and transferred to a volumetric flask of 50 mL; then, 0.5 mL of Carrez Solution I and 0.5 mL of Carrez Solution II were added. The final volume of 50 mL was set with distilled water. The honey solution was filtered, and the first 10 mL of the filtrate was rejected. Lastly, aliquots of 5 mL were pipetted into 2 tubes (reference and sample solution). Then, 5 mL of sodium bisulfite solution 0.2% was added to the reference, and 5 mL of water was added to the sample solution. The absorbance of the reference against the sample solution was determined at 284 and 336 nm with a UV-vis spectrophotometer (Fisher Scientific, Leicestershire, UK).

Diastase activity was determined through the amount of starch converted by a honey solution. The absorbance of yielded blue during the reactions was spectrophotometrically determined at 660 nm with a UV-vis spectrophotometer (Jenway 6305 UV-Visible Spectrophotometer, Staffordshire, UK) at different times until an endpoint of less than 0.235. Diastase activity was calculated as diastase index (DI) or grams of starch hydrolyzed at 40 °C each hour per 100 g of honey.

2.4. Color Determination

Prior to color determination, all samples were decrystallized and left for 20 min in an ultrasound bath to avoid bubbles. Sample color was determined using a HANNA Honey colorimeter (HANNA Instruments, Bedfordshire, UK). This is an instrument that gives the transmittance of honey using glycerol as a reference. Samples were introduced in square optical cuvettes of 10 mm sides, and the color value was taken directly; results are expressed in mm Pfund.

2.5. Determination of Total Polyphenol and Flavonoid Content

The Folin-Ciocalteu spectrophotometric method adapted to honey was used to the quantification of polyphenol content [24]. A UV-vis spectrophotometer (Jenway 6305, Staffordshire, UK) was used for this purpose. Absorbance at 765 nm of a honey solution (0.1 g/mL) that reacted with the Folin-Ciocalteu reactive was determined. Ethanolic solutions of gallic acid in different concentrations (0.01–0.50 mg/mL) were used as a standard to construct the calibration curve. The linearity of the curve was 0.997 (R^2). The polyphenol content of the samples was expressed as gallic acid equivalents in mg/100 g.

Total flavonoid content was measured using a similar spectrophotometric method based on an adaptation of the Dowd method [25]. A solution of aluminum chloride reacted with the flavonoids of the samples, prepared in a concentration of 0.33 g/mL with methanol. Absorbance of yellow yielded by the reaction was measured at 425 nm. Different concentrations of the quercetin flavonoid (0.002–0.01 mg/mL) were employed to construct the calibration curve; linearity was 0.998 (R^2). The flavonoid content of honey samples was expressed as mg equivalent of quercetin per 100 g.

2.6. Radical-Scavenging Activity

The method used to evaluate the antioxidant activity of honey is based on the discoloration of a 2,2-diphenyl-1-picrilhidrazil (DPPH) solution. The methodology measures the radical-scavenging activity of a honey solution against DPPH by spectrophotometry [26]. A solution (0.1 g/mL) of each honey sample in methanol was prepared. Then, 0.3 mL of the honey solution was mixed with 2.7 mL of the DPPH solution (6×10^{-5} M); a blank was also prepared. The honey sample solution and the blank were maintained in the dark at room temperature for 30 min. Then, absorbance at 517 nm was measured with a UV-vis spectrophotometer (Jenway 6305, Staffordshire, UK). Radical Scavenging Activity (RSA) percentage was calculated considering DPPH discoloration for each sample, tested as follows: RSA = [(AbsB − AbsS)/AbsB] × 100, where AbsB is the absorbance of the blank, and AbsS is the absorbance of the honey sample solution.

2.7. Sugar-Composition Analysis

Sugar composition was determined with an ion Dionex ICS-3000 chromatography system (Sunnyvale, CA, USA). The system separated sugars by using an analytical polyvinylidene/polyvinyl benzene CarboPac PA1 column (Dionex 3 × 250 mm) suitable for mono-, di-, and trisaccharides, and oligosaccharide analysis in general, and a pulsed amperometry detector with a gradient of two mobile phases (A and B). Phase A involved ultrapure water, while Phase B involved 200 mM NaOH (HPLC grade, Merck, Kenilworth, NJ, USA). The sugars of honey solutions (10 mg/L) were calculated using the calibration curves of the standard solution for each pure sugar (Sigma, Aldrich, St. Louis, MO, USA). The CHROMELEON Chromatography Management System was used for chromatogram acquisition. The concentrations of the identified sugars (fructose, glucose, sucrose, turanose, trehalose, and maltose) were expressed as g/100 g of honey.

2.8. Statistical Analysis

Multivariate statistical analysis was applied to identify differences and similarities between samples. IBM SPSS statistics software 23.0 (IBM, Armonk, NY, USA) and Statgraphics centurion V18 (The Plains, USA) were used. First, one-way ANOVA was carried out using the bioclimatic area as the factor, and the main pollen grains (those that presented values over 3% of the pollen spectra), physicochemical parameters, bioactive compounds, and sugars as variables. Differences between groups were tested through post hoc comparison using the Bonferroni test. Significance was calculated for $p < 0.05$. Lastly, an exploratory technique like principal-component analysis (PCA) was performed. This technique allows to reduce dimensionality in data, increasing the visibility of the relationship between introduced variables in analysis. The used variables were the same as those mentioned in ANOVA. Samples were considered as cases and marked according to their botanical origin.

3. Results and Discussion

3.1. Pollen Spectra and Content of Honey Samples

The samples were from a large area of northern Algeria (Table 1). We collected 33 samples in the northwest and west of the country, 13 in the north (central part), 14 in the east and northeast, and 2 from southern areas. Sampling covered different bioclimatic regions, the most common being the semiarid region that provided 34 samples, 15 were from Mediterranean subhumid areas, 4 from humid areas (with more than 1000 mm of annual rain), and 9 from arid areas with annual precipitation below 400 mm. Most samples were produced in Tell (42 samples), the most populated region, where agrosystems and forest areas are common. Fifteen were from the steppes of the high plateau, and 5 from the Sahara.

Palynological analysis of the samples showed high diversity in the plants represented in the pollen spectra. Most samples contained more than 20 different pollen grains, and some samples even more than 30.

A total of 104 pollen types belonging to 51 botanical families were identified in the 62 samples. Pollen types *Eucalyptus*, *Olea europaea*, *Brassica napus*, *Echium*, *Ziziphus lotus*, *Papaver rhoeas*, *Genista*, *Foeniculum*, *Hedysarum*, *Tamarix*, *Eryngium campestre*, and *Pimpinella anisum* were presented in more than 50% of the samples. Some of them reached maximal values in pollen spectra higher than 80%. These were *Eucalyptus*, *O. europaea*, *Z. lotus*, and *Foeniculum*. Other common dominant pollen types (≥45% of pollen spectra) were *B. napus* (maximum of 69.0%), *Genista* type (58.8%), *Tamarix* (58.1%), *Eryngium campestre* (52.1%), and *Hedysarum* (45.9%) (Table 2). Sporadically, other pollen types were dominant: *Eriobotrya* (75.4%), *Eruca sativa* (75.1%), *Melilotus* (68.7%), *Punica granatum* (56.2%), and *Erica arborea* (55.1%). The rest of the pollen types were always identified with percentages below 45% of the pollen spectra. Some samples stood out with secondary pollen grains from the Arecaceae family, such as *Chamaerops* (maximum value of 37.3%), *Capparis* (maximum of 36.7%), *Asparagus* (34.3%) or *Paronychia argentea* (maximum of 28.2%).

Quantitative pollen analysis showed 3 samples (pressed honey) with pollen content higher than 100,000 grains of pollen per gram of honey. Two samples had a pollen content between 50,000 and 100,000 pollen grains/g honey (Class IV of Maurizio), 16 were from Class III (10,000–50,000 pollen grain/g honey), 30 had Class II pollen content (2000–10,000 pollen grain/g honey), and 11 a pollen content lower than 2000 pollen grain/g honey (Class I).

3.2. Sample Palynological Profile Regarding Bioclimatic Areas of Origin

Eucalyptus, *Z. lotus* and some *Apiaceae* such as *Foeniculum* or *Eryngium* pollen types were the dominant pollen in many samples from the semiarid and subhumid Tell regions. Other pollen types, such as *Tamarix* or *Hedysarum* were also common in these samples (Figure 2). Less common pollen types, such as *P. granatum* or *Eriobotrya*, were also found as dominant pollen in samples from this origin. The first appeared in three summer honey samples, and the second one in a single sample collected in winter. There stood out 11.1% of samples with high percentages of *O. europaea* pollen, sometimes over 80% of pollen spectra. This was mainly associated with pressed honey and samples that are extraordinarily rich in pollen grains, indicating that this procedure introduces high quantities of pollen grains in honey. Accompanying or important pollen types were found, pollen grains from herbaceous plants common in agrosystems like *Hedysarum*, *Melilotus*, *Onobrychis*, *B. napus*, *Phacelia*, *Echium* or *Melilotus*. Other well-represented pollen types were those from *Apiaceae* such as *Thapsia* or *Pimpinella*. Plants from semiarid and salted lands, named *Tamarix* or *P. argentea*, were frequent in many samples. One of the most representative pollen grains corresponded to *Citrus*, which determined the botanical origin of some samples. In addition to this, *Asparagus* pollen was found with high percentages in samples from Tizi Ouzou and Tlemcen, and *Acacia* in Mostaganem samples. The apicultural value of some *Acacia* species should be studied even though it is considered non-nectariferous. Nectar secretion was found in different *Acacia* species [27] and some *Acacia* spp. honey from different countries were studied [28,29].

Table 2. Main palynological characteristics regarding geographical origin.

Localities	Dominant Pollen (≥45%)	Accompanying Pollen (45–15%)	Important Pollen (15–3%)	Other Pollen (3–1%)
Bechar (n = 1)	Olea (82.8% [1])/1 [2]		Tamarix, Citrus	Sinapis
Blida (n = 2)	Foeniculum (82.8%)/1 Eucalyptus (46.1%)/1	Hedysarum (35.4%)	Pimpinella, Echium, Genista, Punica, Citrus	Paronychia, Brassica
Boumerdes (n = 1)	-	Pimpinella (23.3%), Ziziphus (17.5%)	Thapsia, Apium, Eryngium, Echium, Hedysarum, Eucalyptus, Tamarix	Brassica
El Taref (n = 4)	Eucalyptus (85.7%)/2 Erica (55.1%)/1	Genista (26.3%)	Melilotus, Onobrychis, Malus	Foeniculum, Trifolium. Papaver, Punica, Phacelia
Guelma (n = 5)	Eucalyptus (50.5%)/1	Olea (37.2%), Ziziphus (22.7%), Punica (42.4%), Hedysarum (15.5%), Pimpinella (29.8%), Phacelia (17.6%), Myrtus (17.3%)	Eryngium, Thapsia, Carduus, Borago, Echium, Erica, Genista, Melilotus, Ononis, Trifolium, Papaver, Malus, Ailanthus	Pistacia, Artitalicisia, Brassica, Onobrychis, Peganum, Capparis
Laghouat (n = 1)	Punica (56.2%)/1		Ziziphus, Olea	Acacia, Centaurea, Brassica
M'Sila (n = 2)	Brassica (69.03%)/2	Galega (21.1%)	Artitalicisia, Rosmarinus, Thymus, Capparis	Euphorbia, Ziziphus, Daphne
Mascara (n = 2)	Eucalyptus (69.9%)/1 Olea (56.8%)/1	Tamarix (26.3%)	Genista, Hedysarum, Quercus, Citrus	Pistacia, Thapsia, Melilotus, Chamaerops, Prunus
Médea (n = 7)	Eryngium (52.1%)/1	Ziziphus (41.5%), Melilotus (15.6%), Tamarix (37.0%), Eryngium (27.6%), Hedysarum (19.4%), Pimpinella (16.1%), Eucalyptus (43.0%)	Foeniculum, Pimpinella, Thapsia, Artitalicisia, Carduus, Centaurea, Cichorium, Echium, Brassica, Chenopodium, Convolvulus, Genista, Onobrychis, Quercus, Olea, Papaver, Capparis	Apium, Thymus, Cistus, Cyperus, Euphorbia, Allium, Chamaerops, Phacelia
Mostaganitalic (n = 15)	Eucalyptus (88.9%)/4 Genista (58.8%)/1 Punica (48.6%)/1	Olea (35.7%), Tamarix (19.8%), Melilotus (24.8%), Genista (30.8%), Capparis (24.4%), Paronychia (28.2%), Brassica (15.2%)	Schinus, Foeniculum, Centaurea, Chrysantitalicum, Echium, Sinapis, Buxus, Convolvulus, Acacia, Ceratonia, Hedysarum, Muscari, Chamaerops, Papaver, Ziziphus, Rubus, Citrus, Ailanthus	Melia, Smilax
Naama (n = 3)	Ziziphus (81.5%)/1 Eruca (75.1%)/1	Eucalyptus (32.5%), Melilotus (25.6%)	Pimpinella, Olea, Tamarix	Apium, Eryngium, Acacia, Genista, Chamaerops, Punica, Peganum, Capparis
Relizane (n = 2)	Genista (45.6%)/1	Foeniculum (38.3%), Citrus (16.3%), Ziziphus (38.8%)	Ammi, Eryngium, Sinapis, Olea, Tamarix	Convolvulus, Galega
Saida (n = 3)	Ziziphus (68.9%)/1	Eucalyptus (19.7%), Chamaerops (17.6%), Capparis (36.7%), Echium (18.2%), Centaurea (22.0%)	Tamarix, Papaver, Olea, Asparagus, Hedysarum, Cistus, Echium, Cichorium	Genista, Brassica

Table 2. Cont.

Localities	Dominant Pollen (≥45%)	Accompanying Pollen (45–15%)	Important Pollen (15–3%)	Other Pollen (3–1%)
Sidi Bel Abbés (n = 3)	Eriobotrya (75.4%)/1 Melilotus (68.7%)/1	Globularia (16.5%), Eucalyptus (17.8%), Tamarix (28.9%)	Brassica, Ceratonia, Ziziphus	Artitalicisia, Chrysanthitalicum, Sinapis, Euphorbia, Rosmarinus, Olea
Souk-Ahras (n = 1)	Eucalyptus (76.1%)/1		Melilotus, Ononis	Ammi, Erica, Punica
Tebessá (n = 2)	Hedysarum (45.9%)/1	Brassica (21.3%)	Eryngium, Foeniculum, Sinapis, Lotus, Quercus, Eucalyptus	Apium, Casuarina, Ephedra, Olea, Malus, Citrus, Capparis
Tizi-Ouzou (n = 4)	-	Asparagus (34.3%), Hedysarum (22.2%), Ziziphus (32.1%), Foeniculum (17.9%)	Pimpinella, Carduus, Echium, Thymus, Other Lamiaceae, Olea, Papaver, Rubus, Castanea	Apium, Eryngium, Thapsia, Chrysantitalicum, Cichorium, Brassica, Cistus, Ononis, Quercus
Tlitaliccen (n = 5)	Ziziphus (59.5%)/1 Tamarix (58.1%)/1	Eucalyptus (23.2%), Punica (30.9%), Chamaerops (37.3%)	Schinus, Eryngium, Foeniculum, Pimpinella, Chenopodium, Ceratonia, Olea, Peganum harmala, Capparis	Pistacia, Artitalicisia, Taraxacum, Arctium, Brassica, Cistus, Hedysarum, Muscari, Papaver

[1] The percentage is the maximum value reached for the pollen type. [2] Number of samples in which the pollen type is dominant pollen.

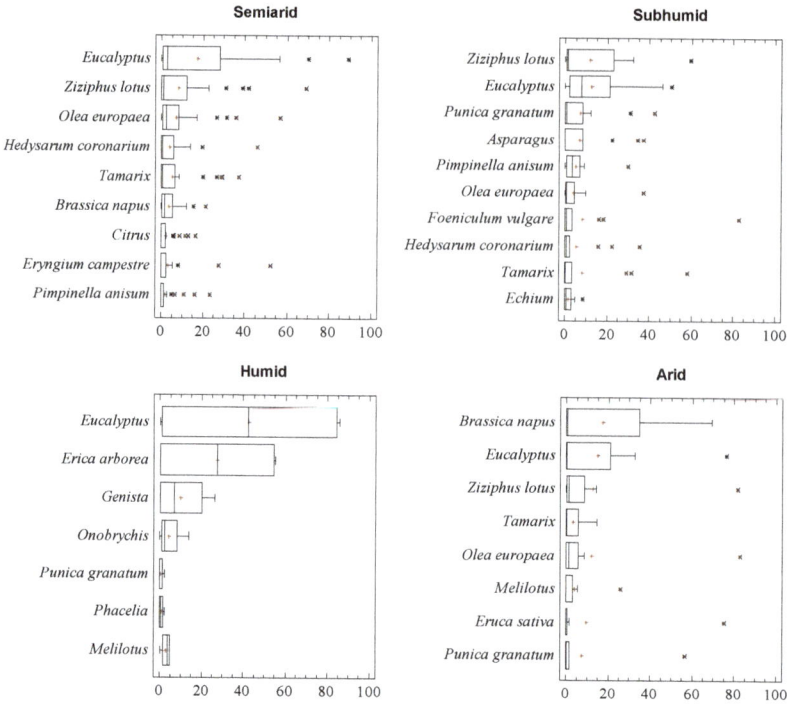

Figure 2. Box and Whisker plot of the main pollen types in each area. * Atypical values, + Centroids.

Regarding honey samples from humid areas, two samples stood out from the El Taref region with *Erica arborea* as the dominant pollen. Samples from the high plateau contained as the dominant pollen Apiaceae-like *Eryngium campestre* type, *Hedysarum* or *Ziziphus lotus*, and, as accompanying or important pollen, a great diversity of *Apiaceae*, such as *Foeniculum*, *Pimpinella*, *Thapsia*, Asteraceae such as *Carduus*, *Centaurea*, or *Cichorium* type. The presence of *Chamaerops* was considered high. Samples from the M'Sila region had *B. napus* as dominant pollen.

Samples from the arid region had as the dominant pollen *P. granatum* (the sample collected in Laghouat) and *Z. lotus* or *Eruca sativa* (honey from the Naama region). The sample from Bechar was considered to be from *Citrus*.

3.3. Quality and Physicochemical Sample Properties

Samples showed important differences in the studied physicochemical parameters. Humidity varied from 14.4% to 22.5%. This maximal value corresponded to a sample collected in El Taref (the humid region). Electrical conductivity was from the 0.133 mS/cm of a sample collected in Sidi bel Abbés to the 1.460 mS/cm of a sample from Naama; pH was from 3.5 (El Taref sample) to 4.7 (Médea sample). Parameters related to sample freshness also had great variation. HMF was higher than 40 mg/kg in two samples, showing the bad quality of these samples. The rest of the samples presented lower values, frequently near 10 mg/kg, with the lowest being 1 mg/kg. Following this trend, diastase activity was low in samples with high HMF content. Excepting these samples, the lowest value was 8.9° and the highest was 40.6°. The high HMF content of some Algerian honey samples (over the limit of 40 mg/kg established in international quality schemes) was noted before [30]. Improvement of the management practices is necessary to avoid this situation. An important parameter for the first differentiation of samples is color, which varied from 13 (Tebessa sample) to 150 mm Pfund (Tizi

Ouzou sample) (samples with higher HMF content were not considered for mean values of color, diastase content, and HMF content).

Most of the samples had a sugar content according to blossom honey, this being a sum of fructose and glucose higher than 60%. The lowest value corresponded to a honeydew sample (52.8%), while the highest value was 75.6%. Fructose content varied from 33.0% in a Tebessa sample to 44.4% in a Guelma sample. Glucose oscillated from 23.1% to 34.1%. The concentration of other sugars was really low. Maltose varied from 0.6% to 3.6%, turanose from 1.0% to 3.1%, and raffinose had a highest value of 3.2%. Lastly, sucrose had the highest value (3.3%) in a *Citrus* honey, but was only found in 20 samples.

Other important compounds, related to healthy properties of honey, were polyphenols and flavonoids. The first varied from 20.0 mg/100 g in a sample of *E. japonica* from Sidi bel Abbes to 182.3 mg/100 g of a honeydew sample from M'Sila. Flavonoid content varied from 1.0 mg/100 g to 12.9 mg/100 g in a sample from Bechar and a sample from Tlemcen, respectively. Moreover, RSA was greatly diverse, this variation being from 10% to 79.5%.

3.4. Influence of Area of Origin on Sample Characteristics

ANOVA was performed to mark differences and similarities between sample characteristics regarding their bioclimatic region. The highest differences were found between samples from humid areas, situated in the NE of the country, and samples collected in semiarid, subhumid, and arid areas. The most significant ($p < 0.05$) were higher electrical conductivity (mean value of 0.910 mS/cm), darker color (mean value of 122 mm Pfund), high water content (mean value of 21.4%), higher flavonoid content, and lower fructose content (mean value) in samples from humid areas in comparison with the others. In ANOVA results, the higher content in *E. arborea* stood out, together with *Malus* and *Onobrychis* types for honey of this origin. Furthermore, samples from semiarid, subhumid, and arid regions did not show significant differences ($p < 0.05$) between them in physicochemical variables except for lower sample humidity from those of the arid region (mean value of 15.7%). With regard to pollen spectra, samples from arid areas had higher content of *B. napus* and *Rosmarinus* than those from semiarid and subhumid regions ($p < 0.05$). However, due to the great variation between samples, more studies are needed.

3.5. Botanical Characterization of Honey Samples

The botanical origin of samples is the most important attribution to valorize local honey production. Sample typification was done with information provided by beekeepers, physicochemical data, and sample pollen profiles. To calculate the percentage of the main pollen type in the samples, nectarless plants were excluded. The sample classification and physicochemical data of each honey type are shown in Table 3.

The main group was polyfloral honey with 26 samples. This type of honey is produced in all regions, presenting significant variations in the studied variables. The pollen spectra had as main pollen types *Eucalyptus* and *Ziziphus*, but Apiaceae (mainly *Foeniculum*, *Eryngium*, *Pimpinella*, and *Thapsia* pollen types), *Hedysarum*, *Melilotus*, *Centaurea*, *B. napus*, or *Echium*) were common. Pollen grains from plants growing in semiarid and arid areas, such as *Tamarix* or *Capparis spinosa*, were also present in relevant amounts. Five samples produced in Medea, Mostaganem, and Tlemcen had a pollen content higher than 30% of *Tamarix* pollen, but their physicochemical properties, and polyphenol and flavonoid content strongly varied between samples. The apicultural value of *Tamarix* is not well-known; while some authors considered this plant important for honey production [31], other authors indicated that the nectar concentration in flowers of *Tamarix* plants is marginal [32]. Regarding *Capparis spinosa*, their pollen grains reached percentages over 20% in one sample from Sidi bel Abbes and another from Medea (36.7% and 24.4%, respectively). Important nectar production for this plant was indicated in a few studies [33,34], but this honey type is little mentioned in the scientific literature [13,29,35]. Another interesting point regards two samples with a high quantity of *Asparagus* pollen type (22.1% and 34.3%); both were from the Tizi Ouzou region and presented similar patterns in the studied variables. The most significant was the low pollen content (less than 2000 pollen grains/g of honey), the color (value close to 90 mm Pfund), diastase content (more than 30° Gothe), and polyphenol (mean value of 100 mg/100 g) and flavonoid content (mean value of 7 mg/100 g). There is little information about the importance of Asparagaceae plants for honey production, but the pollination activity and pollen production in both female and male plants of *Asparagus* was constated [36]. All samples mentioned before were considered polyfloral, but further studies are needed to endorse possible monovarietal honey.

Regarding the number of pollen types, polyfloral samples had a high number with a mean value of 25, but frequently with more than 30 pollen types. The values on physicochemical parameters, sugar profile, and antioxidant components were like those of other polyfloral honey types studied in the country [8,13].

Seven samples were typified as *Eucalyptus* honey, with a mean value of this pollen of 76.7%. Two samples were pressed honey having a high amount of pollen. The rest had a mean value included in Class III of Maurizio. This is slightly high pollen content in comparison with that of *E. globulus* honey from the Atlantic area of Europe [37]. Some of these samples presented high humidity with a mean value of 19.4%. In comparison with *Eucalyptus* honey from other areas, the studied samples presented high electrical conductivity, color, and pH [3,37,38]. This sample type is mainly produced in forests of coastal areas during summer, so some honeydew supports can influence their physicochemical attributes. *Eucalyptus* samples are produced in the Mostaganem region and eastern areas of the country. In Algeria, *Eucalyptus* trees were introduced in 1850, with *E. camaldulensis* and *E. globulus* being the main introduced species. In the middle of S. XX, large areas of the southeast, center, and west, such as Mostaganem, were intensively planted with *Eucalyptus* for wood production. Currently, species such as *E. gomphocephala*, *E. sideroxylon*, *E. robusta*, or *E. viminalis* grow there [39].

Six samples from the regions of Tizi Ouzou, Tlemcen, M'Sila, and Naama were typified as honeydew honey, while beekeepers had named them forest honey. Samples had the highest electrical conductivity, polyphenol content, and dark color with the lowest sum of fructose and glucose, common features to other types of honeydew honey [40–43]. Little information is available about the honeydew production in Algeria, and knowledge about sources of this honey is scarce. However, it was the subject of a great number of recent studies in other territories [40–46]. Considering the interest of this honey type, and that differentiation between different types of honeydew honey or even other dark blossom honey is difficult [47], more studies regarding this honey type could be useful to valorize local honey production.

Another important source for honey production was *Apiaceae* plants. A total of five samples were typified within this group. Three samples had as the main pollen the *F. vulgare* type, one sample *P. anisum*, and another *E. campestre* type. Samples were included in a single group due to some of them sharing high percentages of the mentioned pollen types and other secondary pollen types, such as *Thapsia* or *Apium* showing the importance for the honey production of this botanical family in the area. In general, samples were characterized by electrical conductivity near 0.495 mS/cm, light amber color, and high fructose content; the sample from *P. anisum* had a darker color. These samples had high polyphenol content (among the highest) and with medium flavonoid content. *Apiaceae* honey from North African countries was mentioned in different papers, mainly from plants as *Eryngium campestre*, *Ammi visnaga*, *Ridolfia segetum*, or *Bupleurum spinosum* in Morocco [48,49]. Some *Apiaceae* from Algeria were studied regarding sugar composition or polyphenol content [12,21]. The studied fennel honey had electrical conductivity, diastase content, and sugar content similar to those from Tenerife samples [50]. A high amount of hydroxycinnamic acids, mainly caffeic acid, was mentioned for this honey type [51].

Table 3. Characteristics of the different honey types. Values are expressed as mean and standard deviation.

	Polyfloral (n = 26)	Eucalyptus (n = 7)	Honeydew (n = 6)	Apiaceae (n = 5)	Citrus (n = 5)	Sedra (n = 4)	Punica (n = 3)	Heather (n = 2)	Retama (n = 1)	Rosmarinus (n = 1)	Medlar (n = 1)	Sulla (n = 1)
Main pollen (%)		Eucalyptus (76.7 ± 9.3)		Apiaceae (59.0 ± 19.5)	Citrus (22.0 ± 6.3)	Z. lotus (52.5%)	P. granatum (53.3%)	E. arborea (55.7 ± 0.6)	Genista 79.9%	Rosmarinus 17.0%	Eriobotrya 75.4%	Hedysarum 49.3%
N. Pollen types	25 ± 6	19 ± 6	23 ± 8	23 ± 6	18 ± 6	26 ± 7	22 ± 3	20 ± 2	12	23	13	23
PK (pollen/g)	8526 ± 13,738	27,985 ± 22,930	7433 ± 5757	5155 ± 3674	9935 ± 5787	16,775 ± 10,200	3612 ± 2033	23,750 ± 17,182	21,925	1575	6125	6750
Humidity (%)	16.6 ± 1.3	19.4 ± 2.0	17.2 ± 0.7	16.2 ± 1.0	17.3 ± 1.6	16.6 ± 1.0	14.9 ± 0.2	20.0 ± 0.2	18.9	16	14.6	15.2
EC (mS/cm)	0.481 ± 0.2	0.825 ± 0.2	1.033 ± 0.231	0.495 ± 0.2	0.318 ± 0.1	0.535 ± 0.1	0.580 ± 0.1	0.956 ± 0.2	0.440	0.330	0.135	0.133
pH	3.9 ± 0.2	4.0 ± 0.2	4.2 ± 0.2	4.0 ± 0.3	3.9 ± 0.2	4.4 ± 0.3	4.1 ± 0.1	4.1 ± 0.1	3.6	3.8	3.9	4.0
Color (mm Pfund)	73 ± 22	96 ± 10	125 ± 13	78 ± 19	45 ± 7	79 ± 9	77 ± 7	141 ± 3	67	13	28	34
DI (° Ghote)	20.1 ± 8.9	18.0 ± 6.6	21.5 ± 8.1	25.7 ± 5.0	18.7 ± 6.3	23 ± 3.2	14.1 ± 6.7	14.0 ± 6.6	28.3	6.4	20.6	6.7
HMF (mg/100 g)	1.7 ± 0.9	1.2 ± 0.4	1.7 ± 1.5	1.4 ± 1.1	1.3 ± 0.6	0.9 ± 0.6	1.4 ± 0.4	1.4 ± 1.9	1.4	0.9	0.7	1.0
Fructose (%)	39.9 ± 2.3	37.8 ± 1.9	38.6 ± 0.8	41.5 ± 0.6	40.8 ± 1.7	39.5 ± 0.7	38.5 ± 0.4	37.1 ± 0.2	38.2	38.1	40.3	33
Glucose (%)	29.3 ± 2.7	29.8 ± 1.8	28.0 ± 2.6	29.3 ± 2.3	29.3 ± 2.0	28.8 ± 1.8	28.9 ± 0.4	31.2 ± 3.3	31.6	30.1	34.1	25.9
Sacarose (%)	1.1	nd	nd	nd	2.3	nd	0.1	nd	0.7	nd	0.6	1.3
Maltose (%)	2.1 ± 0.6	2.0 ± 0.4	2.2 ± 0.6	1.3 ± 0.3	2.3 ± 0.7	2.1 ± 0.5	2.4 ± 0.4	1.5 ± 0.1	2.4	3.3	2.1	1.8
Turanose (%)	1.7 ± 0.4	2.0 ± 0.4	1.7 ± 0.5	1.8 ± 0.5	1.7 ± 0.3	2.5 ± 0.5	2.0 ± 0.2	1.3 ± 0.1	1.3	3.1	1.7	1.2
Raffinose (%)	0.4 ± 0.6	1.5 ± 1.3	0.4 ± 0.5	0.2 ± 0.1	0.2 ± 0.2	0.2 ± 0.1	0.3 ± 0.1	0.1 ± 0.0	0.1	0.3	0.1	1.5
Polyphenol (mg/100 g)	67.7 ± 22.2	72.7 ± 16.8	141.2 ± 34.1	104.5 ± 10.3	50.5 ± 31.5	71.6 ± 21.3	95.1 ± 5.4	130.8 ± 10.8	48.1	26.5	20	60.1
Flavonoid (mg/100 g)	4.9 ± 1.8	7.1 ± 1.2	10.6 ± 1.4	5.5 ± 0.8	4.1 ± 2.7	5.6 ± 1.0	5.1 ± 1.0	11.1 ± 0.9	1.4	1.0	1.4	5.9
RSA (%)	30.6 ± 12.2	42.4 ± 10.6	61.5 ± 14.7	28.0 ± 9.28	16.8 ± 8.7	31.1 ± 8.8	33.6 ± 8.3	43.7 ± 5.0	22.4	13.3	14.3	17.0

nd: not detected.

Citrus samples are produced mainly in the Tell region (Mostaganem, Mascara, and Relizane), where these fruit plants are common in agricultural lands, but also in oases of arid areas (Bechar). Five samples were considered from this botanical origin. The pollen of *Citrus* was identified with a mean value of 22.0% (nectarless pollen excluded), less than 15% of pollen spectra. It is common to find this pollen type with other pollen types such as *O. europaea*, *Genista*, or *Tamarix*. Samples had extralight amber color, mean electrical conductivity of 0.318 mS/cm, and high fructose and glucose content besides the highest mean sucrose content. These samples had similarities with other studied samples from Blida region, such as the presence of high percentages of *O. europaea*, and similar values in electrical conductivity, fructose, and glucose content [8,52].

Four samples were considered uniforal sedra honey. The most important species in the area to produce this type of honey is *Z. lotus*. The samples had light amber color, mean electrical conductivity of 0.535 mS/cm, high pH, and medium polyphenol and flavonoid contents. Sedra honey is more known than other monofloral honey types, as consumers appreciate them much due to healthy properties attributed to them [53]. The characteristics of the honey type are in accordance with the results of other publications [19,29].

The typification of three samples as pomegranate samples (*P. granatum*) stands out. The production of this honey type is not mentioned in the country, as it is rare. *P. granatum* pollen was presented in a mean value of 53.3%, and samples had medium-to-low pollen content (Class II of Maurizio). Water content was low (mean of 14.9%), color was extralight amber, the mean of electrical conductivity was of 0.580 mS/cm, and polyphenol and flavonoid contents were medium.

Two samples from El Taref were typified as heather honey, with *E. arborea* having a mean value of 55.7%. Northwestern evergreen forests and high shrubs growing in humid and warm climates facilitate the production of this honey type. Samples presented high humidity content (mean value 20%), high electrical conductivity, and the darkest color. Furthermore, polyphenol content was high, and samples presented major flavonoid content (mean of 11.1 mg/100 g). However, *E. arborea* honey remains produced in the area is poorly described [12].

Lastly, one sample was typified as unifloral from *Retama sphaerocarpa* (*Genista* pollen type was present in 79.9% of the pollen spectra), one sample as *Rosmarinus* (*Rosmarinus* pollen percentage was 17.0%), one sample as medlar honey (*E. japonica* pollen content was 75.4%), and one as sulla honey, with a pollen content of *Hedysarum* of 49.3%. Samples from *Rosmarinus*, *Eriobotrya*, and *Hedysarum* had the clearest color, the first being an extrawhite honey, and the last two white honey. The best-described honey type is *Hedysarum* honey [2,7,12].

As commented before, samples presented a wide variation in physicochemical parameters and pollen content. To reduce the dimensionality of the dataset and increase the interpretability of results, PCA was applied. The first nine components explained 72% of data variance, corresponding the sum of Components 1 and 2 to 30%. The plot of the scores of these two components is included in Figure 3A. Left is flavonoid content, RSA, and color close to electrical conductivity (negative quadrant) and polyphenol content (positive quadrant). *Erica* pollen is situated in the negative quadrant in the left with the electrical conductivity. Right shows *Eriobotrya* near glucose. Some pollen variables, such as *Punica*, *Capparis*, or *Tamarix*, were in the same direction as that of Fructose and *Apiaceae*. *Ziziphus*, *Brassica*, and pH are situated in the same quadrant as that of polyphenol content. On the opposite quadrant are *Citrus* and *Rosmarinus* pollen types, close to Maltose. Lastly, at the bottom, *Fabaceae* close to *Eucalyptus*, humidity, and raffinose sugar. PCA grouped samples from the same botanical origin (Figure 3B). Left are honeydew samples (Hd), *Erica* (Er), and *Eucalyptus* (Eu). This position corresponds to samples with the highest humidity content (*Eucalyptus* and *Erica* samples), major electrical conductivity, and darker color. The influence of polyphenol content determined the situation of *B. napus* in this area, as samples with a high content of this pollen type were considered honeydew honey due to their color, electrical conductivity, and the polyphenol content. In the top of the figure are *Apiaceae* (A) samples and *Z. lotus* (Z) honey, and in the center are *P. granatum* samples (Pu), close to some polyfloral samples. This sample group was separated from other, more disperse groups, situated in the left, which included types

Citrus (C), *Hedysarum* (H), *Rosmarinus* (Ro), and *R. sphaerocarpa* honey. The *Eriobotrya* sample was clearly separated.

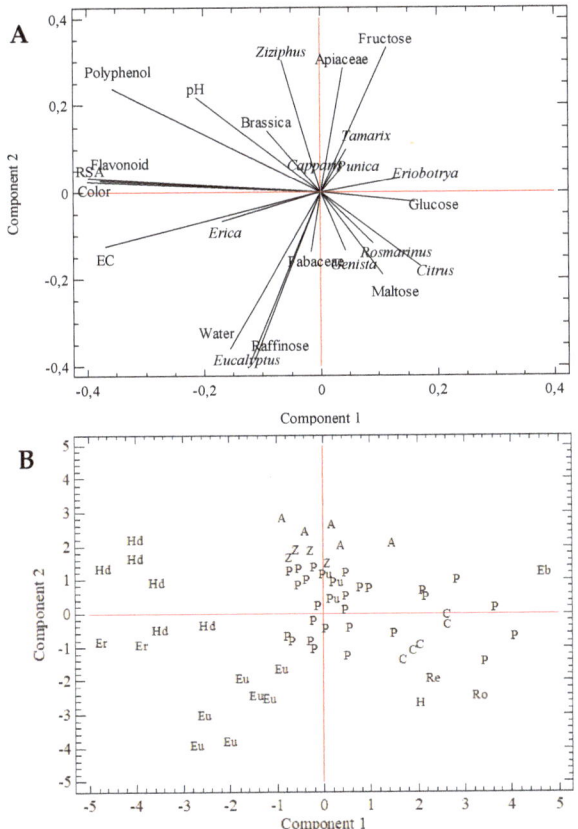

Figure 3. Principal-component analysis. (**A**) Loading plot of variables; (**B**) Score plot of the first two extracted components and sample botanical origin. A, *Apiaceae*; C, *Citrus*; Eb, *Eriobotrya*; Er, *Erica*; Eu, *Eucalyptus*; Hd, honeydew; H, *Hedysarum*; P, polyfloral; Pu, *Punica*; Re, *Retama*; Ro, *Rosmarinus*; Z, *Ziziphus*.

PCA results showed sample similarities classified with the same botanical origin, but also variation in some of these sample groups. This was the case of samples from *Eucalyptus*, honeydew, and, to a lesser extent, *Apiaceae* or *Citrus* samples. On the other hand, common Mediterranean honey types (*Citrus*, *Rosmarinus*, *Hedysarum*, or *Retama*) were close, evidencing the similarities between them.

4. Conclusions

The results of this paper showed the great variability in local honey production of Algerian beekeeping, and the potentiality regarding different honey types that could be obtained. Some of the plant species mentioned in this work, such as *Eucalyptus*, *Brassica napus*, *Hedysarum*, and *Citrus*, are common honey plants in Mediterranean areas. However, others, such as *Capparis spinosa*, *Asparagus*, *Tamarix*, *Ziziphus lotus*, and some *Apiaceae* plants (*Eryngium*, *Thapsia*, *Pimpinella*), even *Acacia*, are representative of the honey of this country, useful as markers to guarantee their authenticity. In any case, there is little information on the apicultural value of some of them and of their honey; thus, more research is needed on these topics.

Author Contributions: Conceptualization, M.H., A.H. and D.F.; methodology, M.H., M.S.R.-F., B.M. and O.E.; formal analysis, M.H., D.F., M.S.R.-F. and O.E., data curation, M.H., M.C.S., M.S.R.-F.; writing—original draft preparation, M.H., M.C.S.; writing—review and editing, O.E., M.S.R.-F.; supervision, A.H., B.M., D.F. All authors have read and agreed to the published version of the manuscript.

Funding: This research received no external funding.

Acknowledgments: We express our deepest gratitude to all the beekeepers for graciously helping us to obtain honey samples.

Conflicts of Interest: The authors declare no conflict of interest

References

1. Hussein, M.H. A review of beekeeping in Arab countries. *Bee World* **2000**, *81*, 56–71. [CrossRef]
2. Makhloufi, C.; Kerkvliet, J.D.; D'albore, G.R.; Choukri, A.; Samar, R. Characterization of Algerian honeys by palynological and physico-chemical methods. *Apidologie* **2010**, *41*, 509–521. [CrossRef]
3. Escuredo, O.; Míguez, M.; Fernández-González, M.; Seijo, M.C. Nutritional value and antioxidant activity of honeys produced in a European Atlantic area. *Food Chem.* **2013**, *138*, 851–856. [CrossRef] [PubMed]
4. Migliore, J. Plant biogeography in the western mediterranean basin: New insights from phylogeographical studies. *BMIB-Bollettino dei Musei e degli Istituti Biologici* **2013**, *75*, 64–68.
5. Médail, F.; Quézel, P. Hot-Spots Analysis for Conservation of Plant Biodiversity in the Mediterranean Basin. *Ann. Mo. Bot. Gard.* **1997**, *84*, 112–127. [CrossRef]
6. Mediouni, K. *Elaboration d'un Bilan et d'une Strategie Nationale de Developpement Durable de la Diversite Biologique*; Projet ALG/97 G31; Tome IX. Editions du Ministere de la Amenagement du territoire et de Environnement; FEM/PNUD: New York, NY, USA, 2004; p. 69.
7. Zerrouk, S.; Boughediri, L.; Seijo, M.C.; Fallico, B.; Arena, E.; Ballistreri, G. Pollen spectrum and physicochemical attributes of sulla (Hedysarum coronarium) honeys of Médéa region (Algeria). *Albanian J. Agric. Sci.* **2013**, *12*, 511–517.
8. Makhloufi, C.; Kerkvliet, J.; Schweitzer, P. Characterisation of some monofloral Algerian honeys by pollen analysis. *Grana* **2015**, *54*, 156–166. [CrossRef]
9. Mekious, S.; Houmani, Z.; Houmani, M. Étude des potentialités mellifères de deux régions du Nord de l'Algérie. *Phytothérapie* **2018**, 1–6. [CrossRef]
10. Zerrouk, S.; Seijo, M.C.; Boughediri, L.; Escuredo, O.; Rodríguez-Flores, M.S. Palynological characterisation of Algerian honeys according to their geographical and botanical origin. *Grana* **2014**, *53*, 147–158. [CrossRef]
11. Latifa, H.; Mouna, B.; Arezki, M. Ziziphus lotus and Euphorbia bupleuroides Algerian honeys. *World Appl. Sci. J.* **2013**, *24*, 1536–1543.
12. Ouchemoukh, S.; Amessis-Ouchemoukh, N.; Romero, M.G.; Aboud, F.; Giuseppe, A.; Gutierrez, A.F.; Segura-Carretero, A. Characterisation of phenolic compounds in Algerian honeys by RP-HPLC coupled to electrospray time-of-flight mass spectrometry. *LWT-Food Sci. Technol.* **2017**, *85*, 460–469. [CrossRef]
13. Boutabia, L.; Telailia, S.; Chefrour, A. Spectre pollinique de miels d'abeille (Apis mellifera L.) de la région d'El Tarf (Nord-Est algérien). *Livest. Res. Rural. Dev.* **2016**, *28*, 1–8.
14. Ouchemoukh, S.; Louaileche, H.; Schweitzer, P. Physicochemical characteristics and pollen spectrum of some Algerian honeys. *Food Control* **2007**, *18*, 52–58. [CrossRef]
15. Azzedine, C.; Marie-José, B.; Yasmina, A.K.; Salima, B.; Ali, T. Melissopalynologic and physicochemical analysis of some north-east Algerian honeys. *Eur. J. Sci. Res.* **2005**, *18*, 389–401.
16. Nair, S.; Maghraoui, N.B. Physicochemical Properties of Honeys Produced in North-West of Algeria. *Adv. Food Sci. Eng.* **2017**, *1*, 123–128. [CrossRef]
17. Mesbahi, M.A.; Ouahrani, M.R.; Rebiai, A.; Amara, D.G.; Chouikh, A.; Mesbahi, M.A.; Ouahrani, M.R.; Rebial, A. Characterization of Zygophyllum album L Monofloral Honey from El-Oued, Algeria. *Curr. Nutr. Food Sci.* **2019**, *15*, 476–483. [CrossRef]
18. Otmani, I.; Abdennour, C.; Dridi, A.; Kahalerras, L.; Halima-Salem, A. Characteristics of the bitter and sweet honey from Algeria Mediterranean coast. *Vet. World* **2019**, *12*, 551–557. [CrossRef]

19. Zerrouk, S.; Seijo, M.C.; Escuredo, O.; Rodríguez-Flores, M.S. Characterization of Ziziphus lotus (jujube) honey produced in Algeria. *J. Apic. Res.* **2017**, *57*, 166–174. [CrossRef]
20. Alzahrani, H.A.; Boukraa, L.; Bellik, Y.; Abdellah, F.; Bakhotmah, B.A.; Kolayli, S.; Sahin, H. Evaluation of the Antioxidant Activity of Three Varieties of Honey from Different Botanical and Geographical Origins. *Glob. J. Health Sci.* **2012**, *4*, 191–196. [CrossRef]
21. Ouchemoukh, S.; Schweitzer, P.; Bey, M.B.; Djoudad-Kadji, H.; Louaileche, H. HPLC sugar profiles of Algerian honeys. *Food Chem.* **2010**, *121*, 561–568. [CrossRef]
22. Neggad, A.; Benkaci-Ali, F.; Alsafra, Z.; Eppe, G. Headspace Solid Phase Microextraction Coupled to GC/MS for the Analysis of Volatiles of Honeys from Arid and Mediterranean Areas of Algeria. *Chem. Biodivers.* **2019**, *16*, e1900267. [CrossRef]
23. Bogdanov, S.; Martin, P.; Lullmann, C. *Harmonised Methods of the International Honey Commission*; Swiss Bee Research Centre: Liebefeld, Switzerland, 2002.
24. Singleton, V.L.; Orthofer, R.; Lamuela-Raventos, R.M. Analysis of total phenols and other oxidation substrates and antioxidants by means of folin-ciocalteu reagent. *Methods Enzymol.* **1999**, *299*, 152–178.
25. Arvouet-Grand, A.; Vennat, B.; Pourrat, A.; Legret, P. Standardization of propolis extract and identification of principal constituents. *J. Pharm. Belg.* **1994**, *49*, 462–468.
26. Brand-Williams, W.; Cuvelier, M.; Berset, C. Use of a free radical method to evaluate antioxidant activity. *LWT-Food Sci. Technol.* **1995**, *28*, 25–30. [CrossRef]
27. Adgaba, N.; Al-Ghamdi, A.; Tadesse, Y.; Getachew, A.; Awad, A.M.; Rana, R.M.; Owayss, A.A.; Mohammed, S.E.A.; AlQarni, A.S. Nectar secretion dynamics and honey production potentials of some major honey plants in Saudi Arabia. *Saudi J. Boil. Sci.* **2016**, *24*, 180–191. [CrossRef] [PubMed]
28. Al-Khalifa, A.S.; Al-Arify, I. Physicochemical characteristics and pollen spectrum of some Saudi honeys. *Food Chem.* **1999**, *67*, 21–25. [CrossRef]
29. Alqarni, A.S.; Owayss, A.A.; Mahmoud, A.A. Physicochemical characteristics, total phenols and pigments of national and international honeys in Saudi Arabia. *Arab. J. Chem.* **2016**, *9*, 114–120. [CrossRef]
30. Achouri, M.; Selka, M.A.; Chenafa, A.; Brahim, S.; Messafeur, M.A.; Toumi, H. Teneur en 5-hydroxyméthylfurfural (HMF) dans les miels du Nord-Ouest de l'Algérie. *Toxicol. Anal. Clin.* **2019**, *31*, 100–105. [CrossRef]
31. Ahmida, M.H.S.; Elwerfali, S.; Agha, A.; Elagori, M. Physicochemical, Heavy Metals and Phenolic Compounds Analysis of Libyan Honey Samples Collected from Benghazi during 2009–2010. *Food Nutr. Sci.* **2013**, *4*, 33–40. [CrossRef]
32. Andersen, D.C.; Nelson, S. Floral ecology and insect visitation in riparian *Tamarix* sp. (saltcedar). *J. Arid. Environ.* **2013**, *94*, 105–112. [CrossRef]
33. Eisikowitch, D.; Ivri, Y.; Dafni, A. Reward partitioning in *Capparis* spp. along ecological gradient. *Oecologia* **1986**, *71*, 47–50. [CrossRef] [PubMed]
34. Petanidou, T.; Van Laere, A.J.; Smets, E. Change in floral nectar components from fresh to senescent flowers of *Capparis spinosa* (*Capparidaceae*), a nocturnally flowering Mediterranean shrub. *Plant. Syst. Evol.* **1996**, *199*, 79–92. [CrossRef]
35. El-Guendouz, S.; Al-Waili, N.; Aazza, S.; Elamine, Y.; Zizi, S.; Al-Waili, T.; Al-Waili, A.; Lyoussi, B. Antioxidant and diuretic activity of co-administration of *Capparis spinosa* honey and propolis in comparison to furosemide. *Asian Pac. J. Trop. Med.* **2017**, *10*, 974–980. [CrossRef] [PubMed]
36. Greco, C.F.; Banks, P.; Kevan, P.G. Foraging behaviour of honeybees (*Apis mellifera*) on asparagus (*Asparagus officinalis*). *Proc. Entomol. Soc. Ont.* **1995**, *126*, 37–43.
37. Flores, M.S.R.; Pérez, O.E.; Rodríguez-Flores, M.S. Characterization of *Eucalyptus globulus* honeys produced in the Eurosiberian Area of the Iberian Peninsula. *Int. J. Food Prop.* **2014**, *17*, 2177–2191. [CrossRef]
38. Karabagias, I.K.; Maia, M.; Karabagias, V.K.; Gatzias, I.; Badeka, A.V. Characterization of Eucalyptus, Chestnut and Heather Honeys from Portugal Using Multi-Parameter Analysis and Chemo-Calculus. *Foods* **2018**, *7*, 194. [CrossRef]
39. Benayache, S.; Benayache, F.; Benyahia, S.; Chalchat, J.-C.; Garry, R.-P. Leaf Oils of some Eucalyptus Species Growing in Algeria. *J. Essent. Oil Res.* **2001**, *13*, 210–213. [CrossRef]
40. Flores, M.S.R.; Escuredo, O.; Seijo, M.C. Assessment of physicochemical and antioxidant characteristics of Quercus pyrenaica honeydew honeys. *Food Chem.* **2015**, *166*, 101–106. [CrossRef]

41. Seijo, M.C.; Escuredo, O.; Rodríguez-Flores, M.S. Physicochemical Properties and Pollen Profile of Oak Honeydew and Evergreen Oak Honeydew Honeys from Spain: A Comparative Study. *Foods* **2019**, *8*, 126. [CrossRef]
42. Jara-Palacios, M.J.; Ávila, F.J.; Escudero-Gilete, M.L.; Pajuelo, A.G.; Heredia, F.J.; Hernanz, D.; Terrab, A. Physicochemical properties, colour, chemical composition, and antioxidant activity of Spanish Quercus honeydew honeys. *Eur. Food Res. Technol.* **2019**, *245*, 2017–2026. [CrossRef]
43. Vasić, V.; Gašić, U.M.; Stanković, D.; Lušić, D.; Vukić-Lušić, D.; Milojković-Opsenica, D.; Tešić, Ž.; Trifković, J. Towards better quality criteria of European honeydew honey: Phenolic profile and antioxidant capacity. *Food Chem.* **2019**, *274*, 629–641. [CrossRef]
44. Karabagias, I.K.; Karabournioti, S.; Karabagias, V.K.; Badeka, A.V. Palynological, physico-chemical and bioactivity parameters determination, of a less common Greek honeydew honey: "dryomelo". *Food Control* **2020**, *109*, 106940. [CrossRef]
45. Nešović, M.; Gašić, U.M.; Tosti, T.; Trifković, J.; Baošić, R.; Blagojević, S.; Ignjatović, L.; Tešić, Ž. Physicochemical analysis and phenolic profile of polyfloral and honeydew honey from Montenegro. *RSC Adv.* **2020**, *10*, 2462–2471.
46. Shaaban, B.; Seeburger, V.C.; Schroeder, A.; Lohaus, G. Sugar, amino acid and inorganic ion profiling of the honeydew from different hemipteran species feeding on *Abies alba* and *Picea abies*. *PLoS ONE* **2020**, *15*, e0228171. [CrossRef]
47. Vasić, V.; Đurđić, S.; Tosti, T.; Radoičić, A.; Lušić, D.; Milojković-Opsenica, D.; Tešić, Ž.; Trifković, J. Two aspects of honeydew honey authenticity: Application of advance analytical methods and chemometrics. *Food Chem.* **2020**, *305*, 125457. [CrossRef] [PubMed]
48. Terrab, A.; Díez, M.J.; Heredia, F.J. Characterisation of Moroccan unifloral honeys by their physicochemical characteristics. *Food Chem.* **2002**, *79*, 373–379. [CrossRef]
49. Elamine, Y.; Aazza, S.; Lyoussi, B.; Antunes, M.D.; Estevinho, L.M.; Anjos, O.; Resende, M.; Faleiro, M.; Miguel, M.G. Preliminary characterization of a Moroccan honey with a predominance of Bupleurum spinosum pollen. *J. Apic. Res.* **2017**, *57*, 153–165. [CrossRef]
50. Manzanares, A.B.; García, Z.H.; Galdón, B.R.; Rodríguez-Rodríguez, E.M.; Romero, C.D. Physicochemical characteristics and pollen spectrum of monofloral honeys from Tenerife, Spain. *Food Chem.* **2017**, *228*, 441–446. [CrossRef] [PubMed]
51. Vella, A.; Cammilleri, G.; Pulvirenti, A.; Galluzzo, F.; Randisi, B.; Giangrosso, G.; Macaluso, A.; Gennaro, S.; Ciaccio, G.; Cicero, N.; et al. High hydroxycinnamic acids contents in fennel honey produced in Southern Italy. *Nat. Prod. Res.* **2020**, 1–6. [CrossRef] [PubMed]
52. Zerrouk, S.; Boughediri, L.; Seijo, M.C.; Fallico, B.; Arena, E.; Ballistreri, G. Palynological and Physicochemical Properties of Citrus and Eucalyptus Honeys Produced in Blida Region (Algeria). *Eur. J. Sci. Res.* **2013**, *104*, 79–90.
53. Masalha, M.; Abu-Lafi, S.; Abu-Farich, B.; Rayan, M.; Issa, N.; Zeidan, M.; Rayan, A. A New Approach for Indexing Honey for Its Heath/Medicinal Benefits: Visualization of the Concept by Indexing Based on Antioxidant and Antibacterial Activities. *Medicines* **2018**, *5*, 135. [CrossRef] [PubMed]

© 2020 by the authors. Licensee MDPI, Basel, Switzerland. This article is an open access article distributed under the terms and conditions of the Creative Commons Attribution (CC BY) license (http://creativecommons.org/licenses/by/4.0/).

Article

Antioxidant Activity, Total Phenolic Content, Individual Phenolics and Physicochemical Parameters Suitability for Romanian Honey Authentication

Daniela Pauliuc, Florina Dranca and Mircea Oroian *

Faculty of Food Engineering, Stefan cel Mare University of Suceava, 720225 Suceava, Romania; daniela_pauliuc@yahoo.com (D.P.); florina.dranca@usm.ro (F.D.)
* Correspondence: m.oroian@fia.usv.ro; Tel.: +40-744-524-872

Received: 4 February 2020; Accepted: 4 March 2020; Published: 8 March 2020

Abstract: The present study aimed to evaluate the physicochemical characteristics of honey (raspberry, mint, rape, sunflower, thyme and polyfloral) produced in Romania. The honey samples were from the 2017 to 2018 harvest and were subjected to melissopalynological analysis, alongside the determination of the following physicochemical parameters: moisture content, pH, free acidity, electrical conductivity (EC), hydroxymethylfurfural (HMF) content, color, total polyphenols content (TPC), flavonoids content (FC), DPPH radical scavenging activity, phenolic acids, flavonols, sugars and organic acids in order to evaluate the usefulness of this parameters for the classification of honey according to botanical origin. The results of the melissopalynological analysis revealed that five types of honey samples had a percentage of pollen grains above the minimum of 45%, which was required in order to classify the samples as monofloral honey. The total polyphenols content reached the maximum value in the case of dark honey such as mint honey, followed by raspberry, thyme and polifloral honey. Fructose, glucose, maltose, sucrose, turanose, trehalose, melesitose, and raffinose were identified and quantified in all samples. Gluconic acid was the main organic acid in the composition of all honey samples. Principal component analysis (PCA) confirmed the possibility of the botanical authentication of honey based on these physicochemical parameters.

Keywords: honey; authentication; physicochemical parameters; PCA

1. Introduction

Honey is used both as medicine and a food source [1] and it is defined, according to Codex Alimentarius and EU Directive 110/2001 [2,3], as a sweet natural substance produced by bees (*Apis melifera*) from nectar or from the secretions of some plants, which is collected by bees and transformed by combining specific substances [4]. Honey is a complex food product, which is derived from nature and is the only natural sweetener that humans can use without processing [5], and therefore is very important economically [6].

Honey has a very complex chemical composition because it contains about 80% sugars, of which an important part is represented by glucose and fructose, 15–17% water, 0.1–0.4% protein and other compounds that are quantified as ash 0.2% [7,8]. In addition, honey also contains, in small quantities, about 200 other constituents, which include amino acids, phenolic compounds, organic acids, vitamins, minerals, and enzymes [9]. This multitude of minor components can be added by bees or comes directly from nectar due to the ripening process [10,11].

The chemical composition depends on the source of honey, which refers to the botanical and geographical origin, as well as the environmental conditions [12]. Monofloral honey is increasingly required on the market and it is necessary to be able to determine some parameters regarding the authentication of the botanical and geographical origin. Monofloral honey is more expensive than polyfloral honey; honey labeled as having a certain floral origin must come entirely or largely from the specific floral source and exhibit the organoleptic, physicochemical and microscopic characteristics of the honey source, as provided in international food standards [2,11].

Considering that bees feed on various plants, pure monofloral honey is generally very rare. The identification of the origin of honey and the proof of its authenticity has become an important problem with the globalization of the honey market, involving about 150 countries [13]. The interest in identifying the floral origin of honey has increased in recent years due to the high preference of consumers for certain types of honey. Consumer preferences often vary depending on different sensory perceptions and medicinal properties. Thus, numerous research has been published to date, which aimed to develop reliable methods for indicating the floral origin of honey [14].

Pollen analysis can be successfully used for the identification of the floral origin of honey. Therefore, melissopalinology should usually be supplemented by physicochemical and organoleptic analysis. Thus, to classify honey by botanical origin, a global interpretation of all results is required [15]. The melissopalynological analysis consists of counting the pollen grains and classify the honey according to its principal pollen grain percentage, for some honey such as sunflower, raspberry, rape and mint the principal pollen must reach at least 45% of the total pollen grains [16] while for thyme honey the *Thymus* spp. pollen grains must be at least 18% of the total pollen grains [17].

Therefore, new analytical methodologies were used to determine the botanical origin; these include the chromatographic, spectroscopic, e-tongue and molecular biological methods [18,19]. Physicochemical parameters (color, moisture, acidity) can vary widely in different types of honey and this contributes, to a certain extent, to their organoleptic characteristics. This is the reason why chromatographic techniques are more eloquent in the classification of honey and special attention should be paid to identifying certain specific minor components [20]. In addition to the classical techniques used to authenticate honey, the use of DNA-based methods for pollen identification has also spread. DNA-based identification has the potential to reduce processing time and increase the level of discriminated species [21]. Soares et al. [22] reported that they extracted the DNA markers and the yield and purity of the extracts were evaluated by UV spectrophotometry; this method was validated successfully with honey of known origins and applied to the entomological authentication of 20 commercial samples from different European countries.

Spectroscopic techniques, such as Fourier transform infrared spectroscopy (FTIR) and Raman spectroscopy, are alternative methods for authenticating honey and these techniques are reliable, practical and not time-consuming. FTIR spectroscopy is sensitive to the chemical composition of the sample, and when coupled with multivariate statistical analysis, it provides accurate results in determining the botanical origin of honey [9]. Svecnjak et al. [23] used FTIR-ATR spectroscopy to confirm the botanical origin of collected honey samples from beekeepers from different Croatian regions. Rheology and electrical tongue are also part of the alternative methods of authentication of honey. The voltammetry technique implies a high sensitivity and the electronic tongue can be regarded as a reference system in honey authentication [24]. Sousa et al. [25] reached a 100% correct classification of chestnut (*Castanea* spp.), lavender (*Lavandula* spp.) and raspberry honey (*Rubus* spp.) with a potentiometric electronic tongue. The exact classification was obtained after honey samples were separated according to their color and then the authentication of each type of honey was done on their botanical origin.

NMR is a fingerprint technique that is used to obtain information about the structure of components [26]. Spiteri et al. [27] analyzed 816 honey samples from 60 different botanical origins by the NMR technique and observed specific profiles for the botanical sources of origin.

In this study, melissopalynological analysis and analysis of physicochemical parameters (moisture content, pH, free acidity, electrical conductivity, hydroxymethylfurfural content, color, total polyphenols content, flavonoids content, DPPH radical scavenging activity, phenolic acids, flavonols, sugar composition, organic acids compositions) was performed to authenticate the botanical origin of sunflower, raspberry, thyme, mint, rape and polyfloral honey from Romania.

2. Materials and Methods

2.1. Honey Samples

Forty-five honey samples from the flowering season of 2017 to 2018 were purchased from beekeepers or apicultural associations from different regions in Romania.

Honey samples were kept away from sunlight at room temperature until the analysis. Thyme (*Thymus* spp.), rape (*Brassica* spp.), mint (*Mentha piperita*), raspberry (*Rubus idaeus*), and sunflower (*Helianthus* spp.) honey were of interest for this research.

2.2. Melissopalynological Analysis

Melissopalynological analysis was carried out according to Louveaux and Vorwohl [28]. The pollen was examined under a microscope using ×40 magnification on a Motic microscope (Motic, Xiamen, China). For achieving the botanical origin at least 800 pollen grains were counted.

2.3. Physicochemical Analysis

2.3.1. Moisture Content, pH, Free Acidity, HMF Content and Electrical Conductivity

Moisture content, free acidity, pH, HMF content and electrical conductivity were determined according to the methods of the International Honey Commission [8,29].

2.3.2. Color

Honey color analysis was performed using the Pfund scale (Pfund HI, Hanna Instruments, USA) and CIEL*a*b* coordinates (portable chromameter CR-400, Konica Minolta, Tokyo, Japan), respectively.

2.3.3. Determination of Total Phenolic Content

The method proposed by Biesaga et al. [30] was used to determine the total phenolic content (TPC) and sample preparation was made, as follows: 1 g of honey sample was extracted with 5 mL of 40% methanol/acidified water (v/v, pH = 2, HCl). Then, the samples were stirred for 15 min with a magnetic stirrer. From the extract, 0.2 mL was mixed with 2 mL of Folin–Ciocalteu reagent 1:10 and 1.8 mL Na_2CO_3 7.5% (w/v). The samples were kept in the dark for 20 min and the absorbance was measured at 750 nm using a UV-NIR spectrometer HR4000CG-UV-NIR (Ocean Optics, St. Petersburg, FL, USA). Gallic acid solutions with concentrations ranging from 0–400 mg·L^{-1} were used to obtain the calibration curve.

2.3.4. Determination of Flavonoids

From the extract prepared as presented in Section 2.3.3, 5 mL were mixed with 300 µL of $NaNO_2$ 5% (w/v) and 300 µL of $AlCl_3$ 10% (w/v) [30]. After 5 min in the dark, the samples were mixed with 2 mL of NaOH 1 N. The samples were kept for 6 more minutes in the dark and then the absorbance of each sample was read at 510 nm with a HR4000CG-UV-NIR spectrometer. Quercetin solutions with concentrations ranging from 0–10 mg·L^{-1} were used to obtain the calibration curve.

2.3.5. DPPH Assay

The determination of 1,1-diphenyl-2-picrylhydrazyl (DPPH) radical scavenging activity required the following sample preparation: 1 g of honey was dissolved in 5 mL of methanol 40% (v/v, with acidified water) and stirred for 15 min with a magnetic stirrer [31]. Then, 35 µL of honey solution was mixed with 250 µL of DPPH. The absorbance was measured at 515 nm using a QE65000 spectrometer (Ocean Optics, St. Petersburg, FL, USA). The results were expressed as % DPPH using the formula in Equation (1):

$$\% \text{ DPPH} = \left(A_0 - \frac{A_1}{A_0}\right) \times 100, \qquad (1)$$

where A_0 is the DPPH absorbance, A_1 is the sample absorbance.

2.3.6. Determination of Sugars Composition

Sugars composition was determined according to the IHC (International Honey Commission) methods [8,29]. The samples were filtered through 0.45 µm PTFE membrane filters prior to the injection in the HPLC instrument (Schimadzu, Kyoto, Japan) equipped with a LC-20 AD liquid chromatograph, SIL-20A auto sampler, CTO-20AC column oven, and RID-10A refractive index detector. The separation was performed on a Phenomenex Luna® Omega 3 µm SUGAR 100 Å HPLC Column 150 × 4.6 mm. Peaks were identified based on their retention times and the determination of sugar content was made according to the external standard method on peak areas or peak heights. The mobile phase was acetonitrile:water (80:20, v/v), with a flow rate of 1.3 mL·min^{-1}; column and detector temperature was 30 °C and the sample volume injection was 10 µL. Standard solutions of fructose, glucose, maltose, sucrose, turanose, trehalose, melesitose, and raffinose were individually injected to calculate the sugar content of each honey sample by using peak areas based on the retention time.

2.3.7. Determination of Polyphenols Composition

Honey solutions were prepared following the steps presented in Section 2.3.3 [30]. The samples were filtered through 0.45 µm PTFE membrane filters and then injected (with a volume of 10 µL) into the HPLC instrument (Schimadzu, Kyoto, Japan) for analysis using an SPD-M-20A diode array detector. The separation was carried out on a Phenomenex Kinetex 2.6 µm Biphenyl 100 Å HPLC Column 150 × 4.6 mm thermostated at 25 °C. Elution was carried out with a solvent system consisting of 0.1% acetic acid in water (solvent A) and acetonitrile (solvent B) as previously described by Palacios et al. [32] with modifications. The solvent flow rate was of 1 mL·min^{-1}. The determined phenolic compounds were gallic acid, vanillic acid, protocatechuic acid and p-hydroxibenzoic acid at 280 nm, and chlorogenic acid, p-coumaric acid, caffeic acid, rosmarinic acid, myricetin, quercetin, luteolin and kaempherol at 320 nm. The obtained standard calibration curves showed high degrees of linearity ($R^2 > 0.99$). Data collection and subsequent processing were performed using the LC solution software version 1.21 (Shimadzu, Kyoto, Japan).

2.3.8. Determination of Organic Acids Composition

The method used to determine the organic acids involved a sample preparation of 0.5 g of honey mixed with 2.5 mL of 4% metaphosphoric acid (w/v), then the samples were vortexed. After, the samples were centrifuged for 5 min at 3500 rpm using a Z216-MK refrigerated centrifuge (Hermle Labortechnik, Wemingen, Germany) [33]. The sample was injected in the HPLC instrument (Schimadzu, Kyoto, Japan) with a diode array detector. The separation was carried out on a Phenomenex Kinetex® 5 µm C18 100 Å HPLC Column 250 × 4.6 mm. The mobile phase used was a mixture of 0.5% metaphosphoric acid and acetonitrile (50/50, v/v) at a flow rate of 0.8 mL·min^{-1}. The volume of injection was 10 µL. The organic acids identification and quantification were carried out at 210 nm. The organic acids that were determined were acetic acid, lactic acid, propionic acid, butyric acid, and gluconic acid. The concentration of organic acids was expressed as mg/L.

2.4. Statistical Analysis

The analysis of variance (ANOVA) (LSD (least significant difference) test and $\alpha = 0.05$ were applied) and principal component analysis (PCA) were used for achieving the suitability of the analyzed parameters for the botanical authentication of honey. ANOVA was carried out using Statgraphics Centurion XVIII software—trial version (Manugistics Corp., Rockville, MD, USA), while PCA was carried out using Unscrambler X version 10.1 (Camo, Norway), respectively.

3. Results

3.1. Melissopalynological Analysis

Melissopalynological analysis is considered a traditional approach for the determination of the botanical origin of honey, and is a method of analysis that involves microscopic examination of pollen grains in order to identify the plants that were visited by bees during honey production [34]. Pollen analysis is a method developed and proposed by the International Bee Botanical Commission (IBBC) in 1970, which was revised in 1978 [28]. A honey sample can be classified as monofloral honey when more than 45% of the pollen grains belong to a single plant species for rape, raspberry, mint and sunflower [16], while thyme honey must be at least 18% total pollen grains [17]; this type of honey is the most preferred by consumers for its specific aroma, taste and biological properties [18,19]. The melissopalynological analysis is presented in Table 1. The raspberry honey had the principal pollen *Rubus idaeus* (49.1–82.3%), rape honey had the principal pollen *Brassica* spp. (50.1–71.1%), sunflower has the principal pollen *Helianthus* spp. (46.5–92.1%) and thyme had the principal pollen *Thymus* spp. (22–45%), respectively.

Table 1. Pollen types in honey samples.

Honey Type	Principal Pollen Type (min.%–max.%)
Raspberry	*Rubus idaeus* (49.1–82.3%)
Rape	*Brassica* spp. (50.1–71.1%)
Sunflower	*Helianthus* spp. (46.5–92.1%)
Mint	*Mentha* spp. (46.5–65.1%)
Thyme	*Thymus* spp. (22–45%)

Based on pollen analysis, the samples were classified according to the botanical origin as raspberry (6 samples), rape (10 samples), sunflower (9 samples), thyme (4 samples), mint (10 samples) and polyfloral honey (6 samples). Of the monofloral honey samples, the highest percentage of pollen grains was found in rape and sunflower honey.

3.2. Moisture Content

The moisture content of honey is dependent on factors such as the relative humidity in the region where honey comes from and the processing and storage conditions [11]. The Codex Alimentarius standard established that the moisture content of honey must be below 20% [2]. Honey samples that do not meet this criterion could become unstable during storage and thus be susceptible to deterioration by fermentation caused by yeast and bacteria naturally found in honey [35]. The moisture content of the analyzed samples ranged between an average value of 17.36% in thyme samples and a maximum average value of 19.60% in the polyfloral honey samples, as shown in Table 1. The botanical origin of honey did not influence the variation of the moisture content ($p > 0.05$), while the year of honey production had some influence ($p < 0.05$) on this parameter. All honey samples had moisture contents in the limits established by legislation. The mint and thyme honey samples had values between 17.36% (thyme honey) and 17.77% (mint honey), while Boussaid et al. [36] reported for mint honey the value of 19.8% and for thyme honey, 18.16%. The results of this analysis were in accordance with the values reported by Karabagias et al. [17] who determined in honey samples a moisture content that ranged from 10.74% (Symi honey sample) to 20.94% (Lakonia honey sample). Mărghitaș et al. [37] reported a variation in the moisture content for the Romanian honey between 16.6% and 20%, while Küçük et al. [38] reported values of moisture content from 19% to 19.7% for Anatolian honey. Escuredo et al. [39] determined a moisture content between 15.5% and 19.8% for honey samples from Spain.

3.3. pH

Honey has a pH that usually varies between 3.5 and 5.5 and is dependent on the compositions of organic acids, which are chemical components that give the aroma of honey and at the same time protect it against microbiological damage. Therefore, the pH can be considered an indicator of potential microbial growth, as a value of 7.2 to 7.4 is optimal for the development of most microorganisms. The average pH values of the honey samples ranged from 3.91 in the case of thyme honey to a maximum of 4.22 in the case of rape honey (Table 2).

Differences in this parameter were determined by the botanical origin of honey ($p < 0.05$), but not by year ($p > 0.05$). Romanian mint and thyme honey samples had pH values similar to Tunisian mint and thyme honey [36]. In Tunisian mint honey samples, the pH value was 4.11 [36], while in Romanian mint honey, it was 4.20. For thyme honey, the pH value was 3.87 (Tunisian honey) and 3.91 (Romanian honey).

Therefore, a pH between 3.2 and 4.5 was considered acceptable for honey samples [11] and the pH values determined for the studied samples were within this range. The pH of honey is of particular importance during the extraction and storage of honey because of its influence on the texture, stability and storage time [5]. The average pH values of honey samples from Vojvodina (Serbia) ranged between 3.88 (sunflower honey) and 3.99 (acacia honey) [35].

Table 2. Physicochemical properties for different types of honey (raspberry, mint, rape, thyme, polyfloral and sunflower). Mean values and standard deviation in brackets.

Parameter	Origin						F Value	Year		F Value
	Mint	Polyfloral	Rape	Raspberry	Sunflower	Thyme		2017	2018	
L*	35.3 (6.24) [bc]	46.1 (3.67) [a]	41.4 (3.48) [a]	34.4 (3.36) [c]	41.05 (6.93) [a]	43.2 (6.31) [ab]	5.05 ***	40.5 (5.58) [a]	39.3 (7.03) [a]	1.67 [ns]
h_{ab}	65.6 (9.38) [d]	84.7 (10) [b]	96.6 (6.03) [a]	73.4 (8.43) [c]	83.4 (3.17) [b]	81.8 (4.76) [bc]	20.30 ***	79.6 (12.56) [a]	82.2 (13.48) [a]	1.56 [ns]
C_{ab}	19.7 (3.42) [c]	26.2 (5.37) [ab]	19.2 (6.66) [c]	23.8 (1.78) [bc]	27.2 (5.81) [a]	29.8 (4.48) [a]	7.71 ***	26.3 (5.43) [a]	21.1 (5.72) [b]	24.85 ***
Pfund (mm Pfund)	74.3 (14.54) [a]	40.9 (20.41) [cd]	29.4 (11.16) [d]	61.4 (14.35) [ab]	37.6 (10.23) [cd]	50.1 (12.29) [bc]	13.28 ***	49.4 (21.19) [a]	48.1 (21.96) [a]	0.01 [ns]
pH	4.20 (0.25) [a]	4.09 (0.24) [ab]	4.22 (0.08) [a]	4.16 (0.12) [a]	3.94 (0.25) [b]	3.91 (0.19) [ab]	2.51 *	4.08 (0.25) [a]	4.1 (0.21) [a]	0.67 [ns]
Free acidity (meq·kg⁻¹)	26.9 (8.89) [a]	23.9 (12.54) [ab]	16 (4.43) [b]	27.3 (7.71) [a]	31.6 (12.20) [a]	22.5 (8.16) [ab]	3.07 **	22.01 (8.69) [a]	26.8 (11.12) [a]	2.28 [ns]
EC (μS·cm⁻¹)	474 (92.76) [a]	354 (242.77) [abc]	162 (38.26) [d]	446 (68.57) [ab]	362 (55.03) [bc]	244 (54.13) [cd]	10.71 ***	310 (154.86) [a]	367 (151.89) [a]	2.28 [ns]
Moisture (%)	17.7 (1.10) [b]	19.6 (1.65) [a]	18.4 (0.86) [ab]	18.3 (1.05) [ab]	18.4 (1.48) [ab]	17.3 (1.95) [b]	1.71 [ns]	17.9 (1.01) [a]	18.6 (1.56) [a]	3.76 *
HMF (mg·kg⁻¹)	29.2 (23.22) [a]	10 (8.84) [b]	13.3 (14.10) [b]	18.7 (16.33) [b]	8.26 (4.49) [b]	30.8 (20.96) [a]	6.24 ***	28.4 (26.31) [a]	20.2 (26.26) [a]	0.72 [ns]
TPC (mg GAE·100 g⁻¹)	23.7 (4.37) [a]	20.3 (7.67) [a]	19.9 (4.83) [a]	19.9 (4.83) [a]	21.1 (7.18) [a]	18.9 (3.82) [a]	0.35 [ns]	21.4 (5.83) [a]	20.5 (5.98) [a]	1.15 [ns]
FC (mg QE·100 g⁻¹)	25.7 (10.55) [b]	24.1 (5.76) [b]	20.2 (12.21) [b]	33.5 (6.62) [a]	22.8 (8.73) [b]	17.4 (9.33) [b]	2.29 [ns]	21.1 (10.42) [a]	26.3 (9.46) [a]	4.26 *
DPPH (%)	74.03 (5.84) [ab]	70.7 (15.90) [ab]	55.4 (6.88) [c]	79.05 (13.51) [a]	68.03 (8.01) [b]	67.3 (9.82) [ab]	5.24 ***	67.3 (13.12) [a]	69.1 (11.38) [a]	0.68 [ns]

[ns] not significant ($p > 0.05$), * $p < 0.05$, ** $p < 0.01$, *** $p < 0.001$, [a–d] different letters in the same row indicate significant differences between samples ($p < 0.001$) according to the LSD test with $\alpha = 0.05$. Pfund—color in Pfund scale, EC—electrical conductivity, HMF—5-hydroxymethylfurfural, TPC—total phenolic content, FC—flavonoids content, DPPH—radical scavenging activity.

3.4. Free Acidity

The free acidity of honey is determined by the presence of organic acids and other compounds such as esters, lactones and inorganic ions found in its composition [11]. Contribution to this parameter also presents the composition of protein, phenolic acids and vitamin C, which are chemical components that act as H$^+$ donors [40]. Determining the acidity helps to appreciate the freshness of the honey. As the composition of honey deteriorates, an increase of free acidity occurs as a result of the fermentation of sugars into organic acids. According to the EU legislation [41], for this parameter a maximum of 50 milliequivalents of acid per 1000 g is allowed [2]. In our study, the highest acidity was determined in sunflower (31.63 meq·kg^{-1}) honey and the lowest (16.01 meq·kg^{-1}) in rape honey (Table 2). The botanical origin of honey had a significant influence on this parameter ($p < 0.01$), while the year of production determined no significant variation between samples ($p > 0.05$). Lazarević et al. [42] observed similar results, they determined the highest free acidity (27.2 meq/kg) in sunflower honey and the lowest values of the parameter (11.6 meq·kg^{-1}) in acacia honey. Significant differences in the function of the botanical origin of honey were also reported for the free acidity of acacia and hay honey [43]. Oroian and Ropciuc [44] reported that free acidity varied between 6.63 meq·kg^{-1} in tilia honey, 13.02 meq·kg^{-1} in sunflower honey and reached the maximum value in the case of polifloral honey (20.83 meq·kg^{-1}).

3.5. HMF Content

The HMF content is a chemical parameter that can be used to study the degree of freshness of honey and consequently its degree of deterioration. The causes of honey deterioration could be due to strong or prolonged thermal treatment and inadequate storage conditions [45]. As seen in Table 2, honey samples had an HMF content between a minimum of 8.26 mg HMF·kg^{-1} (sunflower honey) and a maximum of 50.8 mg HMF·kg^{-1} (thyme honey). Botanical origin had a significant influence ($p < 0.001$) on this parameter. For some of the samples of mint (two samples) and thyme honey (one sample) that were analyzed, the HMF content was above the maximum concentration (40 mg·kg^{-1}) allowed by European legislation [41]. In the case of these samples, it is possible that there was an overheating during processing and/or storage, which might have influenced the HMF content in honey.

Rodríguez et al. [46] observed that the avocado honey had a maximum level of HMF of 27.1 mg·kg^{-1}. Another study, focused on the quality of honey from Rio Grande do Sul State (Brazil), reported values of 0.47–22.72 mg HMF·kg^{-1} of honey, which met the quality requirements established by both Brazilian legislation (upper limit of 60 mg HMF·kg^{-1}) and international standards (Codex, 2001—maximum of 40 mg HMF·kg^{-1}) [47].

3.6. Color

The appearance of honey is very important for consumers. The color of honey is a sensory parameter that varies between different types of honey and is dependent on chemical parameters such as mineral content and polyphenols content [48]. Regarding the mineral composition, it was argued that amber and dark honey have a higher content of certain minerals (Na, K, Ca, Mg, Fe, Cu, Zn, Al, Ni, Cd and Mn) by comparison to light-colored honey [49]. Furthermore, transition metals seem to influence the color of honey through the formation of complexes with some organic compounds. The color of honey can be also affected by both storage and thermal processing which was linked to the formation of Maillard reaction products [11].

As previously mentioned, color is a parameter that is dependent on the botanical origin of honey, as shown by the values in Table 2. The color values presented on the Pfund scale were used to classify honey by color. The color of the analyzed honey samples varied between white (rape honey), extra light amber (sunflower honey, thyme and polyfloral) and light amber (mint and raspberry honey). Manzanares et al. [50] analyzed 85 samples of honey from Tenerife, Spain and reported values between 24 and 150 mm Pfund. The color of honey samples was characterized by red and yellow shades (first quadrant of CIEL a*b* color space), as a* and b* coordinates had positive values. The lightness values

(L *) of the six Tunisian honey samples analyzed by Boussaid et al. [36] ranged from 36.64 to 51.37, while in our honey samples it ranged from 34.4 to 46.1.

3.7. Electrical Conductivity

Another physical parameter that serves as a means to authenticate honey, and particularly the monofloral types, is electrical conductivity. Electrical conductivity is a parameter included in the new international standards regarding the differentiation between honeydew and flower honey. The limits of this parameter that were specified by standards are 500 to 800 $\mu S \cdot cm^{-1}$ for mixed honey and <500 $\mu S \cdot cm^{-1}$ in the case of pure floral honey with some exceptions [51]. Values greater than 800 $\mu S \cdot cm^{-1}$ are specific to honeydew and therefore are not acceptable for floral honey, and can confirm an adulteration with inverted sugar [52–54]. In the work of Kaskoniene et al. [55] it was shown that floral honey has an electrical conductivity that was lower than that of honeydew, confirming that this parameter is a quality indicator that can be used as a means to distinguish honeydew from floral honey [56].

As the values in Table 2 show, the honey samples analyzed had an electrical conductivity of less than 500 $\mu S \cdot cm^{-1}$, so they can be classified as pure floral honey. Mint honey had the highest electrical conductivity (474.05 $\mu S \cdot cm^{-1}$) followed by raspberry honey (446.16 $\mu S \cdot cm^{-1}$). Polyfloral and sunflower honey presented close values of electrical conductivity (354.09 $\mu S \cdot cm^{-1}$ and 362.27 $\mu S \cdot cm^{-1}$) and rape and thyme honey were characterized by the lowest values of electrical conductivity (162.5 $\mu S \cdot cm^{-1}$ in rape honey and 244.28 $\mu S \cdot cm^{-1}$ in thyme honey). Botanical origin had a significant influence ($p < 0.001$) on the variation of this parameter. Boussaid et al. [36] reported for Tunisian mint honey an electrical conductivity of 430 $\mu S \cdot cm^{-1}$, which was similar to our results for mint honey. In the case of thyme honey, the value reported for Tunisian honey was higher than the electrical conductivity measured for the Romanian thyme honey. Oroian and Ropciuc [44] reported, in the case of sunflower honey, an electrical conductivity value of 346.1 $\mu S \cdot cm^{-1}$ and 431.4 $\mu S \cdot cm^{-1}$ for the polyfloral honey. Usually, monofloral rape honey has low electrical conductivity, 130 to 580 $\mu S \cdot cm^{-1}$ [57] and 110–270 $\mu S \cdot cm^{-1}$ [58], which indicates that this type of honey has a lower mineral content [55]. By comparing these reported values to the values we have determined for the electrical conductivity of our rape honey samples, it can be concluded that the samples analyzed in this study were of pure rape honey.

Regarding the influence of other parameters on the electrical conductivity of honey, it was found that the variation of this parameter positively correlated with an increased ash and acid content [8]. The pollen collected by bees is a major source of minerals, and consequently, in the case of monofloral honey the electrical conductivity correlated with the pollen content [55] and may serve as a means to identify the botanical origin of honey [59]. This parameter was included in international standards to replace the ash content determination [2]. The electrical conductivity is a good criterion for identifying the botanical origin of honey and is also used for its routine control [60].

3.8. Total Phenolic Content

The functional properties of honey are related to the number of natural antioxidants from pollen collected by bees and other floral nectars [61]. The antioxidant effects of honey were attributed to the presence of phenolic acids, flavonoids, ascorbic acid, carotenoids, catalase, peroxidase, as well as Maillard reaction products in the composition of honey [62,63]. Table 2 presents the total phenolic content (TPC) of raspberry, mint, thyme, rape, sunflower and polyfloral honey samples. TPC varied between 18.91 mg GAE ·100 g^{-1} (thyme honey) and 23.71 mg GAE ·100 g^{-1} (mint honey); no significant differences were determined by botanical origin and year.

Chua et al. [64] reported in their study that the TPC of the analyzed honey samples ranged from 110.39 to 196.500 mg GAE ·100 g^{-1}. In their study on four types of honey, Marghitaş et al. [37] reported that sunflower honey had the highest value of total polyphenol content (40 mg GAE ·100 g^{-1}), while acacia honey had values between 2 and 39 mg GAE ·100 g^{-1}. The total phenolic content of Indian honey was found in the range of 47 mg GAE ·100 g^{-1} of honey to 98 mg GAE/100 g of honey [51]. The TPC values determined for Romanian monofloral honey in our study were lower than those obtained by other authors when analyzing honey of different origin.

3.9. Flavonoids Content

The flavonoids of honey may originate from pollen, nectar or propolis [65]. In general, the main flavonoids found in honey are pinocembrin, apigenin, campferol, quercetin, pinobanksin, luteolin, galangin, hesperetin, and isorhamnetin [15]. Flavonoids have low molecular weight and are vital components of honey and its antioxidant properties [66]. Table 2 shows the values determined for flavonoids content by botanical origin and year of honey production. As in the case of total polyphenols, in this study the thyme honey samples had the lowest flavonoid content (17.45 mg QE·100 g^{-1}). The highest flavonoid content was identified in raspberry honey (33.58 mg QE·100 g^{-1}) followed by mint honey (25.73 mg QE·100 g^{-1}), polyfloral honey (24.14 mg QE·100 g^{-1}), sunflower honey (22.86 mg QE·100 g^{-1}) and rape honey (20.25 mg QE·100 g^{-1}). The flavonoids content was influenced by year ($p < 0.05$), but not by botanical origin ($p > 0.05$).

Mărghitaş et al. [37] reported that the total flavonoid content of honey samples ranged between 0.91–2.42 mg QE·100 g^{-1} in acacia honey, 4.70–6.98 mg QE·100 g^{-1} in tilia honey, and 11.53–15.33 mg QE·100 g^{-1} in sunflower honey. Boussaid et al. [36] reported higher total flavonoids content in mint honey (22.45 mg QE·100 g^{-1}), and lower in the case of rosemary (16.24 mg QE·100 g^{-1}), thyme (14.77 mg QE·100 g^{-1}), orange (11.12 mg QE·100 g^{-1}), horehound (11.02 mg QE·100 g^{-1}), and eucalyptus (9.58 mg QE·100 g^{-1}) honey.

3.10. DPPH Assay

The DPPH assay was used as a means to determine the free radical-scavenging activity of the honey samples. In this study the highest DPPH radical scavenging activity (Table 2) was identified for raspberry honey (79.05%) and mint honey (74.03%), and the lowest for thyme (63.77%) and rape honey (55.49%). Lachman et al. [67] also determined a lower antioxidant activity from the DPPH assay for rape honey and higher for raspberry honey in a study on honey samples from the Czech Republic. Blasa et al. [68] and Salonen et al. [69] argued that light-colored honey possessed lower antioxidant activity by comparison to darker colored honey, an observation that seems to be accurate for our study, although the differences between the free radical scavenging activities of our honey samples were not as pronounced as in the case of the above-mentioned studies.

DPPH radical scavenging activity is a parameter that varied significantly ($p < 0.001$) depending on the botanical origin of the honey samples analyzed. By comparison to the antioxidant activities reported for honey samples from other geographical regions, which include studies by Ruiz-Navajas et al. [70] who reported values of 33.4–85.5% for honey from Tabasco (Mexico) and Baltrusaityte et al. [71] who reported for honey from Lithuania values between 31.1% and 86.9%, Romanian honey had overall higher antioxidant activities.

3.11. Sugars Composition

Honey contains simple carbohydrates, namely glucose and fructose known as monosaccharides, which represent 65–80% of the total soluble solids, as well as 25% of other oligosaccharides (disaccharides, trisaccharides, tetrasaccharides) [72]. Determination of disaccharide content (mainly maltose and sucrose) is a tool for characterizing honey; maltose content was used to classify Spanish honey and to differentiate Brazilian honey from different geographical regions [72].

In the analyzed honey samples (Table 3), the highest fructose content was identified in thyme honey (36.77%) and the lowest in polyfloral honey (35.15%). Rape honey had the highest glucose content (31.78%) and polyfloral honey had the lowest content (24.95%) of this monosaccharide. Glucose content was the only parameter that varied significantly ($p < 0.01$) depending on the botanical origin of the analyzed samples. Maltose (maximum value of 1.79% in polyfloral honey), trehalose (maximum value of 2.35% in rape honey) and melesitose (maximum value of 1.34% in thyme honey) were sugars that together with fructose and glucose were found in significant concentrations in the analyzed honey samples. Sucrose and rafinose had values between 0.07% (raspberry honey) and 0.73% (polyfloral honey), respectively 0.21% (rape honey) and 0.42% (polyfloral honey). Apart from individual sugars, for all 45 honey samples, the fructose/glucose ratio was also calculated. When the content of fructose is higher than that of glucose honey is fluid, thus, this ratio can be used to identify the crystallization state of honey [73,74]. Suarez et al. [75] reported that the fructose/glucose ratio might also impact the flavor of honey since fructose is sweeter than glucose. All honey samples examined were fluid, as the fructose/glucose ratio was greater than 1 (Table 2).

Some authors argued that the amount and ratio of specific carbohydrates, such as fructose, glucose and oligosaccharides can be used to identify whether honey is monofloral or polyfloral [76]. Kaskonien & Venskutonis [55] considered that the use of carbohydrates as floral markers is not often preferred because of the difficulties encountered in identifying one or more sugars contained by honey. Cotte et al. [76] analyzed authentic monofloral honey samples and found differences in the carbohydrate composition based on botanical origin. Fir honey samples were high in trisaccharides, in particular raffinose (2.1%), melesitose (5.7%) and erlose (2.1%). In contrast, in rape and sunflower honey these trisaccharides were absent, which serves as a way to distinguish them from other botanical varieties. Acacia honey has a high concentration of trisaccharides (1.9%), with erlose being the predominant trisaccharide in this type of honey; lavender and tilia honey were characterized by lower concentrations of erlose (1.4 and 1.0% respectively) [76].

3.12. Polyphenols Composition

Polyphenols are powerful antioxidants that can reach more than 0.8% (by weight) in bee products [77]. Phenolic acids and flavonoids were extensively investigated in honey [78] and were used to evaluate its quality. The correlations between antioxidant activity and total concentration of phenols was confirmed for seven types of honey from Italy [79] and four honey types from Romania [37]. In another study on Portuguese honey, it was shown that polyphenols in honey were responsible for its antimicrobial effects [80]. Phenolic compounds can be used to classify honey according to its botanical origin [78]. The composition of honey in polyphenols was found to be mostly dependent on the botanical origin due to the fact that these compounds mostly originate from the nectar collected by bees; the nature and quantity of phenolic compounds can also vary with the season, climatic conditions and processing factors [63,79].

In the studied samples, 12 polyphenols were analyzed, which were mostly found in all samples in different concentrations (Table 4). Gallic acid was found at a high concentration (1.55 mg·100 g^{-1}) in mint honey, while the lowest value was identified in thyme honey (0.57 mg·100 g^{-1}). The protocatechuic and 4-hydroxybenzoic acids were identified in higher concentrations in mint (2.04 mg protocatechuic acid·100 g^{-1} and 1.20 mg 4-hydroxybenzoic acid·100 g^{-1}) and raspberry honey (2.57 mg protocatechuic acid·100 g^{-1} and 2.33 mg 4-hydroxybenzoic acid·100 g^{-1}). Compared to other honey types, mint also had a high content of vanillic acid (3.03 mg·100 g^{-1}) and chlorogenic acid (1.48 mg·100 g^{-1}). These two phenolic acids were also found in sunflower and thyme honey in large quantities. Caffeic acid predominated in polyfloral honey (1.20 mg·100 g^{-1}) and did not exceed the level of 0.38 mg·100 g^{-1} in other honey samples. Thyme honey had the highest content of p-coumaric acid, while myricetin predominated in rape honey, although it was found in all types of honey. Rosmarinic acid was found only in raspberry honey at a very small concentration of 0.03 mg·100 g^{-1}, while kaempferol was determined only in polyfloral honey in a concentration of 0.38 mg·100 g^{-1}. Quercetin was quantified only in 3 types of honey: mint, polifloral and sunflower honey and luteolin was not determined in any sample.

Gasic et al. [81] observed that quercetin and eriodictiol can be used for sunflower honey authentication and we observed too that quercetin is presented in the sunflower honey analyzed.

3.13. Organic Acids Composition

Organic acids are found in honey in small quantities (<0.5%), but are important chemical components because of their significant contribution to the stability and preservation of the physicochemical and sensory properties of honey [11].

The total acid content increases due to the fermentation phenomena and aging that may occur during storage [82]. Some authors have suggested that organic acid profiles are useful for identifying the botanical and/or geographical origin of honey [83].

As shown in Table 5, the predominant acid in all the honey samples analyzed was gluconic acid. The maximum gluconic acid content was determined in raspberry honey (4.83 g·kg^{-1}) and the lowest value in rape honey (3.59 g·kg^{-1}). Brugnerotto et al. [84] also identified gluconic acid as the predominant acid in all the honey samples that they studied. Gluconic acid is predominant in both honeydew and floral honey and its concentration can be influenced by the botanical source and the pollen and nectar of the flowers collected by bees. In our study, the concentration of gluconic acid was not influenced by botanical origin or year of production ($p > 0.05$).

Romanian mint and thyme honey were also high in propionic acid (2.67 g·kg^{-1} and 2.36 g·kg^{-1}). Honey samples had a succinic acid content that ranged from a minimum value of 0.05 g·kg^{-1} in raspberry honey to a maximum value of 0.13 g·kg^{-1} in mint honey. In the study conducted by Suarez-Luque et al. [85] on 50 honey samples from Galicia (Santiago de Compostela, Spain), the succinic acid content was much higher. Formic, acetic, lactic and butyric acid were determined in low concentrations in all honey samples. The content of honey samples in both propionic and acetic acids was strongly influenced ($p < 0.001$) by botanical origin.

The quantification of malonic and glycolic acids in floral honey was firstly reported in a study by Brugnerotto et al. [84] who determined concentrations of 82.2–134 mg malonic acid·100 g^{-1} and 27.8–43.7 mg glycolic acid·100 g^{-1}. Acetic, lactic, formic, and propionic acids were identified in lower concentrations, while fumaric and tartaric acids were not detected. In their study, citric and malic acid concentrations were of 48.2–506 mg·100 g^{-1} and 19.9–132 mg·100 g^{-1}, respectively [84].

Suarez-Luque et al. [85] also observed variations in the composition of organic acids in honey that were attributed to its botanical origin. The concentration of citric, malic, succinic and fumaric acid was high in chestnut honey and low in eucalyptus honey. Polyfloral honey had a high content of maleic acid, while clover honey did not contain malic and succinic acids.

The concentration and content of organic acids, as well as ketones and benzene compounds such as 2-hydroxy-2-propanone, 2-phenylethanol, butanoic acid or benzyl alcohol, which were identified in fresh honey, increase with temperature and storage time [86].

3.14. Principal Component Analysis (PCA)

Principal component analysis (PCA) is a statistical procedure that is used to perform a comparison of the results of analytical methods applied to a group of samples. In this study, PCA was applied to analyze and identify the honey samples that share similar characteristics from a total number of 45 samples of different honey types from various regions in Romania. The first principal component (PC-1) accounted for 82% of the variance, while the second principal component (PC-2) accounted for 9% of the variance; together, the first two principal components accounted for 91% of the initial variability. The separation of the honey samples according to botanical origin is shown in Figure 1. As seen in Figure 1, there are three ellipses which represent the rape, sunflower and thyme honey, which are not overlapped with other honey samples, except the polyfloral honey. Regarding the mint and raspberry honey, it can be observed that the raspberry honey ellipse is placed in the mint honey ellipse so a clear separation cannot be observed in this sample. Polyfloral honey was not perfectly grouped due to the fact that this honey type has a wide variety of pollen grains.

In Figure 1, the honey types are marked as: RA—rape, T—thyme, P—polyfloral, S—sunflower, M—mint, and R—raspberry honey. In Figure 2, the parameters used for the projection are abbreviated as: Pf—Pfund color, pH, Fa—free acidity, EC—electrical conductivity, Mo—moisture, HMF, TPC—total polyphenols content, TFC—total flavonoids content, DPPH, GA—gallic acid, PA—protocatechuic acid, 4-hA—4-Hydroxybenzoic acid, VA—vanillic acid, CA—chlorogenic acid, CafA—caffeic acid, p-CA—p-coumaric acid, RA—rosmarinic acid, My—miricetin, Qu—quercetin, Lu—luteolin, Ka—kaempferol, F—fructose content, G—glucose, S—sucrose, Tu—turanose, Ma—manose, Tr—trehalose, Me—melesitose, Ra—raffinose, GluA—gluconic acid, ForA—formic acid, AcetA—acetic acid, ProA—propionic acid, LacA—lactic acid, ButA—butyric acid, and SucA—succinic acid.

In Figure 2, the parameters which are in the outer ellipse have a greater contribution to variability than the parameters located in the inner ellipse. The rape honey samples were correlated with L^* values, pH, c^*_{ab}, h^*_{ab}, turanose content, manose content and HMF content. The thyme honey samples were correlated with trehalose content and mint honey with caffeic acid, p-coumaric acid, vanillic acid, rosmarinic acid and chlorogenic acid content. Regarding the physicochemical parameters, it seems that the moisture content was in opposition to the rest of the parameters.

There was a clear differentiation between Mo variable, the variable groups Tu, L^*, Ma, HMF, pH and h^*_{ab} (PC-1 direction) and variable groups TPC and SucA (PC-2 direction). Between variable groups from the PC-1 direction (Tu, L^*, Ma, HMF, pH, h^*_{ab} and M), there was no correlation with variables TPC and SucA. Between Tu, L^*, HMF, pH and h^*_{ab}, and variable Mo, there was negative correlation and the highest fraction of explained variance among these variables was 82%. Furthermore, the small distance between Tu, L^*, HMF, Ma and h^*_{ab} showed a strong correlation between variables.

Table 3. Sugars content for different types of Romanian honey. Mean values and standard deviation in brackets.

Sugars (%)	Origin					F Value	Year		F Value
	Mint	Polyfloral	Rape	Raspberry	Sunflower	Thyme		2017	2018	
Fructose	36.03 (2.33)[a]	35.15 (1.45)[a]	35.26 (1.28)[a]	36.30 (1.43)[a]	36.74 (1.73)[a]	36.77 (3.79)[a]	0.8 [ns]	36.66 (2.27)[a]	35.45 (1.51)[a]	4.08 [ns]
Glucose	27.87 (2.81)[bc]	24.95 (1.27)[c]	31.78 (2.71)[a]	29.00 (2.72)[ab]	28.37 (3.97)[b]	26.86 (2.80)[bc]	4.66 **	28.4 (3.96)[a]	28.6 (3.11)[a]	0.57 [ns]
Sucrose	0.45 (1.09)[a]	0.73 (1.23)[a]	0.08 (0.20)[a]	0.07 (0.07)[a]	0.35 (0.53)[a]	0.49 (0.78)[a]	0.68 [ns]	0.33 (0.74)[a]	0.34 (0.77)[a]	0.13 [ns]
Turanose	0.42 (0.19)[a]	0.2 (0.11)[a]	0.66 (1.24)[a]	0.29 (0.10)[a]	0.38 (0.30)[a]	0.31 (0.26)[a]	0.86 [ns]	0.53 (0.88)[a]	0.31 (0.20)[a]	1.17 [ns]
Maltose	1.44 (0.49)[a]	1.79 (0.40)[a]	1.82 (1.52)[a]	1.32 (0.40)[a]	1.62 (0.93)[a]	1.48 (0.84)[a]	0.42 [ns]	1.76 (1.12)[a]	1.46 (0.66)[a]	1.86 [ns]
Trehalose	1.45 (0.83)[a]	1.87 (0.58)[a]	2.35 (3.22)[a]	1.57 (0.58)[a]	1.92 (1.03)[a]	2.07 (0.73)[a]	0.33 [ns]	2.1 (2.29)[a]	1.68 (0.83)[a]	0.68 [ns]
Melesitose	1.03 (0.31)[a]	1.10 (0.23)[a]	1.08 (0.69)[a]	0.96 (0.28)[a]	1.06 (0.48)[a]	1.34 (0.84)[a]	0.36 [ns]	1.21 (0.59)[a]	0.97 (0.35)[a]	3.47 [ns]
Raffinose	0.31 (0.15)[ab]	0.42 (0.12)[a]	0.21 (0.11)[b]	0.36 (0.21)[ab]	0.40 (0.27)[ab]	0.40 (0.28)[ab]	1.63 [ns]	0.34 (0.21)[a]	0.33 (0.19)[a]	0.36 [ns]
F/G ratio	1.30 (0.14)[a]	1.40 (0.03)[a]	1.11 (0.11)[b]	1.26 (0.14)[ab]	1.33 (0.29)[a]	1.38 (0.22)[a]	3.06 *	1.32 (0.24)[a]	1.25 (0.14)[a]	2.68 [ns]

[ns] not significant ($p > 0.05$), * $p < 0.05$, ** $p < 0.01$, [a–c] different letters in the same row indicate significant differences between samples ($p < 0.01$), according to LSD test with $\alpha = 0.05$. F—fructose, G—glucose.

Table 4. Polyphenols content for different types of Romanian honey. Mean values and standard deviation in brackets.

Polyphenols (mg·100 g^{-1})	Origin						F Value	Year		F Value
	Mint	Polyfloral	Rape	Raspberry	Sunflower	Thyme		2017	2018	
Gallic acid	1.55 (1.71)[a]	1.03 (0.86)[a]	0.65 (0.27)[a]	0.95 (0.45)[a]	0.83 (0.46)[a]	0.57 (0.24)[a]	1.24 [ns]	1.00 (1.31)[a]	0.94 (0.50)[a]	0.14 [ns]
Protocatechuic acid	2.04 (2.52)[ab]	0.71 (1.20)[bc]	0.44 (0.50)[c]	2.57 (1.47)[a]	1.37 (0.41)[abc]	1.77 (1.05)[abc]	2.61 *	1.21 (1.26)[a]	1.58 (1.77)[a]	0.69 [ns]
4-Hydroxybenzoic acid	1.20 (1.03)[ab]	0.52 (0.53)[b]	0.41 (0.21)[b]	2.33 (3.09)[a]	1.15 (0.9)[ab]	0.83 (0.77)[ab]	1.70 [ns]	0.87 (0.82)[a]	1.18 (1.70)[a]	0.96 [ns]
Vanillic acid	3.03 (3.05)[a]	1.24 (2.49)[ab]	0.17 (0.46)[b]	1.83 (3.10)[ab]	2.35 (2.69)[ab]	1.62 (2.72)[ab]	1.05 [ns]	1.56 (2.99)[a]	1.87 (2.24)[a]	0.04 [ns]
Chlorogenic acid	1.48 (2.56)[a]	1.16 (1.94)[a]	0.004 (0.01)[a]	0.45 (0.73)[a]	0.36 (0.67)[a]	1.41 (2.83)[a]	0.93 [ns]	0.31 (1.26)[a]	1.08 (1.90)[a]	1.28 [ns]
Caffeic acid	0.23 (0.24)[a]	1.20 (2.72)[a]	0.18 (0.06)[a]	0.38 (0.49)[a]	0.30 (0.32)[a]	0.22 (0.30)[a]	0.70 [ns]	0.22 (0.25)[a]	0.51 (1.33)[a]	0.85 [ns]
P-coumaric acid	0.61 (0.53)[a]	0.70 (0.68)[a]	0.46 (0.32)[a]	0.74 (0.85)[a]	0.80 (0.41)[a]	1.06 (1.07)[a]	0.49 [ns]	0.60 (0.57)[a]	0.75 (0.61)[a]	0.55 [ns]
Rosmarinic acid	0[a]	0[a]	0[a]	0.03 (0.09)[a]	0[a]	0[a]	1.07 [ns]	0.01 (0.05)[a]	0[a]	0.44 [ns]
Miricetin	1.86 (0.87)[a]	1.73 (1.40)[a]	2.23 (1.03)[a]	1.02 (1.16)[a]	1.37 (1.21)[a]	1.99 (1.58)[a]	1.01 [ns]	1.50 (1.23)[a]	1.90 (1.09)[a]	0.4 [ns]
Quercetin	0.30 (0.41)[a]	0.99 (2.28)[a]	0[a]	0[a]	0.19 (0.30)[a]	0[a]	1 [ns]	0[a]	0.43 (1.13)[a]	1.85 [ns]
Luteolin	0	0	0	0	0	0	–	0	0	–
Kaempferol	0	0.38 (0.94)[a]	0	0	0	0	1.09 [ns]	0	0.09 (0.46)[a]	0.51 [ns]

[ns] not significant ($p > 0.05$), * $p < 0.05$; [a–c] different letters in the same row indicate significant differences between samples ($p < 0.05$), according to the LSD test with $\alpha = 0.05$.

Table 5. Organic acids content for different types of Romanian honey. Mean values and standard deviation in brackets.

Organic Acids (g·kg^{-1})	Origin						F Value	Year		F Value
	Mint	Polyfloral	Rape	Raspberry	Sunflower	Thyme		2017	2018	
Gluconic acid	4.46 (1.53) ab	4.21 (0.48) ab	3.59 (1.12) b	4.83 (0.34) a	4.76 (0.55) a	4.50 (0.49) ab	1.60 ns	4.11 (1.40) a	4.53 (0.53) a	1.57 ns
Formic acid	0.37 (0.43) ab	0.18 (0.09) b	0.21 (0.17) b	0.28 (0.27) ab	0.53 (0.36) ab	0.77 (1.01) a	1.60 ns	0.30 (0.31) a	0.42 (0.49) a	1.15 ns
Acetic acid	0.77 (0.30) a	0.39 (0.22) bc	0.18 (0.05) c	0.58 (0.20) ab	0.4 (0.26) bc	0.3 (0.08) bc	6.99 ***	0.41 (0.25) a	0.47 (0.33) a	0.75 ns
Propionic acid	2.67 (1.52) a	0.72 (0.26) b	0.62 (0.42) b	0.86 (0.49) b	0.79 (0.28) b	2.36 (0.29) a	11.36 ***	1.48 (1.09) a	1.17 (1.20) a	1.11 ns
Lactic acid	0.18 (0.28) b	0.12 (0.22) b	0.14 (0.26) b	0.09 (0.07) b	0.14 (0.20) b	0.59 (0.52) a	1.66 ns	0.18 (0.26) a	0.17 (0.30) a	0 ns
Butyric acid	0.51 (0.42) a	0.87 (1.61) a	0.07 (0.11) a	0.11 (0.16) a	0.32 (0.38) a	0.23 (0.20) a	1.25 ns	0.28 (0.33) a	0.39 (0.84) a	0.18 ns
Succinic acid	0.13 (0.12) a	0.1 (0.19) a	0.09 (0.14) a	0.05 (0.11) a	0.11 (0.09) a	0.08 (0.07) a	0.24 ns	0.09 (0.13) a	0.10 (0.12) a	0.01 ns

ns not significant ($p > 0.05$), *** $p < 0.001$; a-c different letters in the same row indicate significant differences between samples ($p < 0.05$), according to the LSD test with $\alpha = 0.05$.

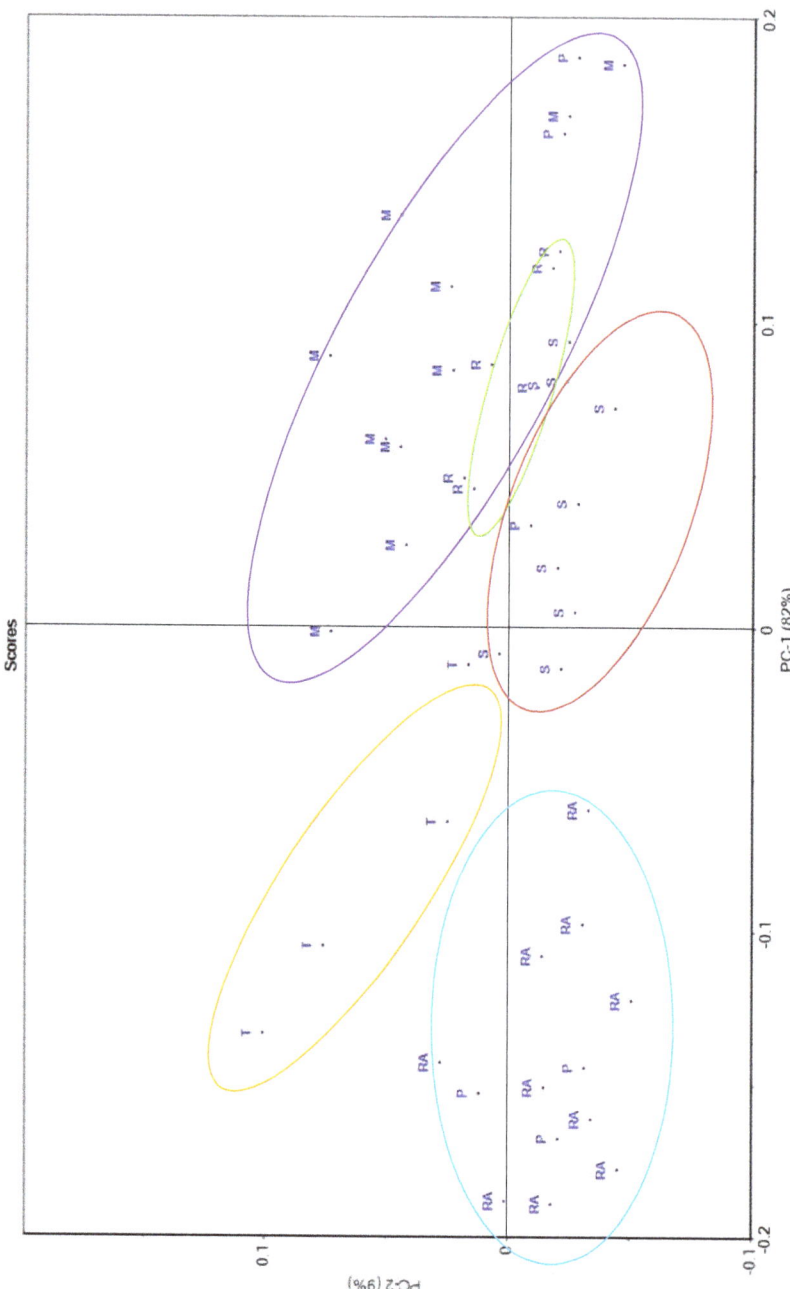

Figure 1. Principal component analysis—scores: RA—rape, T—thyme, P—polyfloral, S—sunflower, M—mint, and R—raspberry honey. Blue ellipse—rape honey group, yellow ellipse—thyme honey group, red ellipse—sunflower honey group, green ellipse—raspberry honey group, and purple ellipse—mint honey group.

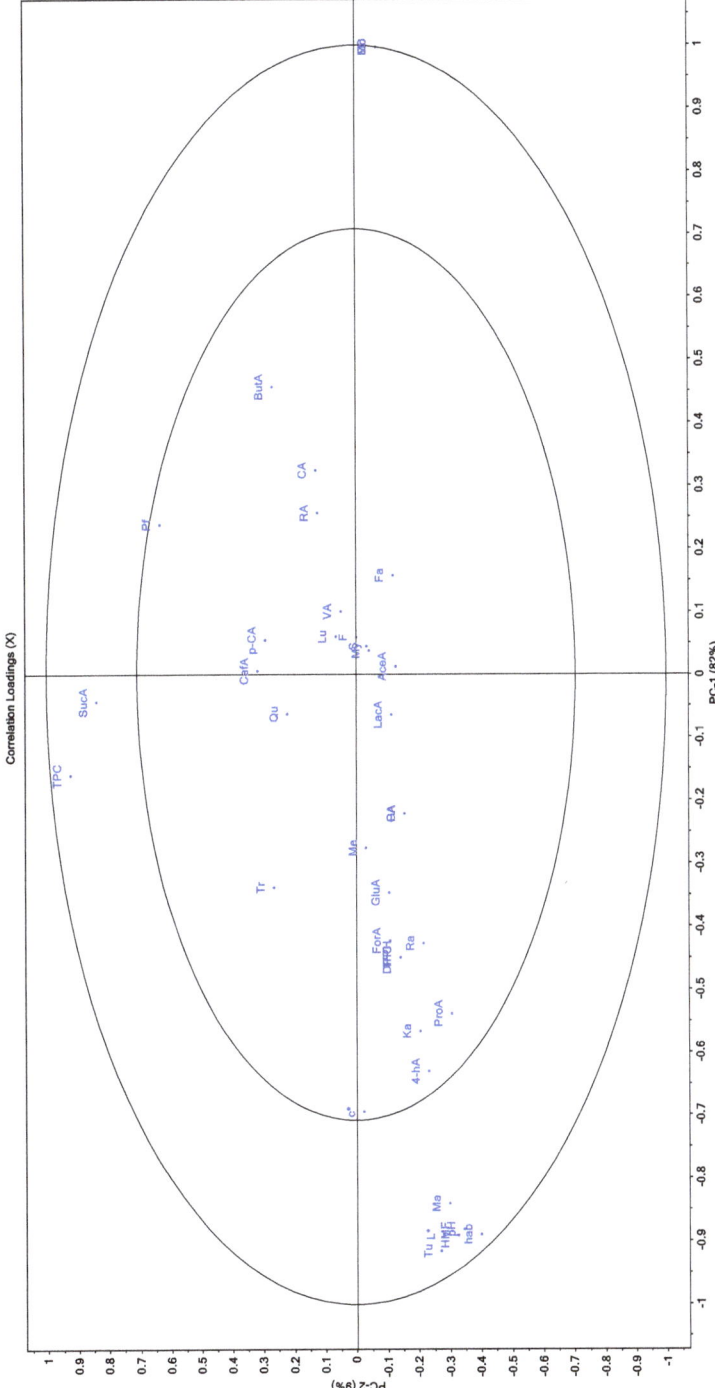

Figure 2. Principal component analysis—loadings: L*, h*ab, c*ab, Pf—Pfund color, pH, Fa—free acidity, EC—electrical conductivity, Mo—moisture, HMF, TPC—total polyphenols content, TFC—total flavonoids content, DPPH, GA—gallic acid, PA—protocatechuic acid, 4-hA—4-Hydroxybenzoic acid, VA—vanillic acid, CA—chlorogenic acid, CafA—caffeic acid, p-CA—p-coumaric acid, RA—rosmarinic acid, My—miricetin, Qu—quercetin, Lu—luteolin, Ka—kaempferol, F—fructose content, G—glucose, S—sucrose, Tu—turanose, Ma—manose, Tr—trehalose, Me—melesitose, Ra—raffinose, GluA—gluconic acid, ForA—formic acid, AcetA—acetic acid, ProA—propionic acid, LacA—lactic acid, ButA—butyric acid, and SucA—succinic acid.

4. Conclusions

The physicochemical parameters of raspberry, mint, sunflower, thyme, rape and polyfloral honey samples from different regions in Romania were analyzed in order to examine their usefulness in the classification of honey according to botanical origin. All honey samples had pH and free acidity values in the limits permitted by quality standards, which confirm the freshness of all honey samples. With the exception of three samples (two samples of mint honey and one sample of thyme honey) that had HMF content above the allowed limit, all the samples that were analyzed in this study met the quality requirements for honey. For most of the physicochemical parameters (color parameters, electrical conductivity, HMF content, DPPH, free acidity and pH) the differences between the measured levels were determined by the botanical origin of honey. The sugar composition, individual phenolic compounds and organic acids composition of honey varied to some extent between samples, however, these parameters were not influenced by the botanical origin of honey. As a consequence, no compound that can be used as a chemical marker was identified. PCA analysis was successful in the rape, sunflower and thyme honey samples while the mint and raspberry honey have not been clearly separated, but for a better classification of unknown honey it is necessary to increase the number of samples analyzed.

Author Contributions: Conceptualization, D.P.; formal analysis, D.P. and F.D.; funding acquisition, M.O.; investigation, D.P.; methodology, D.P., F.D. and M.O.; project administration, M.O.; software, F.D.; supervision, M.O.; validation, D.P.; writing—original draft, D.P.; writing—review and editing, F.D. and M.O. All authors have read and agreed to the published version of the manuscript.

Funding: This work was supported by contract no. 18PFE/16.10.2018 funded by the Ministry of Research and Innovation within Program 1—Development of national research and development system, Subprogram 1.2—Institutional Performance -RDI excellence funding projects.

Conflicts of Interest: The authors declare no conflict of interest.

References

1. Jones, R. Prologue: Honey and healing through the ages. *J. ApiProduct ApiMedical Sci.* **2009**, *1*, 1–5. [CrossRef]
2. Codex Alimentarius Commission. *Codex Alimentarius Commission Standards, Codex Stan 12-1981*; Codex Alimentarius Commission: Rome, Italy, 2001.
3. Communities. Directive 2001/77/Ec. *Off. J. Eur. Communities* **2001**, *6*, 33–40.
4. Oroian, M.; Amariei, S.; Escriche, I.; Leahu, A.; Damian, C.; Gutt, G. Chemical composition and temperature influence on the rheological behaviour of honeys. *Int. J. Food Prop.* **2014**, *17*, 2228–2240. [CrossRef]
5. Terrab, A.; Recamales, A.F.; Hernanz, D.; Heredia, F.J. Characterisation of Spanish thyme honeys by their physicochemical characteristics and mineral contents. *Food Chem.* **2004**, *88*, 537–542. [CrossRef]
6. El Sohaimy, S.A.; Masry, S.H.D.; Shehata, M.G. Physicochemical characteristics of honey from different origins. *Ann. Agric. Sci.* **2015**, *60*, 279–287. [CrossRef]
7. Anjos, O.; Campos, M.G.; Ruiz, P.C.; Antunes, P. Application of FTIR-ATR spectroscopy to the quantification of sugar in honey. *Food Chem.* **2015**, *169*, 218–223. [CrossRef]
8. Bogdanov, S.; Martin, P. Honey authenticity. *Mitteilungen Aus Leb. Und Hyg.* **2002**, *93*, 232–254.
9. Kasprzyk, I.; Depciuch, J.; Grabek-Lejko, D.; Parlinska-Wojtan, M. FTIR-ATR spectroscopy of pollen and honey as a tool for unifloral honey authentication. The case study of rape honey. *Food Control* **2018**, *84*, 33–40. [CrossRef]
10. Anklam, E. A review of the analytical methods to determine the geographical and botanical origin of honey. *Food Chem.* **1998**, *63*, 549–562. [CrossRef]
11. Da Silva, P.M.; Gauche, C.; Gonzaga, L.V.; Costa, A.C.O.; Fett, R. Honey: Chemical composition, stability and authenticity. *Food Chem.* **2016**, *196*, 309–323. [CrossRef]
12. Baroni, M.V.; Podio, N.S.; Badini, R.G.; Inga, M.; Ostera, H.A.; Cagnoni, M.; Gautier, E.A.; García, P.P.; Hoogewerff, J.; Wunderlin, D.A. Linking soil, water, and honey composition to assess the geographical origin of Argentinean honey by multielemental and isotopic analyses. *J. Agric. Food Chem.* **2015**, *63*, 4638–4645. [CrossRef] [PubMed]

13. Kaškoniene, V.; Venskutonis, P.R. Floral markers in honey of various botanical and geographic origins: A review. *Compr. Rev. Food Sci. Food Saf.* **2010**, *9*, 620–634. [CrossRef]
14. Escriche, I.; Kadar, M.; Domenech, E.; Gil-Sánchez, L. A potentiometric electronic tongue for the discrimination of honey according to the botanical origin. Comparison with traditional methodologies: Physicochemical parameters and volatile profile. *J. Food Eng.* **2012**, *109*, 449–456. [CrossRef]
15. Bogdanov, S.; Jurendic, T.; Sieber, R.; Gallmann, P. Honey for nutrition and health: A review. *J. Am. Coll. Nutr.* **2008**, *27*, 677–689. [CrossRef] [PubMed]
16. Siddiqui, A.J.; Musharraf, S.G.; Choudhary, M.I.; Rahman, A.U. Application of analytical methods in authentication and adulteration of honey. *Food Chem.* **2017**, *217*, 687–698. [CrossRef] [PubMed]
17. Karabagias, I.K.; Badeka, A.; Kontakos, S.; Karabournioti, S.; Kontominas, M.G. Characterization and classification of *Thymus capitatus* (L.) honey according to geographical origin based on volatile compounds, physicochemical parameters and chemometrics. *Food Res. Int.* **2014**, *55*, 363–372. [CrossRef]
18. Consonni, R.; Cagliani, L.R. Recent developments in honey characterization. *RSC Adv.* **2015**, *5*, 59696–59714. [CrossRef]
19. Oroian, M.; Paduret, S.; Ropciuc, S. Honey adulteration detection: voltammetric e-tongue versus official methods for physicochemical parameter determination. *J. Sci. Food Agric.* **2018**, *98*, 4304–4311. [CrossRef]
20. Escriche, I.; Sobrino-Gregorio, L.; Conchado, A.; Juan-Borrás, M. Volatile profile in the accurate labelling of monofloral honey. The case of lavender and thyme honey. *Food Chem.* **2017**, *226*, 61–68. [CrossRef]
21. Hawkins, J.; de Vere, N.; Griffith, A.; Ford, C.R.; Allainguillaume, J.; Hegarty, M.J.; Baillie, L.; Adams-Groom, B. Using DNA metabarcoding to identify the floral composition of honey: A new tool for investigating honey bee foraging preferences. *PLoS ONE* **2015**, *10*, e0134735. [CrossRef]
22. Soares, S.; Grazina, L.; Mafra, I.; Costa, J.; Pinto, M.A.; Oliveira, M.B.P.P.; Amaral, J.S. Towards honey authentication: Differentiation of Apis mellifera subspecies in European honeys based on mitochondrial DNA markers. *Food Chem.* **2019**, *283*, 294–301. [CrossRef] [PubMed]
23. Svečnjak, L.; Biliškov, N.; Bubalo, D.; Barišić, D. Application of infrared spectroscopy in honey analysis. *Agric. Conspec. Sci.* **2011**, *76*, 191–195.
24. Elamine, Y.; Inácio, P.M.C.; Lyoussi, B.; Anjos, O.; Estevinho, L.M.; Miguel, M.d.; Gomes, H.L. Insight into the sensing mechanism of an impedance based electronic tongue for honey botanic origin discrimination. *Sens. Actuators B Chem.* **2019**, *285*, 24–33. [CrossRef]
25. Sousa, M.E.B.C.; Dias, L.G.; Veloso, A.C.A.; Estevinho, L.; Peres, A.M.; Machado, A.A.S.C. Practical procedure for discriminating monofloral honey with a broad pollen profile variability using an electronic tongue. *Talanta* **2014**, *128*, 284–292. [CrossRef]
26. Bertram, H.C.; Kristensen, N.B.; Malmendal, A.; Nielsen, N.C.; Bro, R.; Andersen, H.J.; Harmon, D.L. A metabolomic investigation of splanchnic metabolism using 1H NMR spectroscopy of bovine blood plasma. *Anal. Chim. Acta* **2005**, *536*, 1–6. [CrossRef]
27. Spiteri, M.; Jamin, E.; Thomas, F.; Rebours, A.; Lees, M.; Rogers, K.M.; Rutledge, D.N. Fast and global authenticity screening of honey using 1H-NMR profiling. *Food Chem.* **2015**, *189*, 60–66. [CrossRef] [PubMed]
28. Louveaux, J.; Maurizio, A.; Vorwohl, G. Methods of melissopalynology. *Bee World* **1978**, *59*, 139–157. [CrossRef]
29. Bogdanov, S.; Lüllmann, C.; Martin, P.; von der Ohe, W.; Russmann, H.; Vorwohl, G.; Oddo, L.P.; Sabatini, A.-G.; Marcazzan, G.L.; Piro, R.; et al. Honey quality and international regulatory standards: Review by the international honey commission. *Bee World* **1999**, *80*, 61–69. [CrossRef]
30. Biesaga, M.; Pyrzyńska, K. Stability of bioactive polyphenols from honey during different extraction methods. *Food Chem.* **2013**, *136*, 46–54. [CrossRef]
31. Brand-Williams, W.; Cuvelier, M.E.; Berset, C. Use of a free radical method to evaluate antioxidant activity. *LWT-Food Sci. Technol.* **1995**, *28*, 25–30. [CrossRef]
32. Palacios, I.; Lozano, M.; Moro, C.; D'Arrigo, M.; Rostagno, M.A.; Martínez, J.A.; García-Lafuente, A.; Guillamón, E.; Villares, A. Antioxidant properties of phenolic compounds occurring in edible mushrooms. *Food Chem.* **2011**, *128*, 674–678. [CrossRef]
33. Özcelik, S.; Kuley, E.; Özogul, F. Formation of lactic, acetic, succinic, propionic, formic and butyric acid by lactic acid bacteria. *LWT-Food Sci. Technol.* **2016**, *73*, 536–542. [CrossRef]
34. Pires, J.; Estevinho, M.L.; Feás, X.; Cantalapiedra, J.; Iglesias, A. Pollen spectrum and physico-chemical attributes of heather (*Erica* sp.) honeys of north Portugal. *J. Sci. Food Agric.* **2009**, *89*, 1862–1870. [CrossRef]

35. Sakač, M.B.; Jovanov, P.T.; Marić, A.Z.; Pezo, L.L.; Kevrešan, Ž.S.; Novaković, A.R.; Nedeljković, N.M. Physicochemical properties and mineral content of honey samples from Vojvodina (Republic of Serbia). *Food Chem.* **2019**, *276*, 15–21. [CrossRef] [PubMed]
36. Boussaid, A.; Chouaibi, M.; Rezig, L.; Hellal, R.; Donsì, F.; Ferrari, G.; Hamdi, S. Physicochemical and bioactive properties of six honey samples from various floral origins from Tunisia. *Arab. J. Chem.* **2018**, *11*, 265–274. [CrossRef]
37. Al, M.L.; Daniel, D.; Moise, A.; Bobis, O.; Laslo, L.; Bogdanov, S. Physico-chemical and bioactive properties of different floral origin honeys from Romania. *Food Chem.* **2009**, *112*, 863–867. [CrossRef]
38. Küçük, M.; Kolayli, S.; Karaoğlu, Ş.; Ulusoy, E.; Baltaci, C.; Candan, F. Biological activities and chemical composition of three honeys of different types from Anatolia. *Food Chem.* **2007**, *100*, 526–534. [CrossRef]
39. Escuredo, O.; Míguez, M.; Fernández-González, M.; Seijo, M.C. Nutritional value and antioxidant activity of honeys produced in a European Atlantic area. *Food Chem.* **2013**, *138*, 851–856. [CrossRef]
40. Halliwell, B.; Gutteridge, J.M.C. Cellular responses to oxidative stress: Adaptation, damage, repair, senescence and death. In *Free Radicals in Biology and Medicine*; Halliwell, B., Gutteridge, J.M.C., Eds.; Oxford University Press: Oxford, UK, 2007.
41. Council, E.U. Council Directive 2001/110/EC of 20 December 2001 relating to honey. *Communities* **2002**, *10*, 47–52.
42. Lazarević, K.B.; Andrić, F.; Trifković, J.; Tešić, Ž.; Milojković-Opsenica, D. Characterisation of Serbian unifloral honeys according to their physicochemical parameters. *Food Chem.* **2012**, *132*, 2060–2064. [CrossRef]
43. Šarić, G.; Matković, D.; Hruškar, M.; Vahčić, N. Characterisation and classification of Croatian honey by physicochemical parameters. *Food Technol. Biotechnol.* **2008**, *46*, 355–367.
44. Oroian, M.; Ropciuc, S. Honey authentication based on physicochemical parameters and phenolic compounds. *Comput. Electron. Agric.* **2017**, *138*, 148–156. [CrossRef]
45. Önür, İ.; Misra, N.N.; Barba, F.J.; Putnik, P.; Lorenzo, J.M.; Gökmen, V.; Alpas, H. Effects of ultrasound and high pressure on physicochemical properties and HMF formation in Turkish honey types. *J. Food Eng.* **2018**, *219*, 129–136. [CrossRef]
46. Rodríguez, I.; Cámara-Martos, F.; Flores, J.M.; Serrano, S. Spanish avocado (*Persea americana* Mill.) honey: Authentication based on its composition criteria, mineral content and sensory attributes. *LWT* **2019**, *111*, 561–572. [CrossRef]
47. Nascimento, K.S.D.; Sattler, J.A.G.; Macedo, L.F.L.; González, C.V.S.; de Melo, I.L.P.; Araújo, E.d.; Granato, D.; Sattler, A.; de Almeida-Muradian, L.B. Phenolic compounds, antioxidant capacity and physicochemical properties of Brazilian Apis mellifera honeys. *LWT-Food Sci. Technol.* **2018**, *91*, 85–94. [CrossRef]
48. González-Miret, M.L.; Terrab, A.; Hernanz, D.; Fernández-Recamales, M.Á.; Heredia, F.J. Multivariate correlation between color and mineral composition of honeys and by their botanical origin. *J. Agric. Food Chem.* **2005**, *53*, 2574–2580. [CrossRef]
49. Solayman, M.; Islam, M.A.; Paul, S.; Ali, Y.; Khalil, M.I.; Alam, N.; Gan, S.H. Physicochemical properties, minerals, trace elements, and heavy metals in honey of different origins: A comprehensive review. *Compr. Rev. Food Sci. Food Saf.* **2016**, *15*, 219–233. [CrossRef]
50. Manzanares, A.B.; García, Z.H.; Galdón, B.R.; Rodríguez, E.R.; Romero, C.D. Physicochemical characteristics of minor monofloral honeys from Tenerife, Spain. *LWT-Food Sci. Technol.* **2014**, *55*, 572–578. [CrossRef]
51. Saxena, S.; Gautam, S.; Sharma, A. Physical, biochemical and antioxidant properties of some Indian honeys. *Food Chem.* **2010**, *118*, 391–397. [CrossRef]
52. Oroian, M.; Amariei, S.; Leahu, A.; Gutt, G. Multi-element composition of honey as a suitable tool for its authenticity analysis. *Pol. J. Food Nutr. Sci.* **2015**, *65*, 93–100. [CrossRef]
53. Oroian, M.; Ropciuc, S.; Paduret, S.; Todosi, E. Rheological analysis of honeydew honey adulterated with glucose, fructose, inverted sugar, hydrolysed inulin syrup and malt wort. *LWT* **2018**, *95*, 1–8. [CrossRef]
54. Oroian, M.; Ropciuc, S.; Paduret, S. Honey adulteration detection using raman spectroscopy. *Food Anal. Methods* **2017**, *11*, 959–968. [CrossRef]
55. Kaškoniene, V.; Venskutonis, P.R.; Čeksteryte, V. Carbohydrate composition and electrical conductivity of different origin honeys from Lithuania. *LWT-Food Sci. Technol.* **2010**, *43*, 801–807. [CrossRef]
56. Oroian, M.; Amariei, S.; Rosu, A.; Gutt, G. Classification of unifloral honeys using multivariate analysis. *J. Essent. Oil Res.* **2015**, *27*, 533–544. [CrossRef]

57. Piazza, M.G.; Oddo, L.P. Bibliographical review of the main European unifloral honeys. *Apidologie* **2004**, *35*, S94–S111. [CrossRef]
58. Devillers, J.; Morlot, M.; Pham-Delègue, M.H.; Doré, J.C. Classification of monofloral honeys based on their quality control data. *Food Chem.* **2004**, *86*, 305–312. [CrossRef]
59. Terrab, A.; González, A.G.; Díez, M.J.; Heredia, F.J. Characterisation of Moroccan unifloral honeys using multivariate analysis. *Eur. Food Res. Technol.* **2003**, *218*, 88–95. [CrossRef]
60. Yücel, Y.; Sultanoğlu, P. Characterization of Hatay honeys according to their multi-element analysis using ICP-OES combined with chemometrics. *Food Chem.* **2013**, *140*, 231–237. [CrossRef]
61. Schramm, D.D.; Karim, M.; Schrader, H.R.; Holt, R.R.; Cardetti, M.; Keen, C.L. Honey with high levels of antioxidants can provide protection to healthy human subjects. *J. Agric. Food Chem.* **2003**, *51*, 1732–1735. [CrossRef]
62. Sergiel, I.; Pohl, P.; Biesaga, M. Characterisation of honeys according to their content of phenolic compounds using high performance liquid chromatography/tandem mass spectrometry. *Food Chem.* **2014**, *145*, 404–408. [CrossRef]
63. Gheldof, N.; Engeseth, N.J. Antioxidant capacity of honeys from various floral sources based on the determination of oxygen radical absorbance capacity and inhibition of in vitro lipoprotein oxidation in human serum samples. *J. Agric. Food Chem.* **2002**, *50*, 3050–3055. [CrossRef]
64. Chua, L.S.; Rahaman, N.L.A.; Adnan, N.A.; Tan, T.T.E. Antioxidant activity of three honey samples in relation with their biochemical components. *J. Anal. Methods Chem.* **2013**, *2013*, 313798. [CrossRef]
65. Hamdy, A.A.; Ismail, H.M.; Al-Ahwal, A.E.-M.A.; Gomaa, N.F. Determination of flavonoid and phenolic Acid contents of clover, cotton and citrus floral honeys. *J. Egypt. Public Health Assoc.* **2009**, *84*, 245–259. [PubMed]
66. Khalil, M.I.; Moniruzzaman, M.; Boukraâ, L.; Benhanifia, M.; Islam, M.A.; Islam, M.N.; Sulaiman, S.A.; Gan, S.H. Physicochemical and antioxidant properties of algerian honey. *Molecules* **2012**, *17*, 11199–11215. [CrossRef] [PubMed]
67. Lachman, J.; Orsák, M.; Hejtmánková, A.; Kovářová, E. Evaluation of antioxidant activity and total phenolics of selected Czech honeys. *LWT-Food Sci. Technol.* **2010**, *43*, 52–58. [CrossRef]
68. Blasa, M.; Candiracci, M.; Accorsi, A.; Piacentini, M.P.; Albertini, M.C.; Piatti, E. Raw Millefiori honey is packed full of antioxidants. *Food Chem.* **2006**, *97*, 217–222. [CrossRef]
69. Salonen, A.; Virjamo, V.; Tammela, P.; Fauch, L.; Julkunen-Tiitto, R. Screening bioactivity and bioactive constituents of Nordic unifloral honeys. *Food Chem.* **2017**, *237*, 214–224. [CrossRef] [PubMed]
70. Ruiz-Navajas, Y.; Viuda-Martos, M.; Fernández-López, J.; Zaldivar-Cruz, J.M.; Kuri, V.; Pérez-Álvarez, J.Á. Antioxidant activity of artisanal honey from Tabasco, Mexico. *Int. J. Food Prop.* **2011**, *14*, 459–470. [CrossRef]
71. Baltrušaityte, V.; Venskutonis, P.R.; Čeksteryte, V. Radical scavenging activity of different floral origin honey and beebread phenolic extracts. *Food Chem.* **2007**, *101*, 502–514. [CrossRef]
72. Leite, J.M.D.; Trugo, L.C.; Costa, L.S.M.; Quinteiro, L.M.C.; Barth, O.M.; Dutra, V.M.L.; de Maria, C.A.B. Determination of oligosaccharides in Brazilian honeys of different botanical origin. *Food Chem.* **2000**, *70*, 93–98. [CrossRef]
73. Ouchemoukh, S.; Schweitzer, P.; Bey, M.B.; Djoudad-Kadji, H.; Louaileche, H. HPLC sugar profiles of Algerian honeys. *Food Chem.* **2010**, *121*, 561–568. [CrossRef]
74. Venir, E.; Spaziani, M.; Maltini, E. Crystallization in "Tarassaco" Italian honey studied by DSC. *Food Chem.* **2010**, *122*, 410–415. [CrossRef]
75. Alvarez-Suarez, J.M.; Tulipani, S.; Díaz, D.; Estevez, Y.; Romandini, S.; Giampieri, F.; Damiani, E.; Astolfi, P.; Bompadre, S.; Battino, M. Antioxidant and antimicrobial capacity of several monofloral Cuban honeys and their correlation with color, polyphenol content and other chemical compounds. *Food Chem. Toxicol.* **2010**, *48*, 2490–2499. [CrossRef] [PubMed]
76. Cotte, J.F.; Casabianca, H.; Chardon, S.; Lheritier, J.; Grenier-Loustalot, M.F. Chromatographic analysis of sugars applied to the characterisation of monofloral honey. *Anal. Bioanal. Chem.* **2004**, *380*, 698–705. [CrossRef] [PubMed]
77. Kroyer, G.; Hegedus, N. Evaluation of bioactive properties of pollen extracts as functional dietary food supplement. *Innov. Food Sci. Emerg. Technol.* **2001**, *2*, 171–174. [CrossRef]

78. Kečkeš, S.; Gašić, U.; Veličković, T.Ć.; Milojković-Opsenica, D.; Natić, M.; Tešić, Ž. The determination of phenolic profiles of Serbian unifloral honeys using ultra-high-performance liquid chromatography/high resolution accurate mass spectrometry. *Food Chem.* **2013**, *138*, 32–40. [CrossRef] [PubMed]
79. Rosa, A.; Tuberoso, C.I.G.; Atzeri, A.; Melis, M.P.; Bifulco, E.; Dess, M.A. Antioxidant profile of strawberry tree honey and its marker homogentisic acid in several models of oxidative stress. *Food Chem.* **2011**, *129*, 1045–1053. [CrossRef] [PubMed]
80. Estevinho, L.; Pereira, A.P.; Moreira, L.; Dias, L.G.; Pereira, E. Antioxidant and antimicrobial effects of phenolic compounds extracts of Northeast Portugal honey. *Food Chem. Toxicol.* **2008**, *46*, 3774–3779. [CrossRef]
81. Gašić, U.M.; Milojković-Opsenica, D.M.; Tešić, Ž.L. Polyphenols as possible markers of botanical origin of honey. *J. AOAC Int.* **2017**, *100*, 852–861. [CrossRef]
82. Mato, I.; Huidobro, J.F.; Simal-Lozano, J.; Sancho, M.T. Rapid determination of nonaromatic organic acids in honey by capillary zone electrophoresis with direct ultraviolet detection. *J. Agric. Food Chem.* **2006**, *54*, 1541–1550. [CrossRef]
83. Daniele, G.; Maitre, D.; Casabianca, H. Identification, quantification and carbon stable isotopes determinations of organic acids in monofloral honeys. A powerful tool for botanical and authenticity control. *Rapid Commun. Mass Spectrom.* **2012**, *26*, 1993–1998. [CrossRef] [PubMed]
84. Brugnerotto, P.; della Betta, F.; Gonzaga, L.V.; Fett, R.; Costa, A.C.O. A capillary electrophoresis method to determine aliphatic organic acids in bracatinga honeydew honey and floral honey. *J. Food Compos. Anal.* **2019**, *82*, 103243. [CrossRef]
85. Suárez-Luque, S.; Mato, I.; Huidobro, J.F.; Simal-Lozano, J.; Sancho, M.T. Rapid determination of minority organic acids in honey by high-performance liquid chromatography. *J. Chromatogr. A* **2002**, *955*, 207–214. [CrossRef]
86. Barra, M.P.G.; Ponce-Díaz, M.C.; Venegas-Gallegos, C. Volatile compounds in honey produced in the central valley of Ñuble Province, Chile. *Chil. J. Agric. Res.* **2010**, *70*, 75–84. [CrossRef]

© 2020 by the authors. Licensee MDPI, Basel, Switzerland. This article is an open access article distributed under the terms and conditions of the Creative Commons Attribution (CC BY) license (http://creativecommons.org/licenses/by/4.0/).

MDPI
St. Alban-Anlage 66
4052 Basel
Switzerland
Tel. +41 61 683 77 34
Fax +41 61 302 89 18
www.mdpi.com

Foods Editorial Office
E-mail: foods@mdpi.com
www.mdpi.com/journal/foods

www.ingramcontent.com/pod-product-compliance
Lightning Source LLC
LaVergne TN
LVHW070744100526
838202LV00013B/1300